*Drugs and
the Cell Cycle*

CELL BIOLOGY: A Series of Monographs

EDITORS

D. E. BUETOW

Department of Physiology
and Biophysics
University of Illinois
Urbana, Illinois

I. L. CAMERON

Department of Anatomy
University of Texas
Medical School at San Antonio
San Antonio, Texas

G. M. PADILLA

Department of Physiology and Pharmacology
Duke University Medical Center
Durham, North Carolina

G. M. Padilla, G. L. Whitson, and I. L. Cameron (editors). THE CELL CYCLE: Gene-Enzyme Interactions, 1969

A. M. Zimmerman (editor). HIGH PRESSURE EFFECTS ON CELLULAR PROCESSES, 1970

I. L. Cameron and J. D. Thrasher (editors). CELLULAR AND MOLECULAR RENEWAL IN THE MAMMALIAN BODY, 1971

I. L. Cameron, G. M. Padilla, and A. M. Zimmerman (editors). DEVELOPMENTAL ASPECTS OF THE CELL CYCLE, 1971

P. F. Smith. THE BIOLOGY OF MYCOPLASMAS, 1971

Gary L. Whitson (editor). CONCEPTS IN RADIATION CELL BIOLOGY, 1972

Donald L. Hill. THE BIOCHEMISTRY AND PHYSIOLOGY OF *TETRAHYMENA*, 1972

Kwang W. Jeon (editor). THE BIOLOGY OF AMOEBA, 1973

Dean F. Martin and George M. Padilla (editors). MARINE PHARMACOGNOSY: Action of Marine Biotoxins at the Cellular Level, 1973

Joseph A. Erwin (editor). LIPIDS AND BIOMEMBRANES OF EUKARYOTIC MICROORGANISMS, 1973

A. M. Zimmerman, G. M. Padilla, and I. L. Cameron (editors). DRUGS AND THE CELL CYCLE, 1973

In preparation

Stuart Coward (editor). CELL DIFFERENTIATION

Drugs and the Cell Cycle

Edited by

A. M. ZIMMERMAN

Department of Zoology
University of Toronto
Toronto, Canada

G. M. PADILLA

Department of Physiology and Pharmacology
Duke University Medical Center
Durham, North Carolina

I. L. CAMERON

Department of Anatomy
University of Texas
Medical School at San Antonio
San Antonio, Texas

ACADEMIC PRESS New York and London   1973
A Subsidiary of Harcourt Brace Jovanovich, Publishers

COPYRIGHT © 1973, BY ACADEMIC PRESS, INC.
ALL RIGHTS RESERVED.
NO PART OF THIS PUBLICATION MAY BE REPRODUCED OR
TRANSMITTED IN ANY FORM OR BY ANY MEANS, ELECTRONIC
OR MECHANICAL, INCLUDING PHOTOCOPY, RECORDING, OR ANY
INFORMATION STORAGE AND RETRIEVAL SYSTEM, WITHOUT
PERMISSION IN WRITING FROM THE PUBLISHER.

ACADEMIC PRESS, INC.
111 Fifth Avenue, New York, New York 10003

United Kingdom Edition published by
ACADEMIC PRESS, INC. (LONDON) LTD.
24/28 Oval Road, London NW1

Library of Congress Cataloging in Publication Data

Zimmerman, Arthur M   DATE

    (Cell biology)
    Includes bibliographies.
    1. Cells, Effect of drugs on.   2.   Cell Cycle.
I. Padilla, George M., joint author.   II.   Cameron,
Ivan L., joint author.   III.   Title.
[DNLM: 1.   Cell differentiation.   2.   Cell division.
3.   Pharmacology.   QH605 Z72d 1973]
QH650.Z55   615.7   72-7689
ISBN 0-12-781260-1

PRINTED IN THE UNITED STATES OF AMERICA

Contents

LIST OF CONTRIBUTORS	ix
PREFACE	xi

1 Perspectives on Drugs and the Cell Cycle
Ivan L. Cameron

I. Chemotherapy and the Cell Cycle	1
II. The Cell Cycle as a Sensitive Indicator for Drug Analysis	8
References	10

2 Macromolecular Assembly Inhibitors and Their Action on the Cell Cycle
Edwin W. Taylor

I. Introduction	11
II. Colchicine	13
III. Vinblastine	17
IV. Cytochalasin and Microfilaments	20
References	21

3 The Effect of Mercuric Compounds on Dividing Cells
Jack D. Thrasher

I. Introduction	25
II. Mercaptide Formation by Mercurials	26
III. Production of c-Mitotic Figures by Organomercurials	28
IV. Mercuric Compounds and the Generation Time, DNA and Protein Synthesis, the Cell Cycle, and Cilia Regeneration in *Tetrahymena pyriformis*	31

v

V. Distribution of ³H-Methylmercuric Chloride in Fertilized Eggs of
   Lytechinus pictus 40
VI. Observations on the Structural Organization of *Tetrahymena
   pyriformis* Treated with Mercuric Chloride 41
VII. Concluding Remarks 46
   References 46

4 Adrenergic Drugs on the Cell Cycle

E. R. Jakoi and G. M. Padilla

I. Introduction 49
II. Experimental Results 50
III. Discussion and Summary 61
   References 64

5 Action of Narcotic and Hallucinogenic Agents on the Cell Cycle

Arthur M. Zimmerman and Daniel K. McClean

I. Introduction 67
II. The Chemistry of Selected Drugs 69
III. Action of Drugs on Prokaryotic Cells 70
IV. Action of Drugs on Eukaryotic Cells 76
V. Present Studies on *Tetrahymena* 82
VI. Concluding Remarks 90
   References 91

6 Effects of Drugs on Hepatic Cell Proliferation

J. W. Grisham

Abbreviations 95
I. Introduction 96
II. Kinetics of Hepatic Cell Proliferation 97
III. Regulation of Hepatic Cell Proliferation 110
IV. Drugs and Hepatic Cell Proliferation 111
V. Concluding Remarks 126
   References 127

7 Effects of Mitogens on the Mitotic Cycle: A Biochemical Evaluation of Lymphocyte Activation

Herbert L. Cooper

I. Introduction 138
II. Background: Peripheral Blood Lymphocytes 138

III. Nature of Mitogenic Substances	139
IV. General Characteristics of Mitogen-Induced Lymphocyte Growth	145
V. Biochemical Events in Lymphocyte Transformation	148
VI. Discussion and Conclusions	177
References	181

8 Intestinal Cytodynamics: Adductions from Drug Radiation Studies

Ronald F. Hagemann and S. Lesher

I. Introduction	195
II. Nature of the Crypt Progenitor Cell	196
III. Factors Influencing Intestinal Tolerance to Cytotoxins	200
IV. Control Aspects of Intestinal Cell Renewal	205
V. Summary	213
References	215

9 The Effects of Antitumor Drugs on the Cell Cycle

Joseph Hoffman and Joseph Post

I. Introduction	219
II. Antitumor Drugs	221
III. General Discussion	239
IV. Summary	242
References	242

10 Effects of Purines, Pyrimidines, Nucleosides, and Chemically Related Compounds on the Cell Cycle

Glynn P. Wheeler and Linda Simpson-Herren

I. Introduction	250
II. Biochemical Mechanisms of Agents	252
III. Effect of Agents on Cells	255
IV. Correlation of Biochemical Effects and Effects on the Cell Cycle	263
V. Examples of the Logic and Use of These Agents for Cancer Chemotherapy	287
VI. Concluding Comments	292
References	293

AUTHOR INDEX	307
SUBJECT INDEX	324

List of Contributors

Numbers in parentheses indicate the pages on which the authors' contributions begin.

IVAN L. CAMERON (1), Department of Anatomy, University of Texas Medical School at San Antonio, San Antonio, Texas

HERBERT L. COOPER (137), Cell Biology Section, Laboratory of Biochemistry, National Institute of Dental Research, National Institutes of Health, Bethesda, Maryland

J. W. GRISHAM (95), Department of Pathology, Washington University School of Medicine, St. Louis, Missouri

RONALD F. HAGEMANN (195), Cell and Radiation Biology Laboratories, Department of Radiology, Allegheny General Hospital, Pittsburg, Pennsylvania

JOSEPH HOFFMAN (219), New York University School of Medicine, New York University Research Service, Goldwater Memorial Hospital, Welfare Island, New York, New York

E. R. JAKOI (49), Department of Physiology and Pharmacology, Duke University Medical Center, Durham, North Carolina

S. LESHER (195), Cell and Radiation Biology Laboratories, Department of Radiology, Allegheny General Hospital, Pittsburgh, Pennsylvania

DANIEL K. MCCLEAN (67),* Department of Zoology, University of Toronto, Toronto, Canada

---

* Present address: Department of Biochemistry, McMaster University Medical Center, Hamilton, Ontario, Canada.

G. M. PADILLA (49), Department of Physiology and Pharmacology, Duke University Medical Center, Durham, North Carolina

JOSEPH POST (219), New York University School of Medicine, New York University Research Service, Goldwater Memorial Hospital, Welfare Island, New York, New York

LINDA SIMPSON-HERREN (249), Biochemistry Department, Southern Research Institute, Birmingham, Alabama

EDWIN W. TAYLOR (11), Department of Biophysics, University of Chicago, Chicago, Illinois

JACK D. THRASHER (25),* Department of Anatomy, School of Medicine, University of California, Los Angeles, California

GLYNN P. WHEELER (249), Biochemistry Department, Southern Research Institute, Birmingham, Alabama

ARTHUR M. ZIMMERMAN (67), Department of Zoology, University of Toronto, Toronto, Canada

* Present address: 16610 Hart Street, Van Nuyar, California 91406.

Preface

In 1876 Tyndall reported that a species of penicillin exhibited action antagonistic to bacterial growth. It was not until 1940, however, that a preparation of solid penicillin was placed into the therapeutic arsenal of the medical world.

In this book an attempt is made to introduce fundamental principles and studies on the mechanisms of drug action on proliferating cells in an effort to reduce the time lag between observation and practical application. The subject matter reviewed will be of interest to investigators in many disciplines, particularly to physiologists, pharmacologists, and oncologists, as well as to those working in cellular, developmental, and molecular biology. This work should serve to bridge the gap between experimental laboratory observations and their potential relevance to mankind.

This volume is comprised of chapters dealing with plant alkaloids, alkylating agents, mercurials, adrenergic agents, radiomimetics, narcotics, hallucinogens, mitogens, hepatotoxins, antibiotics, and antimetabolites of various types. The drugs used in cancer chemotherapy are given special emphasis. Bacteria, protozoa, sea urchin, and mammalian cell systems are discussed. The mammalian cell studies deal with both *in vitro* and *in vivo* cell systems.

A great deal of information and current concepts are summarized in this book, and it is hoped that it will act as a stimulus for new research.

A. M. ZIMMERMAN
G. M. PADILLA
I. L. CAMERON

1

Perspectives on Drugs and the Cell Cycle

IVAN L. CAMERON

I. Chemotherapy and the Cell Cycle 1
II. The Cell Cycle as a Sensitive Indicator for Drug Analysis 8
    References 10

A dictionary defines drugs as chemical substances that are given to people and animals as medicine. Certainly most of the chemical agents listed and discussed in this book have been tried as potential medicines and may qualify as drugs on this basis alone. This dictionary definition of drugs is inadequate for our discussion because it limits use of the term drugs to medicinal chemicals. The definition ignores the fact that many chemical agents commonly referred to as drugs are exceedingly useful tools and probes for working out metabolic pathways and cellular processes even though they have little or no direct medicinal value. We, therefore, choose to use the word drug in a broader sense.

Several of the chapters in this book bring together new and diversely scattered information about the action of cancer chemotherapeutic drugs on the cell cycle of normal and tumor cell populations. Other contributions illustrate how information on the effects of drugs on the cell cycle can give a better understanding of the sequence of events taking place in the cell cycle. Concurrently, these same cell cycle studies are giving a much better understanding of how a particular drug works.

I. Chemotherapy and the Cell Cycle

Much of what is presented in this book can be related to the concepts of chemotherapy. These concepts were first applied to antimicrobial agents by

Paul Ehrlich, who is recognized as the father of chemotherapy (Franklin and Snow, 1971). During the decade following 1902, Ehrlich established most of the concepts and principles from which subsequent work on chemotherapy has evolved. Although the principles of chemotherapy were well established by Ehrlich, the field had only limited practical success until the introduction of antibiotics in the late 1930's and early 1940's. It is really only since 1959, and the rapid accumulation of information on cell proliferation kinetics, that rational cancer chemotherapy has developed from an art to a science (for general reference, see Elkerbout et al., 1971).

It has been an objective of chemotherapy to describe the mode of action of a specific drug as related to its biological effects on sensitive cells and to describe in molecular detail the interaction between the inhibitor (drug) and its target or receptor within the cell. This principle was stated by Ehrlich himself in 1909: "In order to pursue chemotherapy we must look for substances which possess a high affinity and high lethal potency in relation to the parasites, as selectively as possible. In other words, we must learn to aim and to aim in a chemical sense." (The term parasite in this quotation may refer to viruses, to bacteria, to fungi, or to cancer cells.)

Using these principles, Ehrlich himself had some success in development of chemotherapeutic treatments. Among his therapeutic contributions was the synthesis of several organoarsenical compounds, one of which (Neosalvarsan) was used as the main treatment of syphilis until penicillin was produced. Another drug coming from Ehrlich's work was suramin, produced from trypan red, and used in the treatment of trypanosomiasis. His work also led to the use of a product of methylene blue called mepacrine (atabrine or quinacrine) which has antimalarial value.

The influence of chemotherapy had its most striking manifestations in the antibiotic revolution, which led to the development of penicillin and other antibiotic drugs such as those developed by the soil microbiologist Waksman in the 1940's. The influence of the antimicrobical drugs can be appreciated by an analysis of Table I. Here we see the 10 leading causes of death in the United States in the years 1900 and 1959. It can be seen that bacterial infections are involved in six of the ten main causes of death in 1900. After the development and application of antibiotics, only pneumonia remained as a bacterial infection among the ten leading causes of death.

Table II lists the mode and site of action of some common antibiotics. These antibiotics are isolated as substances elaborated by various microorganisms and, in low concentration, inhibit the growth of other cells or microorganisms. A perusal of the table will indicate that the various agents have rather specific sites of action in cells. Perhaps the best known of the antibiotics on this list are the penicillins. Penicillins interfere with cell wall

TABLE I
THE TEN LEADING CAUSES OF DEATH IN THE UNITED STATES IN 1900 AND 1959[a]

Rank (1900)	Cause of death	Percent of deaths from all causes	Rank (1959)	Cause of death	Percent of deaths from all causes
1	Pneumonia and influenza	11.8	1	Diseases of the heart	38.6
2	Tuberculosis	11.3	2	Cancer and other malignancies	15.7
3	Diarrhea and enteritis	8.3	3	Cerebral hemorrhage	11.5
4	Diseases of the heart	8.0	4	Accidents	5.4
5	Cerebral hemorrhage	6.2	5	Certain diseases of early infancy	4.1
6	Nephritis	5.2	6	Pneumonia and influenza	3.5
7	Accidents	4.2	7	General arteriosclerosis	2.1
8	Cancer	3.7	8	Diabetes mellitus	1.7
9	Diphtheria	2.3	9	Congenital malformations	1.3
10	Meningitis	2.0	10	Cirrhosis of liver	1.2

[a] Modified after Strehler (1962).

TABLE II
MODE AND SITE OF ACTION OF SOME ANTIBIOTICS

Antibiotic	Action
Penicillins and cycloserine	Interferes with cell wall (murein) synthesis
Azaserine and DON	Blocks *de novo* synthesis of purine nucleotides
Mitomycin	Cross-linking of DNA strands
Actinomycin D	Suppress DNA-dependent RNA synthesis probably by binding double-stranded DNA or by intercalating into DNA
Rifamycin	Inhibits bacterial RNA polymerase
Puromycin	Prematurely terminates growing peptide chain on 70 S or 80 S ribosome
Streptomycin	Inhibition of initiation complex formation and transfer RNA-ribosome interaction works specifically on 70 S ribosomes
Tetracyclines	Inhibits binding of aminoacyl-tRNA to acceptor site on both 70 S and 80 S ribosomes
Chloramphenicol and erythromycin	Inhibitors of peptide bond formation and translocation on 70 S ribosomes
Cycloheximide	Specifically inhibits function of 80 S ribosome
Sulfanilamide	Interferes with folic acid synthesis in bacteria; mimics *p*-aminobenzoic acid
Methotrexate	Inhibits folic reductase
Antimycin	Blocks respiratory chain immediately before cytochrome c_1
Oligomycin	Interferes with oxidative phosphorylation

production in bacteria. This inhibition produces a cytostatic effect on growing bacteria.

A good example of the relationship between the action of the drugs and the cell cycle is illustrated by the action that penicillin has on growing bacteria. It is thought that the sole bacterial chromosome replicates during the cell cycle and that the two resulting chromosomes are attached in some way to the cell membrane. The newly replicated bacterial chromosomes are then distributed to daughter cells by means of growth of the cell membrane between the points of attachment of the chromosomes. This separates physically the two bacterial chromosomes and starts the next cell cycle. Penicillin interferes with the growth of the cell wall material and, therefore, interferes with membrane expansion and the separation of the bacterial chromosomes. The drug does not actually kill the cell directly but simply interferes with the formation of the cell wall and cell reproduction.

Animal cells, which do not possess cell walls, are not affected by this drug. This difference between bacteria and animal cells is the basis of the selectivity of penicillin action. Likewise, bacteria in a spore state are not affected by the drug because no new cell wall synthesis is occurring. On the other hand, those bacteria in a growth state will have a weakened cell wall which can then be attacked by cytocidal agents, such as phenol or hypotonic solutions. A weakened cell wall cannot, for instance, keep the cell from swelling and rupturing in hypotonic solutions.

Among the list of compounds in Table II are other antibiotics that possess specificity of action. For example, rifamycin inhibits specifically bacterial RNA polymerase. Streptomycin, chloramphenicol, and erythromycin act to inhibit protein synthesis which specifically involves 70 S ribosomes. Eukaryotic cells having 80 S ribosomes in their cytoplasm are generally not adversely affected by these antibiotics. On the other hand, cycloheximide specifically inhibits the function of 80 S ribosomes in eukaryotic cells. It seems important to mention that mitochondria and chloroplasts of eukaryotic cells also contain 70 S ribosomes, whereas the rest of the eukaryotic cell has 80 S ribosomes. This accounts for the fact that it is possible to selectively inhibit chloroplast reproduction and, therefore, to bleach the chloroplasts from the eukaryotic cell, *Euglena,* by streptomycin treatment. However, *Euglena* is still able to grow and reproduce if additional nutrient supplements are added to the growth media. For similar reasons we may be able to account for the failure of lymphoid cells to produce antibodies when the cells have been treated with chloramphenicol and to explain the observed deficiency of cytochrome c reductase in rat heart cells cultured with chloramphenicol. Presumably, in these latter cases, the drug is preferentially acting on the 70 S ribosomes of mitochondria. Thus, it appears that treatment of bacterial infections with some inhibitors of protein synthesis may depend on the specificity of attack on the 70 S ribosome of bacteria, leaving the 80 S predominant ribosome of the host organism unaffected. In cases of those protein-inhibiting drugs that affect both the 70 S and 80 S ribosomes, such as puromycin and the tetracyclines, a differential permeability of the drugs into the bacteria may account for the specificity of inhibitory action.

Clearly, then, chemotherapy takes advantage of the chemical differences existing between the parasite and the host. Such differences are readily apparent between bacteria and the animal cell, but what are the differences between normal animal cells and cancer cells? Unfortunately, few differences are now known (Elkerbout *et al.,* 1971). Introduction of the tools needed for the study of cell proliferation kinetics *in vivo* and *in vitro* have led us to realize that there are at least some small differences in the cytokinetics of some cancer cell populations in comparison to the cytokinetics in the normal cell populations of the hosts. These cytokinetic studies have led to the recognition that a few specific types of cancer are rapidly proliferating and that most of the cancer cells are in the proliferative state (have a high growth fraction). In man, these rapidly proliferating cancers include choriocarcinoma, Burkett's tumor, Hodgkin's disease, acute lymphocytic leukemia, and Wilm's tumor. A list of some drugs used in cancer chemotherapy is given in Table III. A detailed review of action of some of these drugs can be found in the chapter by Wheeler and Simpson-Herren (this volume). These drugs clearly have selective effectiveness for proliferating cells. Table III

TABLE III

THE ACTION AND CELL CYCLE PHASE SPECIFICITY OF SOME CANCER CHEMOTHERAPEUTIC DRUGS[a]

Drug	Action and end product affected	Cell cycle phase specificity
Cytosine arabinoside	Inhibition of nucleotide reductase (DNA)	S-phase specific
Hydroxyurea	Inhibition of nucleotide reductase (DNA)	S-phase specific
Guanozole	Inhibition of nucleotide reductase (DNA)	S-phase specific
Methotrexate	Inhibits folic reductase (DNA, RNA, protein)	S-phase specific but self-limiting[b]
6-Mercaptopurine	Inhibits PRPP → phosphoribosylamine (DNA, RNA)	S-phase specific but self-limiting[b]
5-Fluorouracil	Inhibits thymidylate synthetase, incorporates into RNA (DNA, RNA)	S-phase specific but self-limiting[b]
Cyclophosphamide and 1,3-bis(2-chlorethyl)-1-nitrosourea (BCNU)	Cross-links DNA strands	Specific for proliferating cells, especially those cells lacking DNA repair enzymes
Vinblastine and vincristine	Inhibits assembly of microtubular proteins into microtubules	Mitosis
Cytochalasin B	Inhibits function of microfilaments	Cytokinesis
Actinomycin D	Inhibits DNA dependent RNA synthesis	Not considered cell cycle specific but may have some selectivity for proliferating cells

[a] Modified after Skipper et al. (1970).
[b] These drugs tend to retard proliferating cells not in S phase at the time of drug application from progressing to S phase.

suggests that an optimal schedule for S-phase specific drugs would have to reach effective serum levels of the drug at intervals of just less than the S-phase duration if all the proliferating tumor cells are to be affected. Skipper *et al.* (1970) have indeed proved this scheduling to be the most effective for the S-phase specific drugs; thus, as these authors point out, optimal drug scheduling can make the difference between failure and success. It is to the credit of modern cancer chemotherapy that the five types of rapidly proliferating human cancer mentioned above are now successfully managed, if not cured.

It is a paradox that among people with cancer those with the five specific types of cancer mentioned above formerly had the worst prognosis for long-term survival, but now they have the best prognosis. It is quite evident that improved quantitation and better fundamental information concerning the pharmacology, the toxicology, the cell-proliferation kinetics, and the cellular response of chemotherapeutic agents will continue to aid clinicians in planning therapeutic regimens.

Just as the drugs listed in Table III inhibit proliferating cells of cancer cell populations they also play havoc with proliferating cell populations of the host tissues. These proliferating cell populations include those of the bone marrow, the lymphatic tissues, the linings of the gastrointestinal tract, the epidermis, and other rapidly proliferating cell systems within the body. Continual use of these chemotherapeutic agents can be expected to lead to such cytostatic side effects as would be predicted by interference with cell reproduction in such cell populations. Some of the most adverse effects are brought about by interfering with megakaryocyte proliferation, which causes loss of blood platelets leading to hemorrhage and hemophilia. The loss of production of red blood cells leads to anemia. Interference with lymphocyte, granulocyte, and plasmocyte production causes suppression of the body's defense mechanisms, which normally operate against the spread of invading microorganisms, and also causes the suppression of the body's immunological system.

These drugs cause loss of the linings of the gastrointestinal tract, resulting in stomatitis, diarrhea, and vomiting. One can also expect that inhibition of cell proliferation in the epidermis will lead to the dermatitis, loss of hair, etc. In males sperm production is disturbed; in women, amenorrhea may result. The rapid rates of cell proliferation in embryonic cell populations make the embryo a prime target for drug action. The result of the use of such chemotherapeutic drugs during pregnancy is teratogenic deformations and embryonic death.

It is, therefore, clear that if we are to use these highly toxic agents we must know and exploit the subtle differences that exist between normal and tumor cytokinetics. Some of the contributions to this book give new and useful information on the action of cancer chemotherapeutic drugs on the mammalian cell cycle of normal and tumor cell populations (see the chapters by Hagemann and Lesher and by Hoffman and Post in this volume).

Looking at the problem of cancer chemotherapy from the cytokinetics point of view leads one to the conclusion that the slow-growing tumor, which contains cells with long cell cycle times and, with many cells no longer in the cell cycle at all, will be difficult or impossible to manage or cure.

We should not, however, be detracted from trying to find other possible differences between normal and neoplastic cells, which can be exploited as a basis for selectivity of drug action. Some of these differences may include (1) sensitization of breast and prostatic cancer by particular sex hormones; (2) the selective accumulation of cytotoxic agents by target tissues, such as occurs in the case of radioiodine in the thyroid; (3) the selective activation of drugs by target tissues, such as the activation of cyclophosphamide by those tissues rich in phosphamidase; (4) taking advantage of the particular differences in metabolism, such as starving some of the asparagine requiring leukemic cancers by asparaginase treatment to remove asparagine from the serum; (5) the use of antimetabolites against specific metabolic requirements; (6) the continued development of combination therapy, which overcomes the problem of the tumor cells developing resistance to the action of one drug; or (7) the introduction of carrier molecules onto carcinostatic agents to increase the permeability and effectiveness of the drug, as in the case of uracil mustards. Of course it also seems probable that some of the tumors caused by viral infections will be subject to direct chemotherapeutic attack or to treatment by established immunization techniques.

More studies need to be conducted to determine means for "priming," "setting up," or "synchronization" of tumor cells *in vivo* for chemotherapeutic attack. Such possibilities include starvation and refeeding experiments, use and enforcement of normal diurnal rhythms of drug susceptability, and surgical removal of a large tumor mass, which would stimulate increased cell proliferation of the tumor cells in metastatic sites. Establishing the existence and isolation of tissue or tumor specific factors, such as the tissue specific chalones, or perhaps introducing specific mitogenic materials to stimulate specific types of tumors may also prove rewarding in priming tumor cells for drug therapy. For a further discussion of the nature and action of such tissue specific mitogenic agents the reader is referred to the chapter by Cooper in this volume.

II. The Cell Cycle as a Sensitive Indicator for Drug Analysis

In the United States we are faced daily with the question of the possible dangerous effects of narcotics and hallucinogenic drugs. We hear politicians speak about the need to establish enlightened legislation to enable further research concerning the effects of such drugs. The chapter by Zimmerman and McClean in this volume reviews the known information on mechanism of action of some of the narcotic and hallucinogenic drugs on pro- and eukaryotic cells. This chapter also illustrates the advantages of using synchronized populations of the ciliated protozoan *Tetrahymena pyriformis* for obtaining new information on these drugs. In another chapter, Padilla and Raff explain that varying the physiological conditions of *Tetrahymena pyriformis* can influence and clarify the action of various adrenergenic drugs.

With the burgeoning list of chemicals and drugs that have been recently introduced, one may be justifiably concerned in seeking means to establish rapidly what effects such agents have on biological systems. The chapter by Thrasher attests to the importance and ease of obtaining valuable information on mechanisms of action that such toxic agents as mercury and the organomercurials have on dividing cells such as *Tetrahymena pyriformis* and sea urchin eggs. His data lead to the conclusion that the organomercurials are interacting specifically with the SH groups of soluble microtubular proteins. This, in turn, causes interference with the assembly of the soluble microtubular proteins into microtubules and therefore interferes with those processes that require formation of microtubules. In *Tetrahymena,* as in mammalian cells, such microtubular-requiring events include division of the nuclear materials, as well as the formation and regeneration of cilia. The findings from *Tetrahymena s*hould therefore be applicable to the microtubule systems of mammalian cells. In a related chapter, Taylor demonstrates that other drugs also interfere with the assembly of microtubules. These include the plant alkaloids: colchicine, vinblastine, and vincristine. It is of some interest that the organomercurials apparently are effective at the same or similar sites as are the plant alkaloids; however, the organomercurials are more potent in their action than the plant alkaloids by several orders of magnitude.

Another system that has been used extensively to study the action of drugs on the cell cycle is the regenerating liver. The liver cell populations can be induced to move from a nonproliferative to a proliferative state by partial hepatectomy. The regenerating liver has proved a useful cell system to study the normal biochemical events that occur in the cell cycle. The liver is also implicated as the organ which is most important in metabolizing and detoxifying the various drugs administered to animals and man. The large and fragmented mass of information on this subject is collected and synthesized in the chapter by Grisham.

A number of the contributors to this book use *Tetrahymena pyriformis* for analyzing the action of drugs on the cell cycle. Much is already known about *Tetrahymena* (Hill, 1972), and this species has already contributed greatly to our knowledge of basic biology and drug action. Among the advantages of the use of this cell are its large size and rapid rate of reproduction, its ability to be cultured in a complex but completely defined medium (*Tetrahymena* has very nearly the same nutrient growth requirements as do mammalian cells), its ease of culture in large numbers, and its ability to be synchronized by a number of different approaches. Much is already known about its cell cycle characteristics. As demonstrated in this book, it continues to prove a useful pharmacological and toxicological tool.

The editors of this book hope to bring together basic information obtained from studies of the effects of drugs on simple cell systems and information on drugs being used in clinical trials. This common awareness of the basic and applied aspects should be of profit both to the basic scientist and to the clinician and can reduce the lag time between discovery of fundamental facts and their application. For instance, it is a fact that there was a 10- to 12-year lag between the time Flemming reported on the action of penicillin and the production of penicillin for human use. Of perhaps even more interest or embarrassment is the fact that in 1876 Tyndall described the antagonistic action of a species of *Penicillium* on bacterial growth (Gabriel and Fogel, 1955). Let us hope that better communications can reduce such lag times in the future. Let us also hope that those who work with the simple cell systems will not be ignored when it comes time for support of their work.

Thus, it can be said that as the advancement of biochemical pathways and molecular biology have enabled us to come closer to our goal of understanding the action of drugs, we are now convinced that cell cycle studies will continue to play a key role in evaluating the action of drugs of all types.

References

Elkerbout, F., Thomas, P., and Zwaveling, A., eds. (1971). "Cancer Chemotherapy." Williams & Wilkins, Baltimore, Maryland.

Franklin, T. J., and Snow, G. A. (1971). "Biochemistry of Antimicrobial Action." Academic Press, New York.

Gabriel, M. L., and Fogel, S., eds. (1955). "Great Experiments in Biology." Prentice-Hall, Englewood Cliffs, New Jersey.

Hill, D. L. (1972). "The Biochemistry and Physiology of *Tetrahymena*." Academic Press, New York.

Skipper, H. E., Schabel, F. M., Jr., Mellett, L. B., Montgomery, J. A., Wilkoff, L. J., Lloyd, H. H., and Brockman, R. W. (1970). Implications of biochemical, cytokinetic, pharmacologic, and toxicologic relationships in the design of optimal therapeutic schedules. *Cancer Chemother. Rep.* **54**, 431–450.

Strehler, B. L. (1962). "Time, Cells, and Aging." Academic Press, New York.

2

Macromolecular Assembly Inhibitors and Their Action on the Cell Cycle

EDWIN W. TAYLOR

I. Introduction	11
II. Colchicine	13
III. Vinblastine	17
IV. Cytochalasin and Microfilaments	20
References	21

I. Introduction

For the present discussion, a drug is defined as a compound that specifically blocks a cellular process. Drug inhibition studies have been widely used in elucidating cellular processes, and it is worth asking whether or not there are any general conclusions to be drawn from drug studies. From the molecular biologist's standpoint, the occurrence of specific inhibitors is not surprising. A lesson which has been learned from studies of cellular mechanisms is that weak interactions between protein molecules and small molecules or other macromolecules are important. By weak interactions, one is referring to electrostatic forces, van der Waals forces, or hydrogen bonds. Summation of these interactions leads to the high specificity of protein-binding sites. A molecule that interacts with the site because its structure mimics a naturally occurring cellular constituent will be able to interfere with a particular process. If the inhibitor is also able to enter into a chemical reaction leading to formation of a covalent bond, so much the better, but this is not necessary. The action of most drugs is reversible. The compound can be washed out of the cell, which may then recover from the effect of the inhibition.

Since most processes depend, at some point, on specific protein interactions, it is possible, in principle, to find a specific drug to block any cellular process. In practice, if the reaction is an enzyme process it is often possible to synthesize a substrate analog that is bound to the enzyme site but does not undergo the catalytic process. In other cases one would need to know the three-dimensional structure of the interaction site in order to design the drug. In general this information is not available, and one has to rely on the substances produced by plants or animals, which have presumably been designed by the trial and error process of natural selection. A large number of compounds have been isolated, particularly from molds, that can block various cellular processes with high specificity (for review, see Beard *et al.*, 1969).

The following list of specific inhibitors is included to give the reader some idea of the variety of ways in which cellular mechanisms can be affected. For example, by binding to DNA, actinomycin D blocks transcription. Rifampicin acts to block initiation by binding to RNA polymerase. Streptomycin acts by affecting the function and specificity of ribosomes; puromycin mediates its action by terminating peptide chain growth. Colicin exerts its action through inhibition of oxidative phosphorylation. Valinomycin increases the permeability of membranes to monovalent ions. Tetrodotoxin acts to block the nerve impulse by binding to the sodium channel. Bungarotoxin acts by blocking receptors involved in synaptic transmission.

The cell cycle can be inhibited by a variety of drugs, but this is in itself a trivial statement. To continue to grow and divide, the cell must carry out the complete spectrum of reactions that constitute the cell's metabolism. Compounds that inhibit the synthesis of macromolecules or necessary small molecules, including ATP, will obviously inhibit the cell cycle and have been discussed in detail by many reviewers.

In this chapter we are concerned with compounds that appear to affect specifically the cell's ability to carry out the steps in mitosis without affecting directly macromolecular synthesis or energy metabolism. At present, only a small group of such compounds are known, but the potential number of mitotic inhibitors is large. Proceeding on the assumption that any specific process can be inhibited, we need only remember that mitosis is a complex process, certainly as complex as the assembly of a T2 phage, which appears to require the participation of about 50 gene products. In comparison with our knowledge of phage assembly, our ignorance of the mitotic process is nearly complete.

Mitosis could be blocked at chromosome condensation, nuclear membrane breakdown, spindle assembly, microtubule–kinetochore interaction, chromosome splitting, anaphase movement, spindle disassembly, or cleavage. We have simply listed the steps one can see by looking in a microscope. Several distinct processes may be required to make each of these steps possible, so each of the above could be blocked in different ways by different drugs.

What kinds of inhibitors should one look for? A specific drug can be extremely useful in solving the problem of the mechanism of a complex process. If compound X could be found, which specifically blocked attachment of chromosomes to the spindle by binding to the kinetochore, we would be in a position to begin to dissect this problem.

A number of compounds are known that appear to block mitosis without directly affecting metabolism, i.e., colchicine, podophyllotoxin, griseofulvin, vinblastine, vincristine, and cytochalasin. The remainder of the chapter will be concerned largely with colchicine, vinblastine, and cytochalasin. We shall develop the thesis that there are a group of drugs that act by blocking the assembly of protein subunits into cellular structures. Such drugs could be classified as macromolecular assembly inhibitors.

II. Colchicine

Colchicine is obtained from the autumn crocus *(Colchicum autumnale)*. It has a long medical history through its use in the treatment of gout. Its antimitotic activity was described by Dustin and collaborators (Eigsti and Dustin, 1955). When added to dividing cells it produces a striking effect: the accumulation of cells in mitosis. The metaphase plate configuration of the chromosomes is lost and a well-defined spindle is absent. The effect of colchicine on spindles was studied by Inoué, 1952, who clearly demonstrated loss of spindle structure by birefringence studies. Colchicine has been used extensively to block cells in mitosis for chromosome studies. From these studies one may draw an interesting conclusion. An extremely wide variety of plant and animal cells are blocked in essentially the same way. Thus the target for colchicine must have a similar structure in all cells. With the wisdom of hindsight, a reasonable guess on the mode of action might have been made. A likely candidate for a mitotic inhibitor is a molecule that binds specifically to the protein subunit of spindle fibers and blocks assembly of the spindle. Furthermore, one might expect to find a class of drugs that act by preventing the assembly of globular proteins into intracellular fibers.

The list of intracellular fibrous proteins is probably not complete but presently includes microtubules, actin filaments (5–7 nm), microfilaments

(4–5 nm), neurofilaments (8–10 nm), tonofibrils (variously reported diameters 5–10 nm), and "10 nm filaments." Of this group, colchicine appears to affect only microtubules. Specific inhibitors of assembly of other filaments have not yet been found with the possible exception of microfilaments (to be discussed below). Nevertheless, further compounds appear to be worth seeking. Blocking of assembly of protein subunits could also affect multisubunit enzymes, lipoproteins (Pollard et al., 1969), virus coat proteins, and even membranes.

Since colchicine binding appears to be the best worked out case of drug action by blocking subunit assembly, we would like to consider the evidence for the mechanism of action of colchicine (which is still incomplete).

At a concentration that blocks mitosis (10^{-6}–10^{-7} M) colchicine does not affect cell metabolism or the rate of arrival of cells at mitosis (Puck and Sheffer, 1963; Taylor, 1965). The latter assay should be quite sensitive to interference with a metabolic process. At high concentrations and over longer time intervals, secondary effects are to be expected (cf. Kuzmich and Zimmerman, 1972). The compound is bound reversibly to a cellular constituent and is not metabolized (Borisy and Taylor, 1967b). Colchicine forms a complex with a soluble protein of the cell *in vivo* or *in vitro*. There is, however, some binding to a vesicle fraction, particularly in brain tissue. It has recently been suggested that there is a binding site on a preparation of nerve endings (synaptosomes) (Feit *et al.*, 1971b).

A number of lines of evidence show that the protein in question is a subunit of microtubules (tubulin) and that tubulin has a single high affinity site for colchicine. By employing ^{3}H-colchicine, a relatively simple assay was developed for colchicine-binding activity. The binding activity was found to correlate in a general way with mitotic activity or with the presence of microtubules (Shelanski and Taylor, 1967; Weisenberg *et al.*, 1968). An essentially pure colchicine-binding protein was obtained by selective dissolution of the central pair of microtubules of sperm tails (Shelanski and Taylor, 1968). The protein was also found in the mitotic apparatus (Borisy and Taylor, 1967b). A protein was subsequently purified from brain using colchicine binding as an assay, which has essentially the same properties as the protein obtained from cilia and flagella tubules (Renaud *et al.*, 1968; Shelanski and Taylor, 1968; Stephens, 1968; Weisenberg *et al.*, 1968). Thus, by a somewhat roundabout procedure, the identity of the colchicine-binding protein with tubulin was established. A similar protein has since been isolated in a number of laboratories using the colchicine-binding assay.

In the earlier studies it was concluded that tubulin is a dimer of molecular weight 110,000, which dissociates into 55,000 molecular weight subunits after reduction in guanidine hydrochloride or sodium dodecyl sulfate

(SDS). The protein has one colchicine-binding site and two GTP-binding sites per dimer unit, but GTP exchange occurs at only one of the sites (Weisenberg et al., 1968). More recent studies (Bryan and Wilson, 1971; Feit et al., 1971a; Olmsted et al., 1971) have established that the subunits are not identical. It is not yet clear whether or not only one subunit has a colchicine-binding site, since the alternate explanation that one site is covered in a head-to-tail dimer has not been ruled out.

Do the chemical studies explain the action of colchicine on mitosis and other microtubule systems? It must be admitted that only a partial, and not entirely satisfactory, explanation can be given. Not all microtubule systems are susceptible to colchicine; the flagellum is a notable example. Although the outer doublet tubules are more stable than single tubules, even the central pair of tubules appears to be unaffected by colchicine, and flagella movement is not inhibited. However, if flagella are amputated they do not regrow in the presence of colchicine (Rosenbaum et al., 1969).

Cells that enter mitosis in the presence of colchicine do not form a spindle, but spindles already present appear to require a high drug concentration to be broken down (Taylor, 1965). The isolated mitotic apparatus is stable on addition of colchicine, though here the problem is presumably an alteration of the protein following isolation, which makes the system relatively insoluble.

There is no conclusive evidence that colchicine binds to stable doublet tubules. Single microtubules are soluble under conditions of pH and ionic strength that allow colchicine binding. They are marginally stable in organic solvent mixtures such as hexylene glycol, which inhibits colchicine binding. Consequently, there is no clear evidence that colchicine binds to a microtubule, only that it binds to tubulin.

We are forced to adopt the negative explanation that colchicine blocks polymerization by binding to the subunit. Once the microtubule is formed it may be insensitive to colchicine or it may not bind colchicine. Stable microtubules, as in flagella outer doublets, axostyles, or even central pair tubule of flagella, may be unaffected because either the binding site is covered or binding does not affect the structure. Relatively unstable tubules, as in the mitotic apparatus, appear to be in a dynamic state, as suggested by cooling or D_2O treatment (Inoué and Sato, 1967; Marsland and Zimmerman, 1965). If a polymer-subunit equilibrium occurs, binding to the subunit would shift the equilibrium in the direction of depolymerization. This is an unsatisfactory explanation, since it is not supported by direct evidence, but it is the best one has to offer at the present time. A major difficulty has been the failure to polymerize subunits into tubules under physiological conditions *in vitro,* since such a reaction should be blocked by colchicine. Some progress has been made by G. G. Borisy (personal communication, 1972), who has demonstrated aggregation of tubulin, in a reaction that requires GTP and is inhibited by colchicine. However, very few microtubules have been formed in this aggregation process. The material consists largely of asymmetric aggregates of undefined structure.*

If we assume provisionally that the action of colchicine requires binding to the tubulin subunit, there are two obvious modes of action. The simplest mechanism is that there are a number (two to six) of regions on the dimer that interact with complementary regions of a second dimer to generate the microtubule. The contact region must be limited in extent, 1 nm would be a reasonable estimate for a protein monomer unit of 4 nm in diameter. If colchicine binds at one such contact site it could prevent or reduce greatly the interaction with a second subunit. Depending on the distribution of the remaining contact regions, colchicine would prevent polymerization or lead to the formation of a different structure. A second possibility is that colchicine acts as an allosteric modifier of protein reactivity. Binding at some site other than the contact region changes the configuration of the protein in a way that reduces the protein–protein interaction. The allosteric type of explanation becomes a possibility with the finding that dimer subunits are nonidentical, and only one need have a colchicine acceptor site. Very little evidence is available on this point, but it is not an unreasonable explanation. Although colchicine is fairly strongly bound (association constant is 10^5–10^6 M^{-1}), the rate constants for binding and for dissociation are quite low in comparison to most reactions involving specific binding of small molecules to proteins. The rate constant for binding is of the order of 10^3 M^{-1} sec^{-1}, which is 10 to 100 times smaller than expected for reactions of this type. It is possible that colchicine binds to and stabilizes a configuration of the protein, which is otherwise energetically unfavorable. Colchicine binding has a sharp pH optimum at 6.5, and circular dichroism of tubulin is markedly pH dependent in this range (Ventilla *et al.*, 1972), again suggesting that binding is dependent on configuration. If the binding of a small molecule at an effector site could alter protein configuration in such a way as to block tubule polymerization, it would be a useful cellular control mechanism. In that case the action of colchicine would derive from its structural resemblance to a molecule that is produced by the cell to control polymerization. At present the results are inconclusive, and positive evidence, such as an effect of colchicine on absorption spectrum, circular dichroism, or pH titration curve, has not been obtained.

If the detailed structure of a microtubule were known, this would help in determining the number of contact regions. At present the proposed struc-

---

* Assembly of tubulin has been obtained by R. Weissenberg; polymerization requires only the removal of Ca ion and is blocked by colchicine.

tures are speculative (Chasey, 1971; Cohen et al., 1971) and the question of dimer orientation is being debated (Bryan and Wilson, 1971; Feit et al., 1971a; Olmsted et al., 1971). In mitotic cells treated with colchicine the observation has been made that while microtubules decrease in numbers, there is an increase in 10 nm filaments. As the time course is unknown it does not follow that these filaments are breakdown products of tubules. They could equally well be an aggregation of subunits in a different configuration in the presence of colchicine. Tubules are often seen to fragment into protofilaments, which are 40–50 Å in diameter, although Olmsted et al. (1971) describe dissolution of outer doublets into pairs of protofilaments (which are thus 10 mm in diameter). The conditions in the latter instance are different and the result may not be comparable.

III. Vinblastine

A number of other compounds are known that have colchicine-like effects, such as podophyllotoxin and griseofulvin. The latter has been shown to compete for the colchicine-binding site, and it is possible that these compounds have a similar mode of action. A second group of antimitotic agents are the *Vinca* alkaloids, which have been used in cancer chemotherapy. Disordered metaphases are produced and the usual spindle structure is absent, and it might first be supposed that their mechanism of action is similar to colchicine. Although microtubule protein does appear to be the target, there are striking differences in the effects. It was shown (Weisenberg et al., 1968) that vinblastine does not compete for the colchicine-binding site, if anything it enhanced colchicine binding (Wilson and Friedkin, 1967) although this may be a secondary effect. An important step was made by Bensch and Malawista (1968) who observed that cells treated with the alkaloid contained crystalline inclusions, which might be described as an array of tubules. Vinblastine precipitates tubulin selectively and quantitatively (Marantz et al., 1969). The effect is not completely specific since some other proteins can also be precipitated (Wilson et al., 1970). This is not too surprising since vinblastine is charged and some salting-out effects might be expected. Nevertheless, some specific interaction is implied by the preferential precipitation of tubulin. Studies of the interaction of vinblastine and tubulin in solution by Weisenberg and Timasheff (1970) indicate that specific aggregates are formed. Whereas the sedimentation constant of the tubulin dimer is 6 S, with increasing vinblastine concentrations a 9 S component (probably a tetramer) and a 27 S component are formed. The formation of discrete aggregates, rather than a continuous spectrum, suggests that the tetramer and 27 S component have higher stability and are intermediate steps in the formation of the paracrystal.

Although these studies are not conclusive proof for a definite vinblastine-binding site, they are most easily interpreted in terms of specific binding. A direct demonstration of the stoichiometry of vinblastine interaction with tubulin has still not been performed.

Shelanski and collaborators (Bensch et al., 1969; Marantz and Shelanski, 1970) have shown that precipitates obtained from purified tubulin have a structure that is identical to the paracrystals formed in cells. Examination of negatively stained material has led them to conclude that the components of the array are not tubules, but a loose helix. The paracrystal is best described as a hexagonal array of bed springs. These studies provide an interesting example of the formation of a protein polymer of reasonably well-defined structure induced by the action of a small molecule, which, presumably, binds to the protein subunit. It would be surprising if similar reactions did not occur naturally within cells.

The antimitotic effect can be explained in the same general way as for colchicine. In this case, tubulin is removed from the soluble pool by formation of paracrystals, which are stable in the cell. Consequently spindles are unable to form because of lack of precursor. A further observation of Shelanski's is perhaps relevant. Colchicine is bound by paracrystals but GTP is released. The interaction site on the molecule, which leads to paracrystal formation, is different from site blocked by colchicine. The release of GTP might indicate competition for the GTP site or a change in configuration induced by vinblastine to a state that does not bind GTP.

The interpretation is still unclear since we do not have binding data. It could be argued that because vinblastine causes precipitation it simply acts as a cross-linking agent. Magnesium, which is also doubly charged, will precipitate tubulin; there is a correlation between vinblastine and divalent ion precipitation of a number of proteins (Wilson, et al., 1970). The helical structure generated by vinblastine precipitation may thus involve different interaction regions than those important in determining tubule structure. However, if both colchicine and vinblastine act according to the simplest mechanism, each blocks a different interaction site on tubulin, and the structures that form in the presence of each compound provide information on the geometry of subunit interaction.

The problem of microtubule structure will be treated elsewhere. Two plausible structures of the microtubule are shown in Fig. 1. The diagram illustrates the surface lattice (the lattice obtained by cutting the tubule parallel to the axis and unveiling the cylinder to form a flat sheet). The structure is based on the following considerations. Optical diffraction and reconstruction of electron micrographs (Chasey, 1971; Grimstone and Klug, 1966) are consistent with the dimer as the structural unit, with the dimer axis tilted

2. MACROMOLECULAR ASSEMBLY INHIBITORS

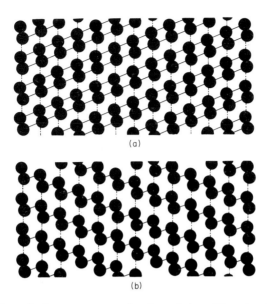

Fig. 1. Surface lattice structures of microtubules. The structures consist of dimer helices with 12 subunits per turn. Dimers are placed on the primitive surface lattice unit [shown as the box in (b)] deduced from X-ray diffraction (Cohen *et al.*, 1971). The two structures correspond to the two possible relative positions of nonidentical monomer subunits for the surface lattice cell. The primitive unit cell dimensions are 4 × 10 nm.

with respect to the tubule axis. The X-ray diffraction evidence (Cohen *et al.*, 1971) suggests a primitive unit cell [shown in Fig. 1(b)], which is rectangular of dimensions 3 × 2.5 nm and a translation per residue along the helix of one-half the dimension of the monomer unit. These dimensions, together with a tubule diameter of 24 nm, require the structure to be made up of three dimer helices. One complete turn of the primary helix is shown in the figure. We have assumed that there are 12 rather than 13 units per turn. Denoting the units of the dimer as x and y, the contacts between adjacent units could be x–x [Fig. 1(a)] or x–y [Fig. 1(b)]. The two structures are actually very similar. If the dimer axes were not tilted they would simply be mirror images corresponding right- and left-handed helices. We have indicated two types of contacts (vertical dot–dash and lateral solid lines). If an inhibitor blocks the vertical contacts, a polymer might still form and would appear as an approximately 10-nm filament, which could form a loose helix of pitch 24 nm. Such a helix would hardly be stable, although it might be favored by making side-to-side contacts in a paracrystal. An inhibitor which blocked the lateral contacts could lead to formation of 5-nm filaments. If more complicated contacts were postulated, other helical structures could be generated.

The models are intended primarily to illustrate how binding of small molecules at critical sites could prevent the formation of microtubules and lead to the generation of new structures. Since the packing of dimers on the microtubule surface lattice has not been established, models A or B do not represent, except by chance, the actual microtubule structure.

IV. Cytochalasin and Microfilaments

The class of compounds termed cytochalasins are produced by a variety of fungi from different genera (Aldrich *et al.*, 1967). Almost immediately, interest was generated by the finding that cytochalasin inhibits cleavage (Carter, 1967), causes extrusion of nuclei, and blocks a variety of motile and secretory processes. A large number of studies have since appeared and were recently reviewed by Wessels and co-workers (1971). It is generally supposed that cytochalasin affects the filament ring of the cleavage furrow (Schroeder, 1970) and that similar filaments are often found near the periphery of the cell and may be involved in other motile processes such as undulating membrane in fibroblasts, growth cone in neurons, and granule movements in cells. A large number of preliminary reports have appeared, and the state of the field is somewhat confused. The importance of microfilaments in all these processes have been stressed, particularly by Wessels, and it is further stated that microfilaments disappear on cytochalasin treatment. Confusion arises as to what is a microfilament and whether or not the different classes of intracellular filaments have been properly distinguished in many of the studies. Microfilaments often appear as a cross-linked array or matting of filaments about 4–5 nm in diameter. Clearly they may be difficult to distinguish, particularly in dense cytoplasm. It has been shown recently that many cell types contain actinlike filaments. Some authors appear to also equate microfilaments and actinlike filaments. In a few cases, purified actin has been isolated from nonmuscle cells, i.e., slime mold and amoebae probably from intestinal brush border and sympathetic ganglia (reviewed by Taylor, 1972). It is safe to assume that an actin–myosin system is responsible for protoplasmic streaming at least in those cells from which it has been isolated. Further, as shown by Ishikawa *et al.* (1969), actin filaments in cells can bind heavy meromyosin (the soluble fragment of the myosin molecule designated HMM). There is strong evidence that this reaction is a specific and general test for actin. However, there is no convincing evidence that microfilaments and actin filaments are identical. Filaments decorated by HMM are often found in cell cortex and for whatever reason they may not be found in the cortex after cytochalasin treatment. However, actin, as seen by electron microscopy, appears much thicker than 40 Å. It generally does not form a branching network and does form parallel filament bundles; i.e., it does not resemble the fine filaments, termed microfilaments. Furthermore, HMM has not been shown to decorate the fine microfilament network. A recent paper by Spooner *et al.* (1971) shows both 5 to 7-nm filament bundles and a 4- to 5-nm meshwork in glial cells, and only the latter disappear on cytochalasin treatment.

Attempts to demonstrate any effect of cytochalasin on G-actin polymerization and F-actin stability have been negative. Preliminary studies have failed to show binding of cytochalasin to actin (Daniels and Shelanski, 1972; Forer *et al.,* 1972). The crucial experiment, demonstrating the binding of cytochalasin to a cellular protein, has yet to be done. Recently progress has been made in preparation of a labeled cytochalasin by Tanenbaum; this point may soon be settled.

What is the mechanism of action of cytochalasin? It would be tempting to suppose that, like colchicine and vinblastine, it interacts with a fibrous protein necessary for some aspect of cell motility. Cytochalasin does not affect microtubules directly nor does it block anaphase. It prevents cleavage, and, in some cases, microfilaments disappear after cytochalasin treatment and reappear when the drug is removed. In addition, a variety of cellular motile processes are blocked. But from this evidence alone it cannot be concluded that cytochalasin interacts directly with microfilament protein or even that it depolymerizes microfilaments. Microfilaments as defined here do not appear to be the same as actin filaments, and failure to show an effect of cytochalasin on actin can probably not be taken as evidence against direct action of cytochalasin on microfilaments.

At present, since none of the crucial binding experiments have been done and since nothing is known about the chemistry of microfilament protein, the primary site of action has not been established. Cytochalasin may turn out to be another example of a fibrous protein inhibitor, but if it does not, its action is no less interesting. Perhaps it should be noted that cytochalasin is highly insoluble, that the cell membrane is often altered by the drug, and that many of its effects on cell motility and secretion could be attributed to a primary action on the membrane. Disappearance of microfilaments from the cortex could be due to blocking of their interaction with the membrane.

References

Adelman, M. R., Borisy, G. G., Shelanski, M. L., Weisenberg, R. C., and Taylor, E. W. (1968). Cytoplasmic filaments and tubules. *Fed. Proc., Fed. Amer. Soc. Exp. Biol.* **27**, 1186–1193.

Aldrich, D. J., Armstrong, J. J., Speake, R. N., and Turner, W. B. (1967). The structure of cytochalasins A and B. *J. Chem. Soc., C* Part 2, pp. 1667–1676.

Beard, N. S., Jr., Armentrout, S. A., and Weisberger, A. S. (1969). Inhibition of mammalian protein synthesis by antibiotics. *Pharmacol. Rev.* **21**, 213–245.

Bensch, K. G., and Malawista, S. E. (1968). Microtubule crystals: A new biophysical phenomenon induced by vinca alkaloids. *Nature (London)* **218**, 1176–1177.

Bensch, K. G., Marantz, R., Wisniewski, H., and Shelanski, M. L. (1969). Induction in vitro of microtubular crystals by vinca alkaloids. *Science* **165**, 495–496.

Borisy, G. G., and Taylor, E. W. (1967a). The mechanism of action of colchicine. Binding of colchicine - ^{3}H to cellular protein. *J. Cell Biol.* **34**, 525–533.

Borisy, G. G., and Taylor, E. W. (1967b). The mechanism of action of colchicine. Colchicine binding to sea urchin eggs and the mitotic apparatus. *J. Cell Biol.* **34**, 535–548.

Bryan, J., and Wilson, L. (1971). Are cytoplasmic microtubules heteropolymers? *Proc. Nat. Acad. Sci. U.S.* **68**, 1762–1766.

Carter, S. B. (1967). Effects of cytochalasins on mammalian cells. *Nature (London)* **213**, 261–264.

Chasey, D. (1971). Ph.D. Thesis, King's College, London.

Cohen, C., Harrison, S. C., and Stephens, R. E. (1971). X-ray diffraction from microtubules. *J. Mol. Biol.* **59**, 375–380.

Daniels, M., and Shelanski, M. L. (1972). In preparation.

Eigsti, O. J., and Dustin, P., Jr. (1955). "Colchicine in Agriculture, Medicine, Biology and Chemistry." Iowa State Coll. Press, Ames.

Feit, H., Slusarek, L., and Shelanski, M. L. (1971a). Heterogeneity of tubulin subunits. *Proc. Nat. Acad. Sci. U. S.* **68**, 2028–2031.

Feit, H., Dutton, G., Barondes, S. H., and Shelanski, M. L. (1971b). *J. Cell Biol.* (in press).

Forer, A., Emmerson, J., and Behnke, O. (1972). Cytochalasin B: Does it affect actin-like filaments? *Science* **175**, 774–776.

Grimstone, A. V., and Klug, A. J. (1966). Observations on the substructure of flagellar fibres. *J. Cell Sci.* **1**, 351–362.

Inoué, S. (1952). The effect of colchicine on the microscopic and submicroscopic structure of the mitotic spindle. *Exp. Cell Res., Suppl.* **2**, 305–318.

Inoué, S., and Sato, H. (1967). Cell motility by labile association of molecules. The nature of mitotic spindle fibres and their role in chromosome movement. *In* "Contractile Processes in Macromolecules," New York Heart Symp., pp. 259–292. Little, Brown, Boston, Massachusetts.

Ishikawa, H., Bischoff, R., and Holtzer, H. (1969). Formation of arrowhead complexes with heavy meromyosin in a variety of cell types. *J. Cell Biol.* **43**, 312–328.

Kuzmich, M. J., and Zimmerman, A. M. (1972). Colcemid action on the division schedule of synchronized *Tetrahymena*. *Exp. Cell Res.* **72**, 441–452.

Marantz, R., and Shelanski, M. L. (1970). Structure of microtubular crystals induced by vinblastine in vitro. *J. Cell Biol.* **44**, 234–238.

Marantz, R., Ventilla, M., and Shelanski, M. L. (1969). Vinblastine-induced precipitation of microtubule protein. *Science* **195**, 498–499.

Marsland, D., and Zimmerman, A. M. (1965). Structural stabilization of the mitotic apparatus by heavy water in the cleaving eggs of *Arbacia punctulata*. *Exp. Cell Res.* **38**, 306–313.

Olmsted, J. B., Witman, G. B., Carlson, K., and Rosenbaum, J. L. (1971). Comparison of the microtubule proteins of neuroblastoma cells, brain, and chlamydomonas flagella. *Proc. Nat. Acad. Sci. U.S.* **68**, 2273–2277.

Pollard, H., Scanu, A. M., and Taylor, E. W. (1969). On the geometrical arrangement of the protein subunits of human serum low-density lipoprotein: evidence for a dodecahedral model. *Proc. Nat. Acad. Sci. U.S.* **64**, 304–310.

Puck, T. T., and Sheffer, J. (1963). Life cycle and analysis of mammalian cells. I. A method for localizing metabolic events within the life cycle, and its application to the action of colcemide and sublethal doses of X-irradiation. *Biophys. J.* **3**, 379–397.

Renaud, F. L., Rowe, A. J., and Gibbons, I. R. (1968). Some properties of the protein forming the outer fibers of cilia. *J. Cell Biol.* **36**, 79–90.

Rosenbaum, J., Moulder, J. E., and Ringo, D. L. (1969). Flagellar elongation and shortening in *Chlamydomonas*. *J. Cell Biol.* **41**, 600–619.

Schroeder, T. (1970). The contractile ring 1. Fine structure of dividing mammalian (HeLa) cells and the effects of cytochalasin B. *Z. Zellforsch. Mikrosk. Anat.* **109**, 431–449.

Shelanski, M. L., and Taylor, E. W. (1967). Isolation of a protein subunit from microtubules. *J. Cell Biol.* **34**, 549–554.

Shelanski, M. L., and Taylor, E. W. (1968). Properties of the protein subunit of central-pair and outer-doublet microtubules of sea urchin flagella. *J. Cell Biol.* **38**, 304–315.

Spooner, B. S., Yamada, K. M., and Wessels, N. K. (1971). Microfilaments and cell locomotion. *J. Cell Biol.* **49**, 595–613.

Stephens, R. E. (1968). On the structural protein of flagellar outer fibres. *J. Mol. Biol.* **32**, 277–283.

Taylor, E. W. (1965). The mechanism of colchicine inhibition of mitosis. I. Kinetics of inhibition and the binding of H^3-colchicine. *J. Cell Biol.* **25**, 145–160.

Taylor, E. W. (1972). Biochemistry of muscle contraction. *Ann. Rev. Biochem.* (in press).

Ventilla, M., Cantor, C. R., and Shelanski, M. L. (1972). Submitted for publication.

Weisenberg, R. C., and Timasheff, S. N. (1970). Aggregation of microtubule subunit protein. Effects of divalent cations, colchicine and vinblastine. *Biochemistry* **9**, 4110–4116.

Weisenberg, R. C., Borisy, G. G., and Taylor, E. W. (1968). The colchicine-binding protein of mammalian brain and its relation to microtubules. *Biochemistry* **7**, 4466–4479.

Wessells, N. K., Spooner, B. S., Ash, J. F., Bradley, M. O., Luduena, M. A., Taylor, E. L., Wrenn, J. T., and Yamada, K. M. (1971). Microfilaments in cellular and developmental process. Contractile microfilament machinery of many cell types is reversibly inhibited by cytochalasin B. *Science* **171**, 135–143.

Wilson, L., and Friedkin, M. (1967). The biochemical events of mitosis. II. The *in vivo* and *in vitro* binding of colchicine in grasshopper embryos and its possible selation to inhibition of mitosis. *Biochemistry* **6**, 3126–3135.

Wilson, L., Bryan, J., Ruby, A., and Mazia, D. (1970). Precipitation of proteins by vinblastine and calcium ions. *Proc. Nat. Acad. Sci. U. S.* **66**, 807–814.

3

The Effects of Mercuric Compounds on Dividing Cells

JACK D. THRASHER

I. Introduction	25
II. Mercaptide Formation by Mercurials	26
A. Affinity of Mercury for Various Ligands	26
B. Organomercurials and Enzymatic Activity	27
III. Production of c-Mitotic Figures by Organomercurials	28
IV. Mercuric Compounds and the Generation Time, DNA and Protein Synthesis, the Cell Cycle, and Cilia Regeneration in *Tetrahymena pyriformis*	31
A. Mercuric Compounds and the Generation Time	31
B. Methylmercuric Chloride and the Cell Cycle, DNA, and Protein Synthesis	33
C. Cilia Regeneration: Inhibition by Methylmercuric Chloride	37
V. Distribution of ^{3}H-Methylmercuric Chloride in Fertilized Eggs of *Lytechinus pictus*	40
VI. Observations on the Structural Organization of *Tetrahymena pyriformis* Treated with Mercuric Chloride	41
VII. Concluding Remarks	46
References	46

I. Introduction

The toxic effects of heavy metals on living cells result from their capacity to bind to ligands. The elucidation of the binding of metals to specific intracellular sites is requisite to an understanding of their toxicity. Metal-binding ligands are found in enzymes, substrates, structural constituents, as well as nucleic acids. In general, ligands are chemical groups that are known to associate with hydrogen ions. They include carboxyl, phosphoryl, enediols, amino, phenolic OH, and sulfhydryl groups (for review, see Passow et al., 1961; Madsen, 1963; Passow, 1969).

Of these heavy metals, mercury and its organo derivatives have been receiving considerable attention. This has resulted because mercuric compounds, and, in particular, methylmercury, accumulate in the food chain. Severe neurological damage, such as mental retardation, blindess, palsy, and atrophy of the gray matter, has been well documented in the case of methylmercury poisoning (Yoshino et al., 1966a,b; Borg et al., 1970; Special Report, 1971). In spite of these observations very little is known about the manner in which organomercurials exert their toxicity on living cells. This chapter will attempt to review briefly what is presently known on the ability of mercurials to form mercaptides. In addition, we will present evidence that the binding of these compounds to sulfhydryl groups of soluble proteins is probably the major mechanism of toxicity to dividing cells.

II. Mercaptide Formation by Mercurials

A. AFFINITY OF MERCURY FOR VARIOUS LIGANDS

The capacity of ionic mercury and various organomercurials to form mercaptides has been reviewed (Hughes, 1957; Madsen, 1963; Passow, 1969). Therefore, only the essentials will be reviewed as they pertain to this chapter.

Table I summarizes the association constants of several heavy metals for

TABLE I

FIRST ASSOCIATION CONSTANTS (LOG k_1) FOR THE COMBINATION OF SOME CATIONS AND SMALL MOLECULES

Cation	Sulfide[a]	RS⁻	NH_3	Imidazole	Acetate	Glycinate
Hg^{2+}	53.5	>20	8.8	—	4.0	10.3
Ag^+	50	15	3.2	—	0.7	3.7
Cu^{2+}	41.5	—	4.2	4.4	2.2	8.2
Pb^{2+}	27.5	11	—	—	2.0	5.5
Cd^{2+}	27.2	8	2.7	2.8	1.3	3.9
Ni^{2+}	27.0	—	2.8	3.3	—	5.8
Co^{2+}	26.7	9	—	—	—	4.6
Zn^{2+}	25.2	7	2.8	2.6	1.0	4.8

[a]Taken from Madsen (1963). Many of the constants, particularly those for the mercaptides, should be regarded as approximations.

various ligands. Hg, Ag, and Cu have a special affinity to SH groups. It is for this reason that these metals, and especially organomercurials, have been used to inhibit active thiol sites in enzymes. In addition, sulfhydryl groups have been measured quantitatively by tritration of proteins against various organomercurials. Moreover, these compounds have been used to measure the effect of mercaptide formation on protein structure (for review, see Madsen, 1963).

B. Organomercurials and Enzymatic Activity

The most commonly used organomercurials used to investigate the role of thiol groups in enzymatic activity are p-chloromercuribenzoate, p-chloromercuriphenyl sulfonate, phenylmercuric hydroxide, various halides of methyl mercury, and, less commonly, mersalyl and 4-(p-dimethylaminobenzeeazo) phenylmercuric acetate (Madsen, 1963). The ability of these compounds to bind to SH groups appears to depend on three factors: First, the relative affinity of a mercurial for a thiol group is modified by the type and size (e.g., alkyl versus aryl) residue bonded to the mercury. Second, the affinity of an organomercurial for sulfhydryl groups is also changed by groups adjoining the SH residues. Third, the degree of charge opposite that of the region surrounding the sulfhydryl groups can either increase or decrease the association constant of the mercurial. Thus, Hughes (1957) has shown that the association constant of methylmercuric iodide for closely related proteins (e.g., human versus bovine serum album and bovine oxyhemoglobin versus carbonylhemoglobin) are different. Therefore, in interpreting inhibitory data one must consider these points. We will return to this later in this chapter.

The active thiol group(s) of several enzymes have been inhibited by various organomercurials. These include trypsin (Kreke, 1969), urease (Sumner and Myrbäck, 1930), phosphorylase (Madsen and Gurd, 1956), yeast invertase (Myrbäck, 1957), cytochrome reductase (Strittmatter, 1959), xanthine oxidase (Green and O'Brien, 1967), and adenosine deaminase (Ronca et al., 1967). More recently, Nishida and Yielding (1970) have shown that ^{14}C-methylmercuric iodide binds to the thiol group of glutamate dehydrogenase at a mole ratio of 1:1. The most interesting aspect of their observations is that the binding caused an increase in the catalytic activity of the enzyme. In addition, methylmercury antagonized the allosteric effects of various reagents. These observations point out that organomercurials alter the enzymatic activity of various enzymes only through the role that a specific thiol group has in the function of the enzyme.

Although these experiments demonstrate that various organomercurials can either inhibit or stimulate enzymatic activity by mercaptide formation, one must be cautious when applying this information to a living cell. The problems in the living state arise from (1) the ability of the cell to take up the organomercurials; (2) the fate of the mercuric compound once it is incorporated, i.e., is it catabolized and into what fraction of the cell (organelles, soluble proteins, substrates, nucleic acids, etc.) does it bind? Part of the answer to this question is that most simple organomercurials (phenyl, methyl, and ethyl) are slowly catabolized, have a long half-life, and are bound in the cell as an organomercurial (Berlin, 1963; Berlin and Gibson, 1963; Berlin and Ullberg, 1963a–c; Platonow, 1968; Norseth and Clarkson, 1970). (3) Once bound in the cell, what is the role of the particular thiol groups in the function of the intracellular constituent? and (4) are there any particular class or classes of proteins that are effected more than others because of either enzymatic or conformational roles of sulfhydryl groups?

III. Production of c-Mitotic Figures by Organomercurials

The cytotoxic effects of inorganic, aryl, and alkyl mercuric compounds have been investigated in *Drosophila melanogaster* (Ramel, 1969b), human leukocytes *in vitro* (Fiskesjö, 1970), root tips of *Allium cepa* (Ramel, 1969a,b), and HeLa cells (Umeda *et al.,* 1969). In addition, Skerfving *et al.* (1970) have shown a positive rank correlation between chromosome breakage in lymphocytes of man exposed to methylmercury through fish consumption. Of these observations those on *Allium cepa* and HeLa cells are the most extensive. For greater detail refer to these articles.

The mercuric compounds tested on HeLa cells are mercuric chloride, phenylmercuric chloride, ethylmercuric chloride, and butylmercuric chloride. The effects can be subdivided into acute cell death or chronic inhibition of growth with lethality. Acute concentrations (mercuric chloride = 3.2 mg/liter and organomercurials = 0.32 mg/liter) are characterized by coagulation necrosis. Chronic inhibition of cell growth by the organomercurials (concentrations between 0.03 to 0.32 mg/liter) produced colchicine-type mitotic figures as well as polynuclear giant cells. The morphology of polynuclear cells are different when the results from aryl versus alkyl mercuric compounds are compared. Phenylmercuric chloride causes polynucleosis with two to three nuclei of similar size and cytology to those of control cells. The alkyl mercuric compounds cause an irregularity in both the size and number of nuclei in polynucleated cells. All mercuric compounds tested caused dissociation of the kinetochores and shortening or fragmentation of chromosomes.

The observations made on the c-mitotic effect of phenyl- and methylmercuric hydroxide on root tips of *Allium cepa* are summarized in Table II.

TABLE II

The Effect on Chromosome Fragmentation after Treatment with Phenyl- and Methylmercuric Hydroxide for 24 Hours, followed by 48 Hours Recovery in Water in Root Tips of *A. cepa*[a]

Compound	Conc. (mole/liter $\times 10^{-6}$)	Bridges	Fragmentation	Bridges plus fragmentation	Anaphases With bridges %	With fragmentation %	Total
Phenylmercuric hydroxide	2.5	78	49	18	19.8	13.8	486
	1.2	84	36	22	13.2	7.2	805
	0.25	33	10	2	2.2	0.7	1627
	0	1	2	0	0.1	0.3	752
Methylmercuric hydroxide	2.5	29	7	2	6.1	1.8	507
	1.2	4	8	0	0.6	1.2	688
	0.25	2	4	0	0.3	0.7	582
	0	4	0	0	0.2	0	1750

[a] From Ramel (1969a,b).

TABLE III
COMPARISON OF THE TOXICITY OF EACH MERCURIAL COMPOUND IN MOLAR CONCENTRATIONS TO *T. pyriformis* [a]

Compound	Increase in generation time (%)			Killed (%)	
	10	50	100	50	100
Mercuric chloride	2.8×10^{-6}	1.1×10^{-5}	1.6×10^{-5} [b]	1.7×10^{-5}	2.9×10^{-5}
Phenylmercuric acetate	3.5×10^{-7}	6.9×10^{-7}	1.3×10^{-6} [b]	1.7×10^{-6}	2.5×10^{-6}
Ethylmercuric chloride	1.5×10^{-7}	3.6×10^{-7}	7.4×10^{-7}	1.4×10^{-6}	1.5×10^{-6}
Methylmercuric chloride	8.0×10^{-8} [b]	2.9×10^{-7}	5.6×10^{-7}	8.4×10^{-7}	1.2×10^{-6}

[a] From Thrasher and Adams (1972a).
[b] Estimated values.

Inspection of the results show that the two compounds produce a different percentage of cells that have anaphase bridges with chromosomal fragmentation. Although both organomercurials produce c-mitosis, they appear to have different affinities for mitotic proteins. Phenylmercury could possibly bind more readily to kinetochore–microtubule proteins, whereas methylmercury appears to effect pole-to-pole microtubules. The difference in association affinities could result either from organic residue on the mercury or by groups adjoining the ligands (SH) of the proteins.

In conclusion, organomercurials appear to affect microtubule proteins of the mitotic apparatus in dividing cells. This could result from the binding of the mercurials to free thiol groups of monomers and, thereby, preventing the assembly of microtubule proteins. In addition, in HeLa cells the difference in nuclear morphology between aryl- and alkyl-treated cells suggest that alkyl compounds might interfere with some aspect of nucleic acid metabolism.

IV. Mercuric Compounds and the Generation Time, DNA and Protein Synthesis, the Cell Cycle, and Cilia Regeneration in *Tetrahymena pyriformis*

A. Mercuric Compounds and the Generation Time

The effects of mercuric chloride, phenylmercuric acetate, ethylmercuric chloride, and methylmercuric chloride on the generation time of *Tetrahymena pyriformis* have been investigated (Thrasher and Adams, 1972a). Figures 1–4 and Tables III and IV summarize these experiments. Mercuric chloride causes an increase in the generation time of 5%–148% over controls between concentrations of 0.78–5.3 mg/liter. Phenylmercuric acetate

TABLE IV

The Effects of 2-Mercaptoethanol on the Toxicity of Phenylmercuric Acetate, Ethylmercuric Choloride, and Methylmercuric Chloride on *T. pyriformis* [a,b]

	Control (hours)	Mercurial (hours)	Mercaptoethanol + mercurial (hours)	Mercaptoethanol (hours)
Phenylmercuric acetate	2.03	4.6	4.4	2.1
Ethylmercuric chloride	2.2	4.6	3.0	2.1
Methylmercuric chloride	2.1	3.7	2.4	2.1

[a] Organomercurials tested at concentration that caused 100% increase in generation time. Mercaptoethanol $= 7.1 \times 10^{-7}\ M$.

[b] From Thrasher and Adams (1972a).

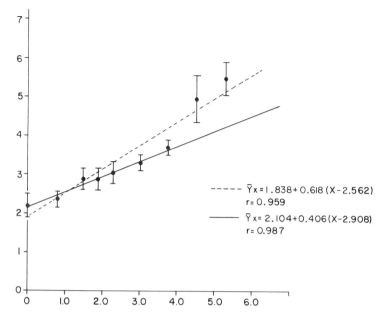

Fig. 1. Effect of $HgCl_2$ on the generation time of *T. pyriformis*. Each point represents the average of three to four experiments (± standard error). Ordinate: generation time in hours; abscissa: mg/liter of $HgCl_2$. (From Thrasher and Adams, 1972a.)

produces an increase from 5% to 168% in concentration ranges of 0.022–0.54 mg/liter. Similiar observations with ethylmercuric chloride and methylmercuric chloride show that they are effective between 0.02 to 0.30 mg/liter and 0.02 to 0.2 mg/liter, respectively.

Table III summarizes the molar concentrations for each mercurial that produce a 10%, 50%, and 100% increase in the generation time. In addition, those concentrations that cause death of 50% and 100% of the cells in 1 hour after exposure are listed. It can be seen that mercuric chloride is the least toxic of the four compounds. Methylmercuric chloride is the most toxic followed by ethylmercuric chloride and phenylmercuric acetate.

Table IV summarizes the data from experiments in which 2-mercaptoethanol was used as an antagonist to the toxicity of methylmercuric chloride, ethylmercuric chloride, and phenylmercuric acetate. Addition of mercaptoethanol to culture conditions containing each mercurial reduces the generation time in the cases of methylmercuric chloride and ethylmercuric chloride. On the other hand, mercaptoethanol does not ameliorate the toxic effects of phenylmercuric acetate on *T. pyriformis*.

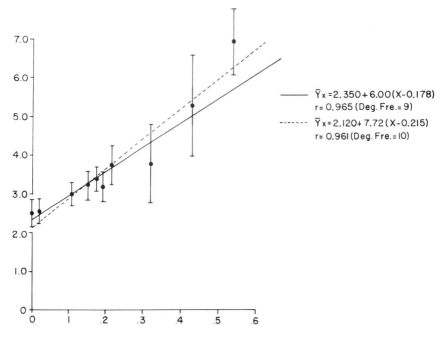

Fig. 2. Effect of phenylmercuric acetate on the generation time of *T. pyriformis*. Symbols, abscissa, and ordinate same as Fig. 1. (From Thrasher and Adams, 1972a.)

Several conclusions can be drawn from these observations. First, *T. pyriformis* is well suited to toxicological investigations with mercuric compounds and possibly other heavy metals. Second, the molar concentrations of each organomercurial are similiar to those that cause c-mitosis in *A. cepa* (Ramel, 1969b,c) and HeLa cells (Umeda *et al.*, 1969). Third, the toxic effects of methylmercuric chloride and ethylmercuric chloride appear to result from mercaptide formation as evidenced by the amelioration by mercaptoethanol. Fourth, the toxicity of phenolmercuric acetate to *T. pyriformis* may result from the fact that arylmercuric compounds have a different association constant for ligands when compared to alkyl derivatives of mercury.

B. Methylmercuric Chloride and the Cell Cycle, DNA, and Protein Synthesis

The effects of methylmercuric chloride on DNA and protein synthesis as well as on the cell cycle of *T. pyriformis* have also been investigated (Thrasher and Adams, 1972b). The results of these experiments are given

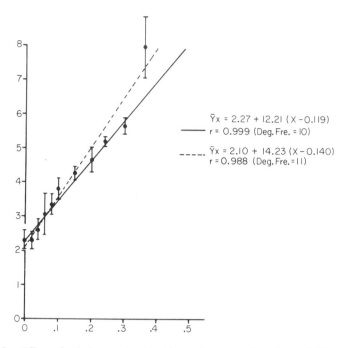

Fig. 3. Effect of ethylmercuric chloride on the generation time of *T. pyriformis*. Symbols, abscissa, and ordinate same as Fig. 1. (From Thrasher and Adams, 1972a.)

in Fig. 5 and 6 and in Table V. Concentrations of methylmercuric chloride that cause 0%, 50%, and 83% increases in the generation time do not affect the rate of incorporation of ^{14}C-leucine into trichloroacetic acid-precipitated proteins (Fig. 5). However, methylmercuric chloride does reduce the incorporation of ^{14}C-thymidine into DNA by 100% at 0.104 mg/liter and 30% at 0.07 mg/liter. The 13% inhibition observed at 0.014 mg/liter could result from normal experimental error.

The effects of methylmercuric chloride on the cell cycle and its phases was investigated under continuous labeling with ^{3}H-thymidine, autoradiography and standard mitotic curve technique; the results are given in Table V. Methylmercuric chloride produces a 62%–70% increase in the duration of $G_2 + \frac{1}{2} D$ (D = division time). The remainder of the cell cycle ($G_1 + S + \frac{1}{2} D$) is increased over control values by 28% at 0.07 mg/liter and 95% with exposure to 0.104 mg/liter. Interestingly, these latter observations correspond to the degree of inhibition of DNA synthesis given in Fig. 6.

From these data it is possible to conclude that one of the major effects of methylmercuric chloride is to inhibit DNA synthesis. This occurs without an

TABLE V
The Effects of Methylmercuric Chloride on the Cell Cycle of T. pyriformis[a]

	GT[b] (minute)	Change (%)	$G_2 + \frac{1}{2} D$[c] (minute)	Change (%)	$G_1 + S + \frac{1}{2} D$ (minute)	Change (%)	S-phase index (at 20 minutes)
Control	120	—	60	—	60	—	0.27
0.014	120	—	60	—	60	—	0.25
0.07	180	50	97	62	83	28	0.21
0.104	220	83	102	70	118	95	0.22

[a] From Thrasher and Adams (1972b).
[b] GT, generation time.

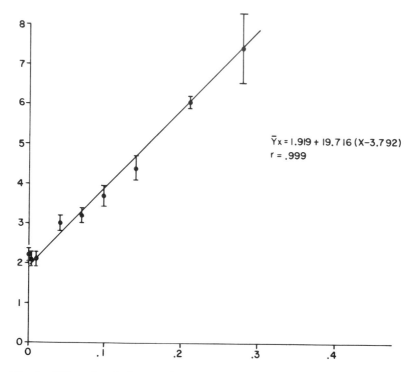

Fig. 4. Effect of methyl mercuric chloride on the generation time of *T. pyriformis*. Symbols, abscissa, and ordinate same as Fig. 1. (From Thrasher and Adams, 1972a.)

alteration in the rate of incorporation of ^{14}C-leucine into proteins. The inhibition of DNA synthesis could result from the binding of methylmercuric chloride to heterocyclic bases of DNA in a manner similar to that of ionic mercury (Davidson *et al.*, 1965; Nandi *et al.*, 1965). However, it has been shown that methylmercuric chloride binds to proteins of the central nervous system and cannot be detected in nucleic acid fractions (Yoshino *et al.*, 1966a,b). At the present time there is no reason to suspect a difference in the intracellular distribution of methylmercuric chloride in *T. pyriformis*.

We now suspect that the inhibition of DNA synthesis results from the mercaptide formation of methylmercuric chloride with soluble proteins responsible for nucleic acid metabolism. Rudick and Cameron (1972) have shown that DNA polymerase appears just prior to DNA synthesis in synchronized *T. pyriformis*. The activity of the enzyme increases fivefold during the S phase. Thus, binding of methylmercuric chloride to this enzyme or others responsible for purine and pyrimidine metabolism could account for the inhibition of DNA synthesis.

3. THE EFFECTS OF MERCURIC COMPOUNDS ON DIVIDING CELLS

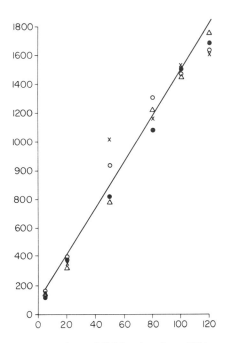

Fig. 5. Rate of incorporation of ^{14}C-leucine into TCA precipitated proteins of *T. pyriformis* under control (o), and methylmercury concentrations of 0.014 (x), 0.07 (●), and 0.104 mg/liter (Δ). Ordinate: dpm/3.2×10^4 cells; abscissa: time in minutes after the addition of ^{14}C-leucine (0.3 μCi/ml) to the growth medium. (From Thrasher and Adams, 1972b.)

It is not surprising that the $G_2 + \frac{1}{2}$ D period of the cell cycle is increased by methylmecuric chloride. Rannestad and Williams (1971) have shown that newly synthesized microtubule proteins are assembled into the oral apparatus of *T. pyriformis* during G_2 and cytokinesis. Since alkylmercuric compounds are known to be mitotic spindle-blocking agents, it is reasonable to assume that methylmercuric chloride inhibits microtubule assembly into the oral apparatus. This probably occurs through the binding of methylmercuric chloride to monomeric units and, thereby, prevents the formation of disulfide linkages.

C. CILIA REGENERATION: INHIBITION BY METHYLMERCURIC CHLORIDE

Cilia of *T. pyriformis* can be easily amputated according to the method of Rosenbaum and Carlson (1969) and as modified in this laboratory (Thrasher *et al.,* 1972a). Regeneration of cilia in the presence of various

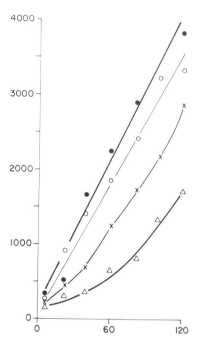

Fig. 6. Rate of incorporation of ^{14}C-thymidine into DNA of *T. pyriformis* in control (●) and under concentrations of methyl mercuric chloride of 0.014 (o), 0.07 (x), and 0.104 mg/liter (△). As in Fig. 5 each symbol represents the average of two determinations per time interval. Ordinate: dpm/3.2×10^4 cells; abscissa: time in minutes after addition of ^{14}C-thymidine (0.5 μCi/ml) to the growth medium. (From Thrasher and Adams, 1972b.)

inhibitors can be determined by enumerating the percentage of cells that have regained motility with time.

Figure 7 demonstrates the regeneration of cilia in the presence of concentrations of methylmercuric chloride that produce 0%, 25%, 50%, and 83% increases in the generation time of *T. pyriformis*. Both controls and cells treated with 0.014 mg/liter of methylmercuric chloride rapidly regain their motility. They reach 50%, 90%, and about 100% motility at 30, 60, and 90 minutes, respectively, after deciliation. It takes cells treated with 0.035 mg/liter 38, 90, and 120 minutes to reach the same percent motility. At the two higher concentrations (0.07 and 0.104 mg/liter) regeneration is much slower. Cells exposed to 0.07 mg/liter reach 50% motility after 45 minutes and 90% after 150 minutes. Values for cells treated with 0.104 mg/liter are 50% at 68 minutes and 88% at 180 minutes.

3. THE EFFECTS OF MERCURIC COMPOUNDS ON DIVIDING CELLS

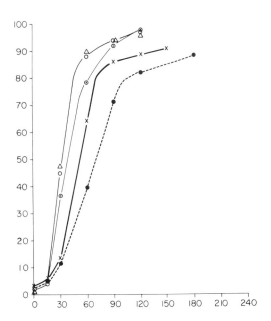

Fig. 7. The rate at which motility returned in *T. pyriformis* following deciliation in controls (o) and methylmercury concentrations of 0.014 (△), 0.035 (⊙), 0.07 (x), and 0.104 mg/liter (●). Each experimental point represents the average of four determinations under each experimental condition. Abscissa: time in minutes after deciliation; ordinate: percentage of motile cells. (From Thrasher et al., 1972a.)

The inhibition of cilia regeneration by methylmercuric chloride is similiar to the effects of colchicine and colcemid (Rosenbaum and Carlson, 1969). The major difference between these two classes of compounds is that methylmercuric chloride is effective at concentrations of 10^{-7}–10^{-8} M, whereas colchicine and Colcemid inhibit cilia regeneration at concentrations which are orders of magnitude greater.

Mechanisms by which methylmercuric chloride inhibit cilia regeneration may be surmized from its ability to form mercaptides (Hughes, 1957). Microtubule proteins of the outer fibers of cilia from *T. pyriformis* contain about 7.5 sulfhydryl groups per 55,000 gm atoms of protein (Renaud et al., 1971). Since methylmercuric chloride does not inhibit gross protein synthesis, it is reasonable to assume that it binds to free thiol groups and other functional ligands of soluble proteins. This binding could prevent disulfide linkages and inhibit the assembly of monomeric units into microtubule proteins of cilia.

TABLE VI

The Specific Activity of Various Fractions of Fertilized Eggs of *L. pictus* after 120 Minutes Labeling with Tritiated Methylmercuric Chloride [a]

	dpm × 10³ per 1000 eggs	Total label (%)	100,000 g Supernatant label (%)
Whole homogenate	5.03	100	—
35,000 g pellet	1.39	27.6	—
100,000 g pellet	0.24	4.8	—
100,000 g supernatant	2.59	51.5	100
Velban precipitate	0.68	13.5	31.4
Velban supernatant	1.58	31.4	61.0

[a] From Adams and Thrasher (1972).

V. Distribution of ³H-Methylmercuric Chloride in Fertilized Eggs of *Lytechinus pictus*

Some preliminary experiments on the incorporation of ³H-methylmercuric chloride into *L. pictus* eggs after 120 minutes of labeling have been carried out (Adams and Thrasher, 1972). Table VI shows the specific activity of ³H-methylmercuric chloride binding to various fractions from *L. pictus* eggs. Approximately 52% of the incorporated label is found in the soluble 100,000 g fraction. Of this, about 14% is precipitated by Velban, while the remainder is found in the Velban supernatant. Translated into the percentage of 100,000 g supernatant label, 31% is precipitated by Velban. Table VII summarizes the results of sodium dodecyl sulfate-gel electropho-

TABLE VII

Molecular Weights of Velban-Precipitated Proteins from 100,000 g Supernatant from Fertilized *L. pictus* Eggs [a]

Peaks		RF	Molecular weights
1	4.14	10.91	180,000*
2	6.10	16.08	148,000
3	12.46	32.85	73,000
4	14.98	39.50	58,000*
5	19.60	51.68	36,400
6	21.75	57.35	29,000*
7	24.50	64.60	22,500
Reference peak	37.92	—	—

[a] From Adams and Thrasher (1972).

resis of the Velban precipitate. At least three of the bands correspond to molecular weights of microtubule proteins. The remaining four proteins are, at present, believed to be hyalin proteins. It is interesting to note that reduction of the Velban precipitate with mercaptoethanol removes all bound label. This suggests that ^{3}H-methylmercuric chloride does bind to free SH groups of soluble proteins. Further work is in progress to examine the incorporation of ^{3}H-methylmercuric chloride at times earlier than 120 minutes.

VI. Observations on the Structural Organization of *Tetrahymena pyriformis* Treated with Mercuric Chloride

Cameron and Tingle (1972) have recently subjected cultures of *T. pyriformis* to mercuric chloride solution in an attempt to find the cellular sites of cytotoxic damage. Dose response curves indicate that *T. pyriformis* are slowly killed in concentrations of mercuric chloride of 1 mg/liter. Therefore, they chose to use 1.25 mg/liter of mercuric chloride for the observations. By using this concentration of mercuric chloride it is possible to maintain all the cells in a viable condition until they are fixed.

T. pyriformis were axenically grown in 2% proteose peptone, 0.1% liver extract. Logarithmically growing cells were centrifuged and washed in distilled water. Half of these was resuspended in 1.25 mg/liter of mercuric chloride, while the other half was resuspended in distilled water. After 8 minutes the cells were centrifuged and fixed in sodium cacodylate buffered solution of 4% glutaraldehyde and 1% osmium tetroxide. Both groups of cells were then embedded in Spurr's media, sectioned on a LKB ultramicrotome, and examined in a Seimens 1A electron microscope.

With phase contrast microscopy, several changes in mercuric chloride-treated cells were observed. The ciliary beating was rapidly retarded and the coordination of the ciliary activity necessary for linear locomotion was lost. The cells then began to spin or rotate around a point. Such cells were somewhat swollen and the contractile vacoule pulsations ceased. No mortality, as determined by cessation of ciliary activity and a change in the refractive index of the cell, occurred in less than 30 minutes. Our laboratory (Thrasher and Adams, 1972a) has made similiar observations at concentrations of mercuric chloride, phenylmercuric acetate, ethylmercuric chloride, and methylmercuric chloride that cause 50% to 100% death of cells in 1 hour after treatment.

Figure 8 is the surface of a cell from the control culture. The anatomy of this cell is intended to serve as a comparison to the mercuric chloride-treated cells. Both longitudinal (LT) and transverse (TT) microtubules can be seen just beneath the cortex. The pellicle of the cell consists of two outer

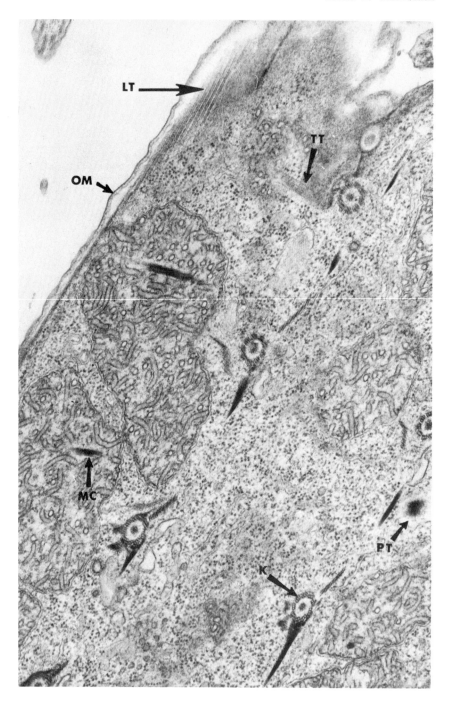

membranes (OM) as well as an inner membrane. The mitochondria contain tubular cristae, considerable ground matrix, and crystalline inclusions. Kinetosomes or basal bodies (K), which are structurally akin to centrioles of mitotically dividing animal cells, are frequently observed. Also indicated are protrichocysts (PT), sometimes referred to as mucigenic bodies or mucocysts.

Figures 9A and 9B represent *T. pyriformis* after 8 minutes exposure to the mercuric chloride solution. A number of cellular alterations can be seen. Perhaps the most striking alteration is in the structure of mitochondria. In many places, the tubular cristae pull away from the outer mitochondrial membrane and become disorganized. The mitochondria also vacuolate (MV) and appear to swell. The ground matrix, visible in normal mitochondria, is significantly reduced. As in the controls the mitochondrial crystalline inclusions are still apparent.

In addition to the mitochondrial changes the treated cells show the release of material from some of the protrichocysts (as is illustrated by the arrows in Fig. 9A), numerous breaks in the outer membrane of the pellicle, the presence of large vacuoles throughout the cytoplasm, condensation of macronuclear chromatin, an increased density of nucleoli, and a loss of intranuclear ground matrix. One often finds small vesicles (SV), primary lysosomes, and myelin figures (MF), both of which suggest that cellular autolysis is occurring. Figure 9B also indicates that mercuric chloride has not caused an alteration of the rows of longitudinally oriented microtubules (LT).

Thus, mercuric chloride-treated cells show a striking change in their mitochondrial morphology, which can be correlated with the observed functional changes of the cell. It seems reasonable to speculate that the observed cytotoxic lesions of the mitochondria cause a decrease in the energy source (ATP) necessary for ciliary activity. This reasoning is strengthened by the observation that the cilia themselves show no obvious cytotoxic damage. The apparent upset in osmoregulation manifested in cellular swelling and cessation of contractile vacuole activity may also be caused by the abrupt change in available energy due to mitochondrial lesions. The condensation of chromatin suggests a direct interaction of ionic mercury with DNA and predicts that the macronucleus has lost some or all its RNA synthetic activity. Although the number of ribosomes and polysomes in mercuric chloride-treated cells seems somewhat less than that of the controls, there are enough polysomes left to suggest that some protein synthesis continues.

Fig. 8. An electron micrograph of *T. pyriformis* cultured under controlled conditions as described in text: 40,800 ×. (From Cameron and Tingle, 1972.)

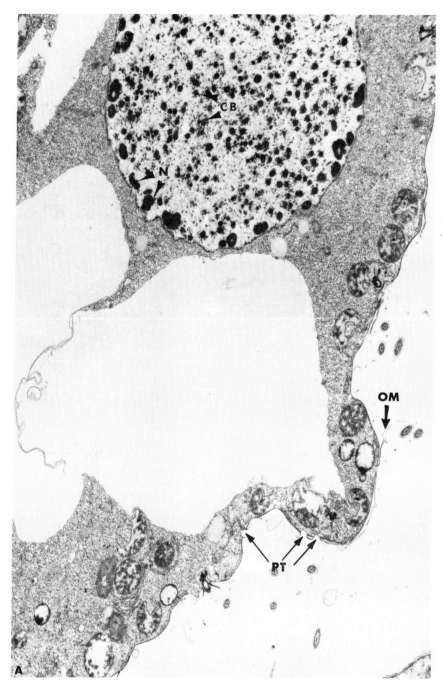

Fig. 9(A). Electron micrographs of *T. pyriformis* exposed to 1.25 mg/liter of mercuric chloride. See text for further description: 7650 ×. (From Cameron and Tingle, 1972.)

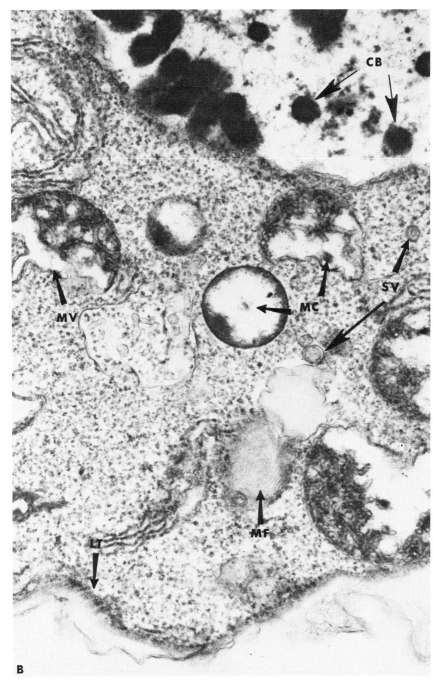

Fig. 9(B). See Fig. 9(A) for explanation: 35,000 ×. (From Cameron and Tingle, 1972.)

Our laboratory (Thrasher *et al.,* 1972b) is currently investigating the cellular site of cytotoxic damage of methylmercuric chloride-treated cells. At the present time we have not observed mitochondrial and nuclear damage at 120 minutes after exposure of *T. pyriformis* to methylmercuric chloride. Thus, it can be concluded that ionic mercury and methylmercuric chloride appear to have different cellular sites of cytotoxic action in *T. pyriformis*. This could result from the relative differences in the affinity of the two compounds for intracellular ligands.

VII. Concluding Remarks

The current available information on the effects of various mercuric compounds on dividing cells has been reviewed. At the present time it appears that the cytotoxicity of these compounds varies according to the organic residue bonded to the mercury. Thus, ionic mercury causes membrane, mitochondrial, nuclear, and cytoplasmic damage. On the other hand, methylmercury appears to bind to SH groups of soluble proteins. This causes either an inhibition or stimulation of catalytic activity of enzymes. In addition, mercaptide formation by methylmercuric chloride may prevent disulfide linkages and, thereby, may interfer with the assembly of microtubule proteins. More information is needed on other organomercurials in order to determine the exact nature of their toxicity.

Acknowledgments

Portions of this work were supported in part by grants from the American Cancer Society Institutional Grant, The Cancer Research Coordinating Committee, The General Research Support Grant, Southern California Edison Company and the National Science Foundation, G.B. 31580.

References

Adams, J. F., and Thrasher, J. D. (1972). Inhibition of cleavage by methyl mercury and the incorporation of ^{3}H-methyl mercury into eggs of *L. pictus*. Unpublished.

Berlin, M. (1963). Renal uptake, excretion and retention of mercury. II. A study in the rabbit during infusion of methyl- and phenylmercuric compounds. *Arch. Environ. Health* **6**, 626–633.

Berlin, M., and Gibson, S. (1963). Renal uptake, excretion and retention of mercury. I. A study in the rabbit during infusion of mercuric chloride. *Arch. Environ. Health* **6**, 617–625.

Berlin, M., and Ullberg, S. (1963a). Accumulation and retention of mercury in the mouse. I. An autoradiographic study after a single intravenous injection of mercuric chloride. *Arch. Environ. Health* **6**, 589–601.

Berlin, M., and Ullberg, S. (1963b). Accumulation and retention of mercury in the mouse. II. An autoradiographic comparison of phenylmercuric acetate and inorganic mercury. *Arch. Environ. Health* **6**, 602–609.

Berlin, M., and Ullberg, S. (1963c). Accumulation and retention of mercury in the mouse. III. An autoradiographic comparison of methylmercuric dicyandiamide with inorganic mercury. *Arch. Environ. Health* **6**, 610–616.

Borg, K., Erne, K., Hanko, E., and Wanntorp, H. (1970). Experimental secondary methyl mercury poisoning in the Goshawk. *Environ. Pollut.* **1**, 91–104.

Cameron, I. L., and Tingle, L. E. (1972). Observations on the structural organization of *Tetrahymena pyriformis* treated with mercuric chloride. Personal communication.

Davidson, N., Widholm, J., Nandi, U. S., Jensen, R., Olivera, B. M., and Wang, J. C. (1965). Preparation and properties of native crab dAT. *Proc. Nat. Acad. Sci. U. S.* **53**, 111–118.

Fiskesjö, G. (1970). The effect of two organic mercury compounds on human leukocytes *in vitro*. *Hereditas* **64**, 142–146.

Green, R. C., and O'Brien, P. J. (1967). Xanthine oxidase inactivation by reagents that modify thiol groups. *Biochem. J.* **105**, 585–589.

Hughes, W. L. (1957). A physicochemical rationale for the biological activity of mercury and its compounds. *Ann. N. Y. Acad. Sci.* **65**, 454–460.

Kreke, C. W. (1969). Inhibition of trypsin sulfhydryl reagents: Selective toxicities of organic mercury compounds. *J. Pharm. Sci.* **58**, 457–459.

Madsen, N. B. (1963). Mercaptide-forming agents. In "Metabolic Inhibitors" (R. M. Hochster and J. H. Quastel, eds.), Vol. 2, pp. 119–143. Academic Press, New York.

Madsen, N. B., and Gurd, F. R. N. (1956). The interaction of muscle phosphorylase with p-chloromercuribenzoate. III. the reversible dissociation of phosphorylase. *J. Biol. Chem.* **223**, 1075–1087.

Myrbäck, K. (1957). Inhibition of yeast invertase (saccharase) by metal ions. V. Inhibition by mercury compounds. *Ark. Kemi* **11**, 471–479.

Nandi, U. S., Wang, J. C., and Davidson, N. (1965). Separation of deoxyribonucleic acids by Hg (II) and Cs_2O_4 density gradient centrifugation. *Biochemistry* **4**, 1687–1702.

Nishida, M., and Yielding, K. L. (1970). Alterations in catalytic and regulatory properties of glutamate dehydrogenase resulting from reaction with one molecule of ^{14}C-labeled methylmercuric iodide. *Arch. Biochem. Biophys.* **141**, 409–415.

Norseth, T., and Clarkson, T. W. (1970). Studies on the biotransformation of ^{203}Hg-labeled methyl mercury chloride in rats. *Arch. Environ. Health* **21**, 717–727.

Passow, H. (1969). The red blood cell: Penetration, distribution and toxic actions of heavy metals. In "Effects of Heavy Metals on Cells, Subcellular Elements and Macromolecules" (J. Maniloff, J. R. Coleman, and M. W. Miller, eds.), pp. 291–340. Springfield, Illinois.

Passow, H., Rothstein, A., and Clarkson, T. W. (1961). The general pharmacology of the heavy metals. *Pharm. Rev.* **13**, 185–224.

Platonow, N. (1968). A study of the metabolic fate of methylmercuric acetate. *Occup. Rev.* **20**, 9–19.

Ramel, C. (1969a). Methylmercury as a mitosis disturbing agent. *Nihonishikai-zasshi* **61**, 1072–1076.

Ramel, C. (1969b). Genetic effects of organic mercury compounds. I. Cytological investigations on allium roots. *Hereditas* **61**, 208–230.

Rannestad, J. and Williams, N. E. (1971). The synthesis of microtubule and other proteins of the oral apparatus in *Tetrahymena pyriformis*. *J. Cell Biol.* **50**, 709–720.

Renaud, F. L., Rowe, A. J., and Gibbons, I. R. (1968). Some properties of the proteins forming the outer fibers of cilia. *J. Cell Biol.* **36**, 79–90.

Ronca, G., Bauer, C., and Rossi, C. A. (1967). Role of sulfhydryl groups in adenosine deaminase. *Eur. J. Biochem.* **1**, 434–438.

Rosenbaum, J. L., and Carlson, K. (1969). Cilia regeneration in *Tetrahymena* and its inhibition by colchicine. *J. Cell Biol.* **40**, 415–425.

Rudick M. J., and Cameron, I. L. (1972). Regulation of DNA synthesis and cell division in starved-refed synchronized *Tetrahymena pyriformis* HSM. *Exp. Cell Res.* **70**, 411–416.

Skerfving, S., Hansson, A., and Lindsten, J. (1970). Chromosome breakage in human subjects exposed to methyl mercury through fish consumption. *Arch. Environ. Health* **21**, 133–139.

Special Report. (1971). Hazards of mercury. Special report to the Secretary's Pesticide Advisory Committee, Department of Health, Education and Welfare. *Environ. Res.* **4**, 1–69.

Strittmatter, P. (1959). The reactive sulfhydryl groups of microsomal cytochrome reductase. *J. Biol. Chem.* **234**, 2661–2664.

Sumner, J. B., and Myrbäck, K. (1930). Über Schwermetall-Inaktivierung hochgereinigter Urease. *Hoppe-Seyler's Z. Physiol. Chem.* **189**, 218–228.

Thrasher, J. D., and Adams, J. F. (1972a). The effects of four mercury compounds on the generation time and cell division in *Tetrahymena pyriformis*, WH14. *Environ. Res.* (in press).

Thrasher, J. D., and Adams, J. F. (1972b). The effects of methylmercuric chloride on the cell cycle of *Tetrahymena pyriformis*, WH14. *Exp. Cell Res.* (in press).

Thrasher, J. D., Maxwell, D. S., and Adams, J. F. (1972a). Cilia regeneration in *Tetrahymena pyriformis:* Inhibition by methylmercuric chloride. *Exp. Cell Res.* (in press).

Thrasher, J. D., Bernard, G., and Maxwell, D. S. (1972b). The ultrastructural changes in *Tetrahymena pyriformis* after exposure to methylmercuric chloride. Unpublished.

Umeda, M., Saito, K., Hirose, K., and Saito, M. (1969). Cytotoxic effects of inorganic, phenyl and alkyl mercuric compounds on HeLa cells. *Jap. J. Exp. Med.* **39**, 47–58.

Yoshino, Y., Mozai, T., and Nakao, K. (1966a). Distribution of mercury in the brain and its subcellular units in experimental organic mercury poisonings. *J. Neurochem.* **13**, 397–406.

Yoshino, Y., Mozai, T., and Nakao, K. (1966b). Biochemical changes in the brain of rats poisoned with an alkylmercury compound, with special reference to the inhibition of protein synthesis in brain cortex slices. *J. Neurochem.* **13**, 1223–1230.

4

Adrenergic Drugs on the Cell Cycle

E. R. JAKOI and G. M. PADILLA

I. Introduction 49
II. Experimental Results 50
III. Discussion and Summary 61
    References 64

I. Introduction

Tricyclic antidepressants such as chlorimipramine (Gyermek, 1966) and desmethylimipramine-HCl (Salama *et al.*, 1971) have been shown to block the catecholamine re-uptake across nerve terminal membrane. Recently, however, Linstead and Wilkie (1971) suggested that chlorimipramine bound to yeast mitochondrial membranes and thus potentially decreased the efficiency of ATP production. Further support for this theory was obtained by Wilkie and Delhanty (1970) who showed that cultured human fibroblasts were arrested in metaphase when exposed to chlorimipramine. These cells exhibited an immediate decrease in oxygen consumption and died ultimately. A partial reversal of death was possible by the addition of exogenous ATP, but no resumption of cell proliferation was observed. In the subsequent work by Mittwoch and Wilkie (1971) on partially synchronized human fibroblasts they again described a chlorimipramine-induced block at metaphase and also a block at the initiation of DNA replication, both of which were partially alleviated by the addition of ATP.

Desmethylimipramine (DMI), an analog of chlorimipramine, has been employed by Connett and Blum (1971) in their work on glycogen metabolism in *Tetrahymena pyriformis* HSM. They reported that in conjunction

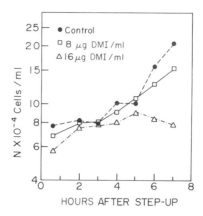

Fig. 1. Effect of DMI concentration on cell density of *Tetrahymena* following a temperature-nutritional step-up.

with a reduced oxygen consumption a general metabolic depression and growth inhibition occurred in cells in the early stationary phase of growth following an overnight exposure to 25 μM DMI. Because of the apparent close interrelationship between nucleic acid synthesis and cell division (Cummins, 1969; Jeon and Lorch, 1971; Tsien and Wattiaux, 1971; Rudick and Cameron, 1972) further elucidation of DMI inhibition of growth, together with an investigation of the effect of this drug on nuclear metabolism, seemed warranted. Thus, a study was undertaken to see if desmethylimipramine interfered with RNA metabolism and cell division in synchronized cell systems. Two methods were employed: the temperature-cycling method of Padilla and Cameron (1964) and the method of starvation–refeeding of *Tetrahymena* developed by Cameron and Jeter (1970). The later method is useful because the cells display clearly defined phases of cell growth and division.

II. Experimental Results

The first series of experiments were conducted with *Tetrahymena* that had been grown into the stationary phase at 15° for 2 days and then diluted with fresh proteose–peptone medium as the temperature was raised to 28°. Figure 1 shows that without the addition of DMI the cell number did not increase for about 3 hours. At this time the cells began to divide and the population density rose from 8×10^4 to 21×10^4 cells/ml. Cells exposed to 8 μg DMI (21 μM) showed a similar pattern of cell division with the exception that population density rose from 6.8×10^4 to 16×10^4 cells/ml.

4. ADRENERGIC DRUGS ON THE CELL CYCLE

This represents a population increase that is 83% of the control cells. When the concentration of DMI was increased to 16 μg (53 μM) the treated cells initially increased in number during the first 2 hours after the temperature step-up, but showed essentially no increase in population density during the next 5 hours. The total increase in cell number was 50% of the control value. This is in agreement with the results of Blum (1968) who found that exponentially growing *Tetrahymena* exposed to 50 μM DMI showed a 43% inhibition as compared to the control cells. As we will discuss later, starved–refed *Tetrahymena* undergo a comparable growth inhibition by DMI at these concentrations.

Since the cells were exposed continuously to DMI but showed a 50% inhibition, even at the highest concentration used, it became interesting to see whether or not the time of addition of DMI was a factor in the inhibition of cell division. We therefore exposed cultures of *Tetrahymena,* treated as in the previous experiment, to DMI (53 μM) at 0, 1, and 2 hours after dilution and temperature step-up. The results of such an experiment are shown in Fig. 2. As in the previous experiment, the control cells began to divide 3 hours after the step-up. Cells essentially double in cell number. If DMI is added at 0 or 1 hour after the step-up, inhibition is such that the treated cells reach a population density that was 61% of the control value. However, when DMI is added at 2 hours after the step-up (C) the treated cells show only a 26% inhibition with respect to the control cells. This suggested that the DMI depression of cytokinesis in these cells was either a cycle-dependent effect or a cycle-independent effect that summates throughout the time of exposure to the drug.

Figures 3 and 4 show the results of experiments in which the effect of DMI on the RNA content of the cells was examined under a similar experimental design. Figure 3 shows that when cells are exposed to DMI at a time

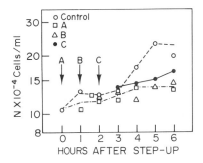

Fig. 2. Effect of the time of addition of DMI on cell division of *Tetrahymena* following a single temperature-nutritional step-up.

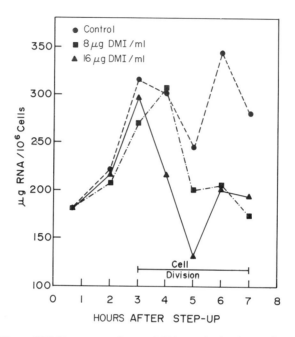

Fig. 3. Effect of DMI concentration on RNA synthesis of *Tetrahymena* following a temperature-nutritional step-up.

of dilution and temperature shift the RNA content increases during the first 3 hours when there is little or no cell division; it decreases markedly in the next 2 hours when the cells begin to divide but fails to increase beyond 200 μg RNA/10^6 cells during the next 2 hours. In the control cells the level increases to approximately 350 μg/10^6 cells. This experiment shows that the initial increase in RNA is insensitive to DMI at these concentrations and it is only after cell division that the cells are inhibited as compared to the control cells. Unfortunately, when temperature-cycling shifts are used to achieve the phase cycle of *Tetrahymena* close alignment between the patterns of cytokinesis and macromolecular synthesis was not always achieved, and thus a concomitant difference in the response of the cell to the drug ensued from one experiment to the next. We thus employed in subsequent studies, to be discussed below, the starvation–refeeding technique of Cameron and Jeter (1970) to analyze further the effect of DMI on the RNA synthesis of *Tetrahymena*.

To summarize the studies up to this point, DMI appeared to affect cells differentially as they proceeded toward cytokinesis and on stimulation to divide by dilution and temperature step-up. The differential sensitivity, moreover, was expressed as a blockage of RNA synthesis as the cells begin to divide. Since the cells were not synchronized fully, it is difficult to assess whether or not the drug-sensitive cells are those cells in the post synthetic, predivision phase of the cell cycle at the time of drug exposure.

A clear temporal alignment of nucleic acid synthesis and cell division in *Tetrahymena* was obtained with the Cameron and Jeter (1970) starvation–refeeding method of synchrony. Prior to starvation, cell stocks were grown over 24 hours to a cell density of 400,000–500,000 cells/ml in a 28° rotary shaker bath. An aliquot of 200,000 cells was harvested, washed, and resuspended in 100 ml of sterile phosphate buffer without shaking at 28° in 1-liter Erlenmeyer flasks. After 24 hours, an equal volume of 3% proteose peptone with 2% liver extract was added to each flask to attain a final cell density of approximately 100,000 cells/ml. The cell cultures were then grown at 28° in a rotary shaker bath until the untreated cell cultures had effectively doubled in cell number.

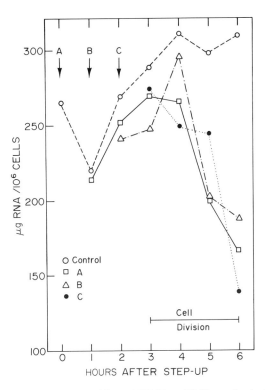

Fig. 4. Effect of the time of addition of DMI on RNA synthesis in *Tetrahymena* following a temperature-nutritional step-up.

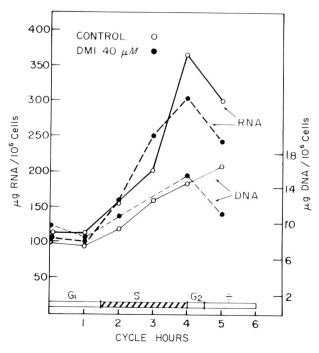

Fig. 5. Effect of 40 μM DMI on the temporal pattern and the extent of nucleic acid synthesis during the starved–refed cell cycle. Operational cell cycle markers are relative to pre- and post-DNA replication and pre-cell division.

Figure 5 shows the time sequence of RNA and DNA synthesis of starved–refed cells under the influence of 40 μM DMI. Control cells are also shown. The drug was added at the time of refeeding (0 hours). During the first hour after refeeding there is no synthesis of either RNA or DNA. Subsequently both nucleic acids increase simultaneously. RNA reaches its maximum level in both the treated and untreated cells by 4 hours after refeeding. The DNA content of the drug exposed cells doubles by the fourth hour, whereas the control cells continue to synthesize DNA until the fifth hour. In other experiments (Table I) the control cells also double their DNA content by the fourth hour. The onset of DNA replication may be delayed by 30 minutes, i.e., the initiation of DNA synthesis occurs between 1.5 to 2 hours after refeeding. In no instance have we found DNA synthesis to begin beyond the second hour after refeeding. This is at variance with the results previously reported by Cameron and Jeter (1970) and Rudick and Cameron (1972) who showed an initial increase of DNA at the third hour.

On the basis of experiments shown in Fig. 5, it was possible to assign operational cell cycle markers to this system: the G_1 period is the interval from 0–1.5 hours; the S period, 1.5–4 hours; G_2 period, 4–4.5 hours; and the division extends from 4.5–6 hours.

Variation of the DMI concentration did not produce either a significant decrease in the percent doubling of DNA throughout the S period or change the time of onset of DNA replication (Table I). RNA synthesis was decreased by desmethylimipramine, but this inhibition was also independent of an increase in DMI concentration from 21 to 40 μM or to 53 μM (Table I). Dependent on the period of the cell cycle considered, these drug concentrations reduced the accumulation of RNA by 15% to 34% of the control cells RNA (Table I).

Although the reduction of net RNA accumulation was independent of DMI concentration, a cycle dependence was apparent. For all the drug exposed cell cultures, the change in RNA content was lowest between the third to the fourth hours after refeeding. This possible cycle dependency was fur-

TABLE I

Effect of DMI Concentration on Nucleic Acid Content of Starved–Refed *Tetrahymena pyriformis* HSM

Treatment	DNA content (μg DNA/10^6 cells) during cycle intervals				Percent of doubling by fourth hour
	0+1 hour	2 hour	3 hour	4 hour	
0 μM DMI (control)	8.86	10.5	13.3	17.6	99.3
21 μM DMI	8.3	11.0	13.9	15.8	95.2
40 μM DMI	8.5	10.3	12.2	17.2	101.2
53 μM DMI	8.12	10.05	12.6	16.0	98.5

Treatment	Change in RNA content (μg RNA/10^6 cells) during cycle interval			
	Δ1–3	Δ2–3	Δ3–4	Δ1–4
0 μM DMI (control)	124.1	80.3	91.5	215.6
21 μM DMI	92.9 (0.75)[a]	60.2 (0.75)	60.4 (0.66)	152.9 (0.71)
40 μM DMI	105.4 (0.85)	67.6 (0.84)	71.2 (0.78)	176.6 (0.82)
53 μM DMI	102.8 (0.83)	62.7 (0.78)	54.2 (0.59)	157.0 (0.73)

[a] In parentheses: the ratio of the change in the experimental RNA content to the change in the control RNA content during the stated cycle interval.

ther pursued in experiments with delayed DMI addition (Jakoi and Padilla, 1973). Forty micromoles DMI given at 0, 1, and 2 hours into the cell cycle elicited a reduced RNA accumulation comparable to that of Table I. However, when the cell cultures were initially exposed to 40 μM DMI at 220 minutes into the cell cycle, no decrease in RNA content occurred (Jakoi and Padilla, 1973). Accumulation of RNA of the starved-refed cell cultures was reduced by 40 μM DMI only if the drug is present prior to the late portion of the S period. Addition of the drug during the late S period does not impair the accumulation of RNA during the end of the S period and in the G_2 period.

Cell division in the starved–refed cell cultures was inhibited by desmethylimipramine-HCl. Twenty-one, 40, and 53 μM DMI caused, respectively, 40%, 26%, and 36% inhibition of cell number doubling in these drug-treated cell populations by the sixth cell cycle hour (Table II), whereas the control cells showed an 8% inhibition of cell number doubling at this hour. Therefore, no correlation existed between an increase in DMI concentrations and an increase in nondividers.

Delayed addition of 40 μM DMI at 0, 1, and 2 hours after refeeding did not change the number of nondividers in the drug-treated cell cultures. However, an addition of 40 μM DMI at 220 minutes after refeeding was ineffective in preventing cell division (Jakoi and Padilla, 1973). In no instance did the drug-treated cells begin division later than 4.5 hours after refeeding, the cell cycle time at which the control cells begin to divide (Fig. 5).

In order to determine the rate of loss of prelabeled RNA during the

TABLE II

Effect of DMI Concentrations on Cell Density of Starved–Refed *Tetrahymena pyriformis* HSM

Treatment	No. 1[a]	N_5/No. 1[b]	N_6/No. 1[c]	Experimental/control	
				N_5/No. 1	N_6/No. 1
0 μM DMI (control)	0.12	1.47	1.82	1.00	1.00
21 μM DMI	0.11	1.11	1.20	0.76	0.66
40 μM DMI	0.13	1.31	1.49	0.89	0.82
53 μM DMI	0.11	1.10	1.28	0.75	0.70

[a] No. 1: average cell number $\times$ 10^6/ml determined at refeeding or at first hour after refeeding

[b] N_5/No. 1: ratio of average cell number $\times$ 10^6/ml at fifth cycle hour to average cell number $\times$ 10^6/ml at refeeding or first cycle hour

[c] N_6/No. 1: ratio of average cell number $\times$ 10^6/ml at sixth cycle hour to average cell number $\times$ 10^6 at refeeding or first cycle hour

starved–refed cell cycle and to examine whether or not DMI had any effect on this loss of RNA, the following experiment was conducted. The logarithmically growing cell stocks of *Tetrahymena* were exposed to 0.2 mCi of ^{3}H-uracil prior to starvation. These cells were transferred subsequently to 100 ml of inorganic phosphate buffer for starvation (28° with no shaking). Refeeding occurred 24 hours later as described previously. Aliquots were removed to determine the specific activity of counts per minute of ^{3}H-uracil incorporated per microgram of RNA for both the drug-treated and untreated cell cultures throughout the cell cycle (Jakoi and Padilla, 1973; see Fig. 6 for experimental methods). The curve in Fig. 6 gives the results of one of two such experiments during the several hours after switching *Tetrahymena* to unlabeled food. The amount of labeled RNA dropped rapidly, reflecting identical intracellular breakdown of RNA in both the 40 μM DMI-treated and control cells. Thus the lower accumulation of RNA in the drug-treated cells was most likely not due to more active ribonucleases in these cells, unless a specific degradation of RNA synthesized only during the starved–refed cycle occurred.

The improbability of RNA degradation in the drug-exposed cell cultures by ribonucleases directed us to consider either a reduced transcription of RNA in these cells or an inherent instability of the product RNA. Riboso-

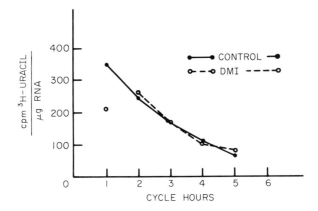

Fig. 6. Intracellular breakdown of prelabeled RNA during the starved–refed cell cycle of *Tetrahymena*. Prestarved, logarithmically growing cell cultures were exposed to 0.2 mCi of ^{3}H-uracil for 24 hours. Subsequently these cell cultures were starved and after 24 hours refed with cold twice-concentrated media. Forty micromolars DMI was added at refeeding to the experimental cell cultures. Tritium incorporation was measured on triplicate 5% TCA pellets. Total micrograms of RNA was determined in triplicate by a modified Schmidt–Thannhauser assay for nucleic acids (De Deken-Grenson and De Deken, 1959).

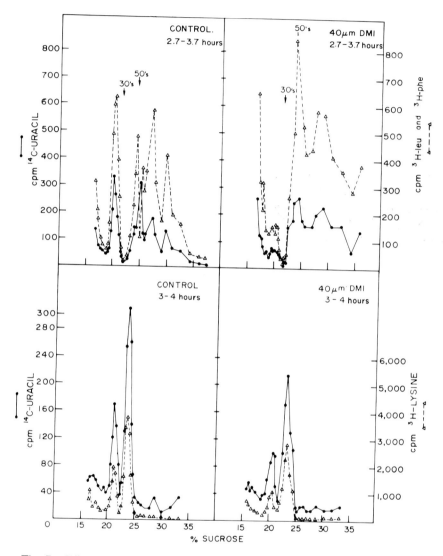

Fig. 7. Effect of 40 µM DMI on the rate sedimentation centrifugation of EDTA-derived ribosomal subunits of starved–refed *Tetahymena*. Starved–refed *Tetrahymena* were exposed at refeeding to 40 µM DMI. Both the drug-treated and control cells were exposed for 1 hour to 0.1 mCi ^{14}C-uracil, 0.2 mCi ^{3}H-leucine, and 0.2 mCi ^{3}H-phenylalanine per 400 ml of cells at 2.7 cycle hours or to 0.02 mCi ^{14}C-uracil and 0.1 mCi ^{3}H-lysine per 400 ml of cells at 3 cycle hours. Incorporation of label was stopped by pouring cell cultures over frozen media containing cold uracil and amino acids at either 3.7 cycle hours or at 4 cycle hours. Ribosomes were isolated from indole lysed cells at 100,000 g in 0.02 M triethanolamine-HCl, pH 7.2 with 0.05

mal RNA was assayed throughout the starved–refed cell cycle as a convenient stable marker of RNA synthesis. During the starvation period of *Tetrahymena,* RNA synthesis is slowed because of nucleolar fusion (Nilsson and Leick, 1970; Cameron *et al.,* 1971). Thus as active protein synthesis increases after refeeding and through cell division (Cameron *et al.,* 1971) an increase in ribosome utilization and, most likely, quantity is required. Furthermore, nascent synthesis of rRNA and ribosomal subunits must occur during the starved–refed cell cycle, since ribosomal subunits are not recycled or stored in cytoplasmic pools in *Tetrahymena* (Leick *et al.,* 1970).

The fidelity of ribosomal RNA synthesis in DMI-treated cells was approached by two experimental methods: (1) zonal rate sedimentation of polysomes isolated from drug-treated and untreated starved-refed *Tetrahymena* and (2) rate sedimentation on linear sucrose gradients of EDTA-derived ribosomal subunits, again from the DMI-treated and control cells. The first series of experiments cosedimented the DMI-treated polysomes with control cell polysomes using a dual label of ^{3}H-uracil and H_3 $^{32}PO_4$ (Jakoi *et al.,* 1973). Ribosomes were isolated from indole lysed cells after 30 minutes exposure to label during the late S and G_2 periods. A reduced incorporation of label was apparent for the polysomes isolated from 40 μM DMI-treated cells. Furthermore, the monosome peak from the drug-exposed cells appeared to sediment more rapidly than that isolated from the control cells. The monosome peak did not accumulate in the drug-treated cells in excess of that found in the control cells. In the second series of experiments, EDTA-derived ribosomal subunits were isolated from the early and late S cycle periods. Carbon-14-uracil was given 1 hour previous to isolation of the subunits (see Fig. 7 legend for experimental technique). The subunit profiles taken at the second hour after refeeding from the drug-treated and untreated cells showed no difference between them (Jakoi *et al.,* 1973). However, the ribosomal subunits isolated from 40 μM DMI-treated cells labeled with ^{14}C-uracil during 2.7 to 3.7 hours after refeeding showed a decreased incorporation of label into both the apparent 50 S and 30 S subunits (Fig. 7, top). The 50 S of the DMI-treated cells incorporated 91% of the ^{14}C-uracil taken up by the 50 S of the control cells; whereas the 30 S of the

---

M KCl, 1 mM MgCl$_2$ and 40% sucrose. Isolated ribosomes were treated for 10 minutes with 500 μl of 1 mM EDTA in triethanolamine buffer, pH 7.2. The ribosomal subunits were then layered on 15%–30% linear sucrose gradient, spun at 21,000 rpm for 12.5 hours in the SW 25.1 Spinco rotor (Leick *et al.,* 1970). Twenty-five drop fractions were collected from the bottom of each gradient. Counts were corrected for background, efficiency, and spillover. Sedimentation markers were ribosomal subunits obtained from *E. coli* K12 exposed to H_3 $^{32}PO_4$ for 24 hours and centrifuged on a separate 15%–30% linear sucrose gradient.

drug-treated cells had only 14% of the carbon label present in the control 30 S (Fig. 7). On the other hand, the incorporation of ^{3}H-leucine and ^{3}H-phenylalanine into the derived 50 S subunit of the 40 μM DMI-treated cells had 180% of that incorporated into the 50 S of the control cells. The 30 S isolated from DMI cells had 23% of the tritium label incorporated into the small subunit from the control cells. When the initial exposure to ^{14}C-uracil was delayed by 20 minutes, the profile of label incorporation into the EDTA derived subunits is shown in the lower portion of Fig. 7. Again a lower amount of ^{14}C-uracil and ^{3}H-lysine was taken up into the 50 S and 30 S ribosomal subunits of the 40 μM *DMI-treated cells*. The 50 S of these cells had incorporated 73% of both the carbon and tritium labels present in the control cells 50 S. The 30 S from the DMI-treated cells had taken up 56% of the carbon label and 59% of the tritium label found in the 30 S from the control cells at this time (Fig. 7, lower portion). Thus DMI altered both the RNA and the protein incorporation into ribosomal subunits produced during the late portions of the S period but did not change those synthesized early in the S period.

Two separate but coordinate inputs are required for the assembly of ribosomal subunits: (1) the transcription of rRNA and (2) the maturation of this RNA and its associated ribosomal proteins (Willems *et al.*, 1969; Kumar, 1970; Craig, 1971). In order to determine if the transcription of rRNA was involved in the reduction of rRNA synthesis in cells exposed to DMI, the following experiment was conducted. Macronuclei of drug-treated and untreated starved–refed *Tetrahymena* were isolated by the method of Gorovsky (1970) at different times during the cell cycle. These nuclei were assayed for the Mg^{2+}-activated DNA-dependent RNA polymerase (Jakoi *et al.*, 1973; Fig. 8). Although the Mg^{2+} activated enzyme had not been shown conclusively to be the ribosomal RNA polymerase in *Tetrahymena*, it has been verified that the higher eukaryotes do possess at least two polymerases: the nucleolar or Mg^{2+}-activated enzyme and the nucleoplasmic or Mn^{2+}-activated enzyme (Roeder and Rutter, 1970). Furthermore, Lee's (1969) work on isolated macronuclei from logarithmically growing and heat-shocked *Tetrahymena* suggests the presence of two distinct polymerases specific for either Mg^{2+} or Mn^{2+}. Figure 8 shows the initial rate of the Mg^{2+}-dependent RNA polymerase during the starved–refed cell cycle for both the 40 μM DMI-treated and untreated cells. Most striking is the presence of only 3% of the control cells activity in the 40 μM DMI-treated macronuclei at the beginning of the third cycle hour. This denotes a preferential cycle-dependent inhibition of the Mg^{2+}-activated DNA-dependent RNA polymerase in those cells treated with DMI.

Similar experiments have been conducted for the Mn^{2+}-activated enzyme (Jakoi *et al.*, 1973). Again a cycle dependence was apparent in that

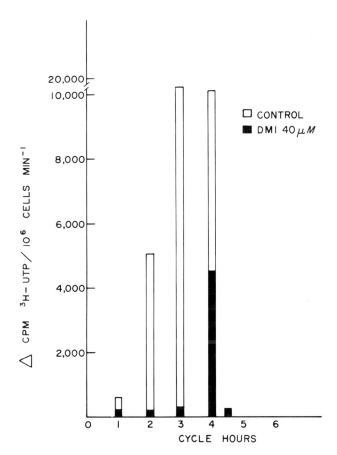

Fig. 8. Effect of 40 μM DMI on the Mg^{2+}-activated DNA-dependent RNA polymerase of isolated whole macronuclei from starved–refed *Tetrahymena* during the cell cycle (Jakoi *et al.*, 1973). The rates presented are calculated from the initial 0- to 1-minute linear incorporation. Later time points of the assay show a nonlinear incorporation of ^{3}H-UTP which approaches an asymptote by 10 minutes of incubation (data not shown).

this enzyme was also more depressed in the mid and late S and G_2 periods in the 40 μM DMI-treated cells.

III. Discussion and Summary

Desmethylimipramine-HCl inhibited cell division and RNA levels in synchronized *Tetrahymena pyriformis* HSM. This inhibition occured in both methods of synchrony used: the temperature step-up of Padilla and Cameron (1964) and the starvation–refeeding method of Cameron and Jeter (1970). However, the inhibition of cell division was dependent on the concentration of DMI present in the media for those cells subjected to a temperature step-up but not for those synchronized by the starvation–refeeding method. Thus, either a random or a nonsaturated specific response to DMI had occurred in the temperature-cycled cells such that an increase in the drug concentration given increased the occurrence of nondividing cells. The starved–refed cells, on the other hand, when exposed to increased concentrations of DMI, did not show a greater incidence of nondividing cells in the cell population. Furthermore, the inhibition of cell division was not saturated at the concentrations of DMI given to these cells because 21 and 53 μM DMI-treated cells had 40% and 36%, respectively, of the cell population not dividing, whereas the 40 μM DMI-treated cells had 26%.

The time of addition of DMI was critical for blocking cell division. In both synchronized cell cultures, the number of nondividing cells was reduced by exposing the cells to DMI 1 hour before the onset of cell division. Therefore, the required protein(s) or RNA(s) necessary for the initiation of cell division in *Tetrahymena* must be made at least 1 hour before the onset of cell fission. This is consistent with the interpretation of Tobey and co-workers (1966) and Cummins (1969) that a protein critical for the initiation of cell division is made at least 1 hour before cell division begins.

DMI also reduced the accumulation of RNA in *Tetrahymena* synchronized by temperature-recycling (Padilla and Cameron, 1964) and starvation–refeeding methods (Cameron and Jeter, 1970) of synchrony. This reduction occurred after the first cell division of those cells synchronized by a temperature step-up but before cell division during the late S and G_2 periods of the starved–refed cells. No correlation between an increase in drug concentration and a decrease in RNA was evident in either synchronized cell population. Thus the reduction in RNA content was apparently random.

However, a distinct cycle dependence was present in both synchronized cell populations. The temperature-cycled cells initially exposed to DMI at the first hour of the cell cycle contained comparatively more RNA during the hour before cell division than did those cells exposed to DMI at 0 hour or at the second hour of the cell cycle. Similarly, an initial addition of DMI at 220 minutes after refeeding no longer reduced RNA accumulation during the late S and G_2 periods of the starved–refed cells. Thus, although random, the accumulation of RNA in synchronized *Tetrahymena* was inhibited by exposure of the cells to DMI either at 0 and 2 hours into the temperature step-up cell cycle or during the G_1 and early S periods of the starved–refed cell cycle.

4. ADRENERGIC DRUGS ON THE CELL CYCLE 63

If the inhibition of cell division in the drug-treated cell cultures resulted from the absence of protein(s) or RNA(s) required for the initiation of cell division, then a random nondiscriminate reduction of RNA is sufficient to block cell division in these cells. Furthermore, this effect is independent of the method of synchrony used to align the nucleic acid metabolism of the cell population, yet is cycle dependent, provided that the synthesis of these division-related products are made at specific times in the cell cycle. Several aspects of RNA metabolism may be abnormal in the DMI-treated cells. Of these possibilities, an enhanced degradation of RNA by ribonucleases is not present. However, the following areas of nucleic acid metabolism potentially are produced in the drug-treated cell cultures: (1) DMI occludes portions of the template DNA, possibly by intercalation within the minor groove, thereby preventing RNA transcription; (2) DMI modifies both the Mg^{2+}- and Mn^{2+}-activated DNA-dependent RNA polymerases by altering the initiation and/or completion of reading; (3) DMI reduces protein translation because of the lower production of active ribosomal subunits; and (4) DMI degrades proteins nonspecifically by increasing pronase activity.

A long-term exposure to DMI would conceivably reduce the RNA levels indiscriminately in these cell cultures and subsequently cellular metabolism in general. Accumulation of protein and RNA deficiencies in the drug-treated cells would block eventually not only RNA and protein synthesis but also DNA replication, cytokinesis, and the production of energy; cellular death would follow. A partial restoration of some cellular metabolism may result from exogenously supplying ATP; however, those cellular processes blocked because of the absence of required RNA(s) or protein(s) would remain inhibited. Such a situation has been described by Wilkie and co-workers (Wilkie and Delhanty, 1970; Mittwoch and Wilkie, 1971) in human fibroblast cell cultures exposed over several hours to chlorimipramine, an analog of DMI. Thus, nondiscriminate inhibition of RNA transcription and protein synthesis is sufficient to explain both the chlorimipramine and the DMI effects on cellular metabolism of two unrelated eukaryotes.

Furthermore, in man, the physiological dosage of DMI administered is comparable to that given to the synchronized *Tetrahymena* cultures in the above experiments. However, in man, desmethylimipramine is given each day for 7–13 days. Recently, administration of DMI given to women in the first trimester of pregnancy has been correlated with a number of deformed births (McBride, 1972) in which the limb bud development had been arrested. This is consistent with a nondiscriminate inhibition of RNA and protein synthesis blocking cell proliferation. This inhibition would be most apparent in those cells highly active in cell growth and division.

References

Blum, J. J. (1968). Effect of adrenergically reactive drugs on peroxisomal enzymes in *Tetrahymena*. *Mol. Pharmacol.* **4**, 247–257.
Cameron, I. L., and Jeter, J. (1970). Synchronization of the cell cycle of *Tetrahymena* by starvation and refeeding. *J. Protozool.* **17**, 429–431.
Cameron, I. L., Griffin, E. E., and Rudick, M. J. (1971). Macromolecular events following refeeding of starved *Tetrahymena*. *Exp. Cell Res.* **65**, 265–272.
Connett, R. J., and Blum, J. J. (1971). Metabolic pathways in *Tetrahymena*: Distribution of carbon label by reactions of the tricarboxylic acid and glyoxylate cycles in normal and desmethylimipramine-treated cells. *Biochemistry* **10**, 3299–3309.
Craig, N. C. (1971). On the regulation of the synthesis of ribosomal proteins in L-cells. *J. Mol. Biol.* **55**, 129–134.
Cummins, J. E. (1969). Nuclear DNA replication and transcription during the cell cycle of *physarum*. In "The Cell Cycle: Gene Enzyme Interactions" (G. M. Padilla, I. L. Cameron, and G. L. Whitson, eds.), pp. 141–158. Academic Press, New York.
De Deken-Grenson, M., and De Deken, R. H. (1959). Elimination of substances interfering with nucleic acids estimation. *Biochim. Biophys. Acta* **31**, 195–207.
Gorovsky, M. A. (1970). Studies on nuclear structure and function in *Tetrahymena pyriformis*. II. Isolation of macro- and micronuclei. *J. Cell Biol.* **47**, 619–630.
Gyermek, L. (1966). Effects of imipramine-like anti-depressant agents on the autonomic nervous system. In "Workshop on Antidepressant Drugs of Non-MAO Inhibitor Type" (D. Efron and S. S. Ketz, eds.), No. 1, pp. 41–62. National Institute of Mental Health, Bethesda, Maryland.
Jakoi, E. R., and Padilla, G. M. (1973). In preparation.
Jakoi, E. R., Elrod, H., and Padilla, G. M. (1973). In preparation.
Jeon, K. W., and Lorch, I. J. (1971). Mitotic inhibition and nucleic acid synthesis in amoeba nuclei. *Nature (London), New Biol.* **231**, 91–93.
Kumar, A. (1970). Ribosome synthesis in *Tetrahymena pyriformis*. *J. Cell Biol.* **45**, 623–634.
Lee, Y. C. (1969). Nuclear RNA metabolism in normal and division synchronized *Tetrahymena*. Ph.D. Thesis, University of California, Los Angeles.
Leick, V., Engberg, J., and Emmersen, J. (1970). Nascent subribosomal particles in *Tetrahymena pyriformis*. *Eur. J. Biochem.* **13**, 238–246.
Linstead, D., and Wilkie, D. (1971). A comparative study of in vivo inhibition of mitochondrial function in saccharomyces cerevisiae by tricyclic and other centrally-acting drugs. *Biochem. Pharmacol.* **20**, 839–846.
McBride, W. G. (1972). Limb deformities associated with iminodibenzyl hydrochloride. *Med. J. Aust.* **1**, No. 10, 492.
Mittwoch, U., and Wilkie, D. (1971). The effect of chlorimipramine on DNA synthesis and mitosis in cultured human cells. *Brit. J. Exp. Pathol.* **52**, 186–191.
Nilsson, J. R., and Leick, V. (1970). Nucleolar organization and ribosome formation in *Tetrahymena pyriformis* GL. *Exp. Cell Res.* **60**, 361–372.
Padilla, G. M., and Cameron, I. L. (1964). Synchronization of cell division in *Tetrahymena pyriformis* by a repetitive temperature cycle. *J. Cell. Comp. Physiol.* **64**, 303–308.
Roeder, R. G., and Rutter, W. J. (1970). Specific nucleolar and nucleoplasmic RNA polymerases. *Proc. Nat. Acad. Sci. U. S.* **65**, 675–682.

Rudick, M. J., and Cameron, I. L. (1972). Regulation of DNA synthesis and cell division in starved-refed synchronized *Tetrahymena pyriformis* HSM. *Exp. Cell Res.* **70**, 411–416.

Salama, A. I., Insalaco, J. R., and Maxwell, R. A. (1971). Concerning the molecular requirements for the inhibition of the uptake of racemic ^{3}H norepinephrine into rat cerebral cortex slices by tricyclic antidepressants and related compounds. *J. Pharmacol. Exp. Ther.* **178**, 474–481.

Tobey, R., Anderson, E. C., and Petersen, D. F. (1966). RNA stability and protein synthesis in relation to the division of mammalian cells. *Proc. Nat. Acad. Sci. U. S.* **56**, 1520–1527.

Tsien, H. C., and Wattiaux, J. M. (1971). Effect of maternal age on DNA and RNA control of drosophila eggs. *Nature (London), New Biol.* **230**, 147–148.

Wilkie, D., and Delhanty, J. (1970). Effects of chlorimipramine on human cells in tissue culture. *Brit J. Exp. Biol.* **51**, 507–511.

Willems, M., Penman, M., and Penman, S. (1969). The regulation of RNA synthesis and processing in the nucleolus during inhibition of protein synthesis. *J. Cell Biol.* **41**, 177–187.

5

Action of Narcotic and Hallucinogenic Agents on the Cell Cycle

ARTHUR M. ZIMMERMAN
and DANIEL K. McCLEAN

I. Introduction	67
II. The Chemistry of Selected Drugs	69
III. Action of Drugs on Prokaryotic Cells	70
A. Cell Division and Growth	70
B. Drug Tolerance	71
C. Biochemical Studies	72
IV. Action of Drugs on Eukaryotic Cells	76
A. Cell Division and Growth	76
B. Drug Tolerance	79
C. Biochemical Studies	80
V. Present Studies on *Tetrahymena*	82
A. Introduction	82
B. Growth Rate, Cell Division, and Drug Tolerance	82
C. Macromolecular Synthesis	85
D. Nature of RNA Inhibition	87
VI. Concluding Remarks	90
References	91

I. Introduction

Considering the extensive medical and nonmedical use of narcotic and hallucinogenic drugs, relatively little is known about the action of these agents on cellular systems. Recently, Hollister (1971) remarked that "marijuana may be unique among drugs, in that more experimentation has been accomplished in man than in animals," furthermore, he states, "it may be necessary to look to additional animal studies to provide leads for pertinent future studies in man." This statement bears eloquent witness to the fact that most research on drugs has been concerned with "social behavior" and the biochemical mechanisms responsible for physiological activity have been relatively unexplored. The purpose of this chapter is to review the action of selected drugs on single cellular systems with an aim to evaluate their effectiveness on growth, cell division, and cellular biosynthesis.

Growth and cell division are sensitive parameters for evaluating drug action. Information from such studies can be best evaluated when combined with biochemical studies that show specifically which systems are affected. During the last several years there have been numerous reports indicating that simple cellular systems are useful in analyzing the effects of drugs. The advantages of employing unicellular systems are associated with the facility of obtaining large amounts of material necessary for biochemical research, the homogeneity of the systems, and the relative simplicity of these systems compared to the complexity of multicellular organisms.

The action of narcotic agents on single cell systems has been studied since the earliest reports over a century ago. Among the first cellular systems investigated were protozoa. Binz (1867), Neresheimer (1903), and Pick and Wasicky (1915) reported the action of morphine on *Amoeba* and *Stentor*. Since then, little has been done on the effects of opium alkaloids on protozoa. A thorough review on the effects of opium alkaloids in a variety of cell types can be found in the work of Krueger *et al.* (1941). Since then there have been several reviews concerning the pharmacology of narcotic drugs (see, for example, Seevers and Deneau, 1963) and biochemical responses to narcotic drugs (Clouet, 1968, 1971b). A useful treatise on the chemistry and pharmacology of hashish has been edited by Wolstenholme and Knight (1965). More recently the proceedings of a conference organized by The Institute for the Study of Drug Dependence has been assembled and edited by Joyce and Curry (1970). The report is primarily concerned with the botany, chemistry, and pharmacology of cannabis. Additional reviews on the action of cannabis also appear in the current literature (Cotten, 1971; Singer, 1972; Le Dain *et al.,* 1972). For a comprehensive review of the action of narcotic drugs the reader is referred to the recently published monograph "Narcotic Drugs: Biochemical Pharmacology" (Clouet, 1971a).

The authors wish to emphasize at the onset of this review that caution should be exercised by the reader in evaluating data obtained from studying the action of drugs on cellular systems and in extrapolating these data to interpret the action of narcotic agents on man.

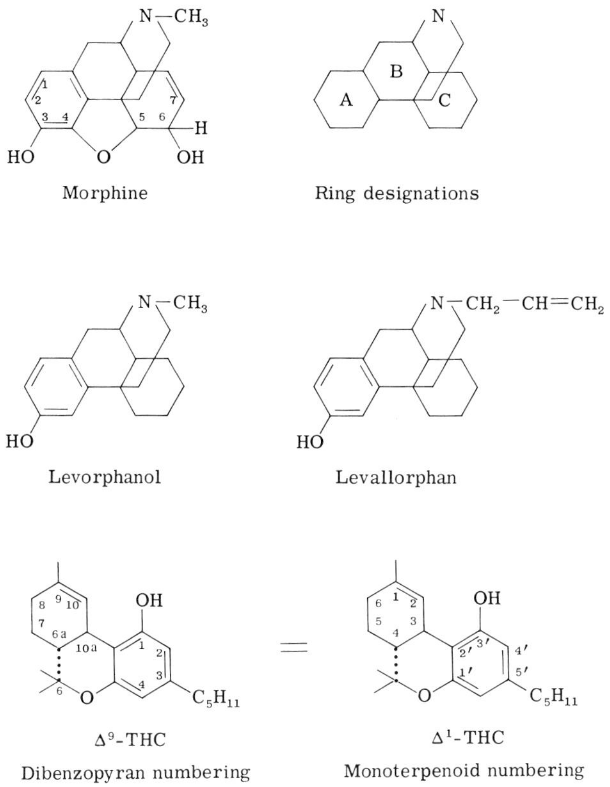

5. ACTION OF NARCOTIC AND HALLUCINOGENIC AGENTS

Fig. 1. Structural diagrams of several typical drugs discussed in this review.

II. The Chemistry of Selected Drugs

In the present review the authors have attempted to restrict the discussion to four drugs: morphine, levorphanol, levallorphan, and Δ^9-tetrahydrocannabinol (Δ^9-THC). Although other agents are included in this review, most of the research conducted in the authors' laboratory is associated with the above compounds (Fig. 1).

Morphine is one of the chief alkaloids of opium. The opium alkaloids are classified into two groups—phenanthrene and benzylisoquinoline; morphine belongs to the phenanthrene class of drugs (Goodman and Gilman, 1970). The structure–activity relationships of morphine have been extensively investigated. From the ring designations one can see that morphine may be considered as a derivative of phenanthrene, bridged across by oxygen and having a nitrogen-containing chain (ethanamine). It has been determined that the analgesic action of morphine is dependent on the phenolic hydroxyl group on the 3 position. Masking this group reduces the analgesic effects. Moreover, by masking or removing the alcoholic hydroxyl group, on the 6 position, the narcotic and respiratory depressant properties are intensified. There are several analogs of morphine in which certain properties are enhanced. For example, acetylation of the 3- and 6-hydroxyl groups of morphine yields diacetylmorphine (heroin), which is known for its euphoric effects.

A derived morphinan, which bears a close resemblance to morphine, is levorphanol (1-N-methyl-3-hydroxymorphinan). The structure–activity relationships of levorphanol are similar to those of morphine (Casy, 1971). It has been reported that in man, levorphanol is more potent than morphine; 2 or 3 mg of levorphanol are equal to approximately 10 mg of morphine (Casy, 1971). Replacement of the N-methyl group by a N-allyl group forms a potent morphine antagonist called levallorphan. It is interesting that levallorphan (N-allyl-3-hydroxymorphinan) is not classified as a narcotic agent and thus it is more readily available for investigative use.

In the last several years there has been renewed interest in the chemistry and pharmacology of marijuana (Mechoulam, 1970; Mechoulam et al., 1970; Dewey et al., 1970; Joyce and Curry, 1970; Neumeyer and Shagoury, 1971). It has been established that the major psychomimetically active compound found in marijuana is Δ^9-tetrahydrocannabinol (Δ^9-THC). Although most of the cannabinoids have been isolated and characterized, there has been no comparable advance in pharmacological and clinical areas of research (Mechoulam, 1970). The nomenclature of the tetrahydrocannabinols is rather confusing; two nomenclature systems are in use today. In one system of nomenclature, the cannabinols are regarded as substituted monoterpenoids. In the second systems, the rules for numbering of pyran-type compounds are used. Thus Δ^9-THC would correspond to the pyran-type nomenclature (dibenzopyran numbering) which is equivalent to Δ^1-THC which is representative of the substituted monoterpenoids nomenclature. In the present review the pyran-type nomenclature is used.

III. Action of Drugs on Prokaryotic Cells

A. Cell Division and Growth

For evaluating the action of drugs on growth and cell division the prokaryotic cells, *Escherichia coli* and *Staphylococcus aureus* have been favored as experimental organisms by Simon and co-workers (Simon, 1963, 1964; Simon and Van Praag, 1964a,b; Simon et al., 1966) and Gale (1970a,b).

Simon (1964) reported the relative potencies of several morphine congeners on bacterial growth. Levorphanol was found to be less effective in inhibiting growth in *E. coli* than its non-narcotic *N*-allyl analog, levallorphan. The order of effectiveness of the series of morphinans tested was as follows: levallorphan > levorphanol > dextrophan > *N*-allylnormorphine > morphine. The study clearly demonstrated that morphine was the least effective growth inhibitor. For example, 3 mM levorphanol caused complete inhibition of the growth of *E. coli* (strain W); however, morphine at the same concentration had no effect. Growth inhibition due to levorphanol was reversed when the drug was removed from the medium.

Simon (1964) also demonstrated that the effectiveness of levorphanol was strongly dependent on pH of the incubation medium. At a pH near neutrality, marked growth inhibition was demonstrated with 3 to 4 mM levorphanol. Below pH 6 there was no inhibition of growth; however, when the pH of the medium was raised to 8.5, inhibition of growth was manifest at 0.1 mM and complete at 1.0 mM.

The K-12 and K-15 strains of *E. coli* reacted similarly to levorphanol with slight variations in sensitivity. Growth of *Diplococcus pneumoniae* type II was inhibited at 0.45 mM levorphanol (pH 7.3–7.8), and at concentrations greater than 1.3 mM the organisms were killed. This is in contrast to the cultures of *E. coli* in which reversible growth inhibition was found at similar concentrations. Other bacteria such as *Bacillus subtilus* (Simon, 1964), *Bacillus megaterium,* and *Micrococcus leysodeikticus* (Simon, 1971) are also quite sensitive to growth inhibition by levorphanol.

Levorphanol and levallorphan are approximately ten times more effective than heroin (the synthetic diacetyl derivative of morphine) in preventing growth of *Staphylococcus aureus*. Gale (1970a) reported no visible growth of cell cultures incubated for 24 hours in 5 mM levorphanol or levallorphan. In contrast to this low growth-inhibiting concentration, a heroin concentration of 50 mM was necessary to prevent growth.

B. Drug Tolerance

Bacterial growth rate is a useful parameter for determining drug tolerance. Gale (1970a) has been able to culture *Staphylococcus aureus* in 20 mM heroin. By serial subculturing cells into media containing higher concentrations of heroin, a culture was developed that, after ten passages, would grow in 200 mM heroin. He reported that under these conditions organisms could then grow in 30 mM levorphanol, a concentration sixfold higher than that which is necessary to prevent growth of control cultures. Thus, under these conditions the cells developed a tolerance to heroin as well as to levorphanol.

As early as 1964, Simon attempted to isolate drug-resistant mutants of *E. coli* for use as an experimental model system. Two mutants, S-3 and S-6, were developed which displayed resistance to high concentrations of levorphanol.

C. Biochemical Studies

A comprehensive review concerning the action of narcotic analgesic drugs on bacteria has been reported recently by Simon (1971). For the past decade he and his collaborators have been investigating the action of morphinans on macromolecular synthesis in microorganisms. Using *E. coli* as a model system, he has been able to study a variety of biochemical events.

1. Nucleic Acid and Protein Synthesis

In a series of articles, Simon and co-workers (Simon, 1963; Simon and Van Praag, 1964a,b; Röschenthaler *et al.*, 1969), Greene and Magasanik (1967), and Gale (1970a) have established that levorphanol and levallorphan affect RNA, DNA, and protein synthesis. The early studies (Simon, 1963) with *E. coli* established that the effect of levorphanol was primarily on RNA synthesis. Subsequent investigations (Simon and Van Praag, 1964a,b) have shown that levorphanol displays a selected inhibition of ribosomal RNA. A levorphanol concentration (1.35 mM) that exerted 90%–95% inhibition of ribosomal RNA (rRNA) synthesis nevertheless permitted the synthesis of messenger RNA (mRNA) and transfer RNA (tRNA) to proceed at about one-half the rate found in a growing control culture.

Simon and Van Praag (1964a) reported that RNA synthesis was decidedly more sensitive to levorphanol than either protein or DNA synthesis. Whereas 80%–90% of the RNA synthesis was inhibited at a concentration of 1 mM levorphanol, DNA synthesis continued essentially undiminished for at least one generation and protein synthesis continued at about one-half the rate of a growing control culture. Röschenthaler *et al.* (1969) working with *E. coli* found similar effects with levallorphan. They reported that 1.54 mM levallorphan inhibits RNA biosynthesis by 70%, whereas protein synthesis was only inhibited by 14%.

In contrast to these studies, Greene and Magasanik (1967), who also work with *E. coli*, reported that fivefold higher concentrations of levallorphan (5 mM) inhibited RNA, DNA, and protein synthesis more than 90%. The action of levallorphan on the inhibition of DNA was shown to be very rapid. The incorporation of radioactive thymidine was inhibited 4 minutes after the addition of 5 mM levallorphan. Levorphanol (5 mM) was found to be a slightly less potent inhibitor of RNA and protein synthesis.

Recently Gale (1970a), studying the action of heroin in *S. aureus,* reported that concentrations of 30–60 mM inhibited protein and nucleic acid synthesis by 40%–50%; however, these same concentrations had a negligible effect on amino acid incorporation in an *in vitro* synthesizing system. Both levorphanol and levallorphan are approximately ten times more effective (on a molar basis) than heroin as inhibitors of protein and nucleic acid synthesis. Whereas 30–60 mM heroin is required to show significant protein and nucleic acid inhibition, only 1–5 mM levorphanol or levallorphan is necessary to show comparable effects. It should be noted that some of the morphinans affect transport of amino acids. Levorphanol, levallorphan, and heroin cause an inhibition of the transport of some amino acids and stimulation of others.

2. *The Nature of RNA Inhibition*

Since the earliest studies had shown that RNA synthesis was markedly inhibited by both levorphanol and levallorphan, investigation into the nature of the inhibition was a natural consequence. Simon and Van Praag (1964b) established that 1.35 mM levorphanol, which inhibited ribosomal RNA synthesis by 90%–95%, allowed continuous synthesis of mRNA and tRNA at about one-half the rate of that found in a growing culture. Sucrose density gradient fractionation of RNA obtained from cells incubated in the presence of levorphanol (1.35 mM) and ^{14}C-uracil for 60 minutes, followed by a 15 minute chase, displayed little radioactivity in the 16 S and 23 S fractions, which was about 6%–8% of the control. However, tRNA from these levorphanol-treated cells showed radioactivity that was about 35%–40% of the control value.

In order to elucidate the action of levallorphan and levorphanol on mRNA synthesis, Simon and Van Praag (1964b) designed two series of experiments in which they used bacterial phage replication and induction of β-galactosidase synthesis. Working on the assumption that mRNA carries the genetic information for the production of phage specific proteins, they investigated the effect of levallorphan on the production of phages T6 and gamma in *E. coli.* Levallorphan (1.3 mM) had little effect on inhibiting DNA bacterial phage production. Thus it was proposed that this agent (which markedly inhibits RNA synthesis) had little effect on the synthesis of mRNA. In their other experiments, a system involving the induction of β galactosidase was used. The induced synthesis of β-galactosidase in *E. coli* provides an unique system for studying transcription and translation. They found that induction of β-galactosidase continued for 60 minutes in the presence of levorphanol at about one-half the rate of the control culture; at the same time only 10%–15% of normal RNA synthesis was taking place.

Röschenthaler et al. (1969), studying the effect of levallorphan on RNA synthesis in E. coli, reported that a concentration of 1.54 mM inhibited total RNA synthesis by 70% and protein synthesis by 14%. By fractionating the RNA from drug-treated cells they found that the synthesis of high molecular weight rRNA (16 S and 23 S) was inhibited more than 4 S RNA, and mRNA was only slightly inhibited. Moreover, the 5 S RNA synthesis was inhibited as much as the high molecular weight RNA.

Greene and Magasanik (1967), using a higher concentration of levallorphan (5 mM), reported essentially complete inhibition of mRNA in E. coli as determined by sucrose density gradient analysis and by β-galactosidase induction studies. These workers also found that levallorphan acted to stabilize preexisting mRNA. Bacillus subtilis was chosen as an organism for investigation because of its sensitivity to actinomycin D and levallorphan (e.g., 5 mM levallorphan caused more than 90% inhibition of protein and RNA synthesis). A culture was given a 30-second pulse of ^{14}C-uracil and treated with actinomycin D in the presence and absence of levallorphan. The radioactivity (of the acid insoluble fraction) decreased more than 80% within 6 minutes in the actinomycin-D-treated cultures. However, the cultures treated with levallorphan plus the actinomycin D showed a reduction of radioactivity of less than 20%. These studies clearly demonstrated that levallorphan protects mRNA from decay.

There appears to be a discrepancy in the studies of Greene and Magasanik (1967) and Simon and co-workers (Simon and Van Praag, 1964a,b; Röschenthaler et al., 1969; Simon, 1971) concerning the mechanisms of action of levorphanol and levallorphan. Greene and Magasanik employed 5 mM concentrations of drug and reported inhibition of all RNA species. They emphasized that Simon and co-workers obtained selective rRNA inhibition (with 1–2 mM concentrations of drug) due to a "shift-down" in bacterial metabolism. In other culture systems it has been shown that when a culture is shifted from a good growth condition to a minimal growth condition ("shift down") rRNA synthesis is selectively inhibited. Thus it is possible that the selectivity reported by Simon and Van Praag (1964b) reflects the shift-down in bacterial metabolism caused by a marginal concentration of the drug.

The action of levorphanol has also been investigated on RNA phage replication and macromolecular synthesis. Simon et al. (1970a) reported that levorphanol (1.6 mM) almost completely inhibited the synthesis of phage RNA and protein as well as the replication of RNA phage (MS-2 and Q_β).

5. ACTION OF NARCOTIC AND HALLUCINOGENIC AGENTS

They determined that levorphanol markedly reduced the replication of MS-2 at concentrations that had a minimal effect on the bacterial host (*E. coli* Q13).

3. *Drugs on ATP and Oxygen Consumption*

Although Greene and Magasanik (1967) demonstrated that RNA and protein synthesis decreased in levallorphan-treated cells, they consider that the major action of levallorphan is on the cellular ATPase system. In their studies with *E. coli* they reported a rapid decrease in the ATP content of the cell, which was followed by a leakage of ATP into the medium. The reduction of cellular ATP was accompanied by an increase in ADP and AMP, which were also seen to leak into the surrounding medium. An investigation of the guanine nucleotides in levallorphan-treated cells showed that GTP decreased 10 minutes after treatment; GDP was also found in the medium. The authors suggested that the change of GTP reflected an event secondary to ATP loss or to the general destruction of nucleotide triphosphates. Levallorphan neither reduced the oxygen consumption nor did it cause an uncoupling of oxidative phosphorylation. The inability to demonstrate an effect of levallorphan on the cellular energy production was the basis by which the authors proposed that levallorphan acted through an activation of cellular ATPase, which in turn resulted in a reduction of ATP.

4. *Action of Drugs on Transport*

Since Greene and Magasanik (1967) have shown that adenine and guanine nucleotides leak out of levallorphan-treated cells, an obvious question remaining to be answered pertained to the effect of this agent on cellular permeability. They reported that levallorphan neither increased nor decreased the rate of entry of galactosides into either permease positive or negative *E. coli* cells. In addition, levallorphan did not cause radioactively labeled proteins to leak out of cells into the surrounding medium.

In order to elucidate further the mechanism of action of levorphanol, Simon and co-workers (1966, 1970b) investigated the relationship between polyamines and levorphanol. Previous reports have suggested that polyamines are related to the regulation of ribosomal RNA synthesis in *E. coli* (Raina and Cohen, 1966; Raina *et al.*, 1967). Recently, Simon *et al.* (1970b) reported on the effect of levorphanol on the transport of polyamines (putrescine and spermidine) in *E. Coli*. They determined that levorphanol stimulated the efflux of the polyamines. Levorphanol inhibited the uptake of putrescine at a drug concentration (0.22 mM) that did not reduce RNA synthesis and the ATP concentration of the cells. The authors concluded that inhibition of uptake and stimulation of efflux of putrescine reflects two separate effects of levorphanol. They also studied the effects of levorphanol on amino acid transport. They reported that transport of radioactive amino acids (leucine, serine, tryptophan, and an algal hydrolyzate) into the acid-soluble pool of chloramphenicol-treated *E. coli* continued for 10 minutes in the presence of levorphanol; this was followed by a sharp and rapid loss of radioactivity.

The transport of amino acids into *Staphylococcus aureus* is altered markedly by levorphanol, levallorphan, and heroin (Gale, 1970a). The activities of these agents on amino acid transport were similar to their effects on growth, i.e., levorphanol and levallorphan were about ten times as effective as heroin in affecting transport. These morphine analogs increased the rate of accumulation of aspartate, glutamate, and alanine in the free amino acid pool; however, the transport of lysine and proline were inhibited. While heroin acted to stimulate the transport of some amino acids and inhibit others, valine, arginine, histidine, leucine, and glycine were not affected by heroin (50 mM). Gale (1970b) reported that heroin stimulated the increase of glycerol into the phospholipid fraction of cells at concentrations that inhibited the transport of lysine. Similar results were found with levorphanol, levallorphan, naloxone, and dextrorphan. However, it has recently been shown that the incorporation of radioactive precursors into total bacterial lipids of *E. coli* was reduced following incubation with levorphanol (Wurster *et al.*, 1971a).

Löser and co-workers (1971) have established that the transfer of an R factor from *Proteus rettgeri* to *E. coli* (W677) during mating was inhibited at concentrations of levallorphan that did not affect the growth of the recipient or donor cells. They proposed that levallorphan caused damage to the sex-pili and to the cell membrane as a whole. In addition, they have reported that Mg^{2+} acts to reverse this phenomenon.

IV. Action of Drugs on Eukaryotic Cells

Our understanding of the action of drugs on eukaryotic cells arises primarily from investigations on cultured HeLa cells. These studies have been supported by the early work with embryonic chick material and to a smaller degree with work on protozoa.

A. Cell Division and Growth

Although it is the intention of the authors to comment on the recent literature published during the last decade, it would be unfortunate if one did not mention briefly the early investigations that laid the foundation for the

5. ACTION OF NARCOTIC AND HALLUCINOGENIC AGENTS

more current studies. Painter et al. (1949) and Pomerat and Leake (1954) investigated the effects of numerous drugs on chick embryos explants. Their criteria consisted of determining the minimum concentration of drug that would cause a total inhibition of tissue growth. In addition they established the "least injurious dose, that is a concentration of a chemical that induced the slightest demonstrable effect" on tissue growth.

A more recent evaluation of the action of selected narcotic agents on mammalian cells can be found in the reports of Noteboom and Mueller (1966, 1969). Using HeLa cells as a test system, they investigated cellular growth as affected by a large series of morphinans and related compounds. Studying HeLa cells grown on monolayers, they determined the percent inhibition of growth at three specified drug concentrations (1.0, 0.1, and 0.01 mM). Of the drugs tested, the N-allyl derivatives (levallorphan and N-allyl-3-methoxymorphinan) were slightly more active than the N-methyl narcotic analogs (levorphanol and N-methyl-3-methoxymorphinan). Since masking the phenolic function with a methoxy group in either levorphanol or levallorphan did not affect the activity, they suggested that the phenolic group was not involved in the growth-inhibitory action. However, the C ring of the molecules appears to play an important role in the inhibitory action of these compounds. Morphine, codeine, and dihydrocodeine and dihydrocodeinone, which have substitutions on the C ring, were strikingly less effective growth inhibitors than N-allylnormorphine, levorphanol, levallorphan, N-methyl-3-methoxymorphinan, and N-allyl-3-methoxymorphinan. The authors concluded that the C-D rings may be concerned specifically with the inhibitory action, but one must also consider the effects of charge and polarity of the compounds. Such structure-activity relationship studies are essential and provide insight for evaluating these drugs on cellular systems.

Simon (1971) has investigated the effects of morphine and related compounds on the cloning of HeLa cells. The inhibition of clone formation by the morphinans was found to be a tenfold more sensitive method for assaying activity than the inhibition of growth in monolayer cultures. By determining the dose that would reduce the clone number by 50% of the control value, a sensitive tool was available for evaluating drug action. Simon reported the relative toxicity of levallorphan > levorphanol > dextrorphan > demerol > N-allylnormorphine > morphine.

Recently the action of Δ^8- and Δ^9-tetrahydrocannabinol (THC), which are the only active levorotatory forms found in natural marijuana, have been investigated in order to determine if these agents would induce chromosomal aberrations when added to cultured human leukocytes (Neu et al., 1970). When the cells were treated with Δ^8-THC (at several concentrations between 30–45 μg/ml) there was a decrease in the mitotic index, which corresponded to the increasing drug dosage. However, no structural rearrangements were observed in the Δ^8-THC-treated cultures, and there was no increase in the number of breaks or gaps from that found in nontreated control cells. Preliminary studies with Δ^9-THC gave results similar to those found with the Δ^8 compound. It is interesting that the results from this study were similar to that found by Tjio et al. (1969) who reported that the hallucinogen, lysergic acid diethylamide (LSD), did not produce damage to human lymphocyte chromosomes.

Although protozoa offer useful model systems for studying drug kinetics, they have been neglected as an experimental tool. Using the large one-celled protozoan, *Amoeba proteus*, the effects of morphine on sol–gel transformation were investigated by Zimmerman (1967). It has been well established that the formation and maintenance of pseudopodia are dependent on the structural characteristics of the cortical plasmagel (ectoplasm). When the gelated ectoplasm is transformed into a sol by low temperature, chemical agents, or by pressure the cell rounds up due to the action of surface forces. The pseudopodial instability, which results from the application of hydrostatic pressure, has been treated quantitatively and has been shown to be due to a shift in sol–gel equilibrium within the cell toward the structurally weaker sol (Landau et al., 1954; Zimmerman and Zimmerman, 1970). Thus the solating action of pressure can be used to evaluate the action of narcotic agents on pseudopodial stability. When amoebae were exposed systematically to increasing concentrations of morphine (0.01 to 2 mM) under standardized compression–duration treatment, there was a corresponding increase in pseudopodial stability. At the highest concentration studied, (2 mM) following a drug treatment of 60 minutes, the difference in stability of the morphine-treated cells compared to nontreated control cells was equivalent to a pressure differential, which ranged from 500–1000 psi. Extending the duration of exposure of the amoebae to morphine did not augment the stability characteristics of the cells. Amoebae displayed some degree of recovery 30 minutes after removal of the morphine; extending the recovery time up to 2 hours produced a somewhat greater recovery. However, even after 3 days the cells did not return to the control value. When N-allylnormorphine was used in combination with the morphine, the stabilizing influence of morphine was reversed (Fig. 2). This study suggests that morphine affects the sol–gel transformation in the amoeba and that the antagonistic action of N-allylnormorphine is exhibited through competition with morphine for the same receptor site.

In addition to the activity of the morphinans on growth inhibition and cell division, it has recently been demonstrated that levorphanol and levallor-

5. ACTION OF NARCOTIC AND HALLUCINOGENIC AGENTS

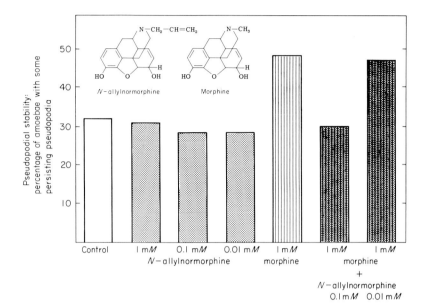

Fig. 2. The action of morphine and *N*-allylnormorphine, separately and in combination, on pseudopodial stability of *Amoeba proteus*. The percentage of amoebae that display some pseudopodia in the presence of these drugs following a standardized treatment of 5000 psi for 20 minutes at 20° is shown. Amoebae treated with 1 mM morphine display increased stability; when *N*-allylnormorphine (0.1 mM) is used in combination with morphine (1 mM), the morphine stabilizing action is counteracted. (From the work of Zimmerman, 1967.)

phan reversibly blocked conduction in the giant axon of the squid. Very low concentrations of levorphanol (0.05 mM) blocked repetitive or spontaneous activity induced by decreasing divalent cations. Moreover, levorphanol did not alter the penetration of ^{14}C-acetylcholine into nerve tissue (Simon and Rosenberg, 1970).

B. Drug Tolerance

There have been several attempts to develop cell cultures that display a tolerance for narcotic drugs. One of the earliest reports was by Sasaki (1938) in which he was able to show morphine tolerance in chick fibroblasts. Heubner *et al.* (1952) were also able to develop tolerance to morphine in muscle explants from chick embryos. More recently, Corssen and Skora (1964) provided evidence of tolerance and physical dependence in cultured human cells derived from a cervical carcinoma. It is interesting to note that most reports in the literature reflect studies in which tolerance was developed, and there appear to be no published reports of "failure to demonstrate tolerance or physical dependence in cultured cells exposed to morphine" (Ruffin et al., 1969). Ruffin and co-workers (1969) studied the effects of prolonged morphine exposure to human heteroploid epithelial-like cell line H. Ep. 2 derived from carcinoma larynx. They reported no evidence of tolerance or physical dependence to morphine.

Cox and Osman (1970) have reported that rats did not display tolerance to the analgesic effects of morphine when they were treated with RNA or protein inhibitors (e.g., actinomycin D, 5-fluorouracil, cycloheximide, and puromycin).

Development of tolerance to morphine has been shown for some protozoans. Zimmerman (1967) was able to demonstrate that *Amoeba proteus* develop tolerance to lethal morphine concentrations. Cells were cultured in a low nonlethal concentration of morphine and then transferred to more lethal concentrations. *Amoeba proteus* were grown in 2 mM morphine for 5–7 days and then transferred to test concentrations of morphine (3 and 5 mM). Control cells were exposed to test concentrations without any morphine pretreatment. When the nonpretreated cells were transferred to 5 mM morphine, 57% and 89% were moribund at 5 and 24 hours, respectively. However, in cells that were pretreated with morphine 13% and 61% were moribund at 5 and 24 hours, respectively. At the lower morphine concentration (3 mM) tolerance to morphine was also evident.

It is of interest to note that in the authors' laboratory it has not been possible to demonstrate tolerance in the ciliate protozoan, *T. pyriformis* (see Section V,B).

C. Biochemical Studies

The effects of levorphanol and levallorphan on the biosynthetic activity of HeLa cells has been investigated by Noteboom and Mueller (1966). These agents caused a marked reduction of RNA and protein synthesis; however, they had no appreciable effect on DNA and phospholipid synthesis. The inhibitory action of the agents seems to be related to the reduction in the protein-synthesizing capacity of the cell. The authors proposed that the reduction of protein synthesis resulted from interference with the utilization of mRNA. This idea was supported by experiments in which it was shown that ribosomes isolated from levallorphan-treated cells did not act in a cell free protein-synthesizing system unless synthetic messenger (poly U) was added to the system. Additional support for this hypothesis arises from sucrose gradient profiles, which showed a reduction in the amount of polysomal material that was recovered from levallorphan-treated cells. In a more recent report, the authors (Noteboom and Mueller, 1969) studied a large series of morphinans and related compounds; their results confirmed and elaborated on their earlier biochemical findings.

It has recently been reported that levorphanol reduces the bactericidal capacity of leukocytes. Levorphanol decreased both phagocytosis and metabolism of the granulocytes (Zucker-Franklin *et al.*, 1971; Wurster *et al.*, 1971b).

Brdar and Fromageot (1970) have found that levallorphan (1 mM) completely suppressed RNA and DNA synthesis in cultured mouse fibroblast (strain L-929) cells. Levallorphan at concentrations of 0.5 to 0.75 mM did not affect the phosphorylation of adenosine or the relative amount of the various adenine nucleotides in the acid soluble pool. However, at those concentrations levallorphan markedly inhibited or suppressed the synthesis of viral (Mengovirus) RNA. The authors suggested that the comparable sensitivities of viral and cellular RNA synthesis is a secondary manifestation of some other effect on the cell, as yet undetermined.

Rossman *et al.* (1971) found that levorphanol inhibited the replication of Sindbis virus as well as synthesis of RNA and protein in chick embryo cells and in rabbit kidney cells. Levorphanol also displayed cytotoxic activity. They suggested that levorphanol exerts its effect through alteration of the cell membrane, either directly or as a result of a metabolic alteration.

It is interesting to note that Sakiyama *et al.* (1969) have reported an inhibitory action of levorphanol on ribosomal RNA synthesis in nucleoli isolated from rat liver. These studies showed that levorphanol inhibited specifically nucleolar ribosomal RNA synthesis; however, it had no effect on messenger or transfer RNA. Not only does morphine reduce the rRNA synthesis in rat liver, but ribosomes isolated from the liver or morphine-treated rats displayed a reduced capacity to incorporate ^{14}C-leucine into protein. Moreover, ribosomes isolated from the brain of rats treated with morphine *in vivo* also exhibited reduced protein synthesis *in vitro* (Clouet and Ratner, 1968). Kuschinsky (1971) has found that the incorporation of ^{14}C-leucine into the proteins of synaptosomes of mouse brain was decreased by 20% following the administration of morphine; the uptake of ^{14}C-leucine into mitochondria was not affected. It was proposed that the uptake of ^{14}C-leucine into synaptosomes is due to inhibition of axoplasmic transport of proteins. The inhibitory action of the drugs was also found to modify the specific activity of cellular nucleases. Datta and Antopol (1971) have recently reported that the chronic administration of morphine to mice caused a decrease in the specific activities of ribonuclease and deoxyribonuclease in liver and brain cells as well as in subcellular fractions (nuclei and mitochondria).

The mechanism of action of the tetrahydrocannabinols have not been as extensively investigated as the morphinans. Recently it has been shown that Δ^9-THC combines with hepatic microsomes, *in vitro,* and is an effective inhibitor of microsomal drug metabolism (Cohen *et al.,* 1971). It has also been demonstrated that Δ^9-THC protects erythrocytes against hemolysis (Chari-Bitron, 1971).

V. Present Studies on *Tetrahymena*

A. Introduction

Synchronized *Tetrahymena* provide excellent material for studying cell division, and, consequently, they have been used in our laboratory for evaluating the action of chemical agents on cell division. One of the most effective methods for synchronizing populations of *Tetrahymena* was described by Scherbaum and Zeuthen (1954). The technique for synchronization consists of subjecting log phase cells to a total of eight heat shocks at 34° for 30 minute durations, altered by 30 minute intervals at 28°. At 70–75 minutes after the last heat shock (designated EH) about 90% of the cells divide synchronously. These cells provide excellent models for studying the action of chemical agents on cell division, not only because of the ease in handling and maintaining the cultures but also because of the extensive biochemical and physiological studies that have been conducted with these cells which offer the investigator a firm foundation for further analysis (see, for example, Zeuthen, 1964).

The heat shock treatment results in a very high degree of division synchrony in which about 90% of the population divides over a 15–20 minute time interval. Macromolecular synthesis continues during the synchronizing treatment; this results in large cells that contain two to three times the amount of RNA, DNA, and protein as that found in log cells. Therefore, in order to investigate drug action on growth it is advantageous to study log growth-phase cultures as well as heat synchronized cells.

B. Growth Rate, Cell Division, and Drug Tolerance

The growth kinetics of log growth-phase cultures of *Tetrahymena* were determined in systematically varied concentrations of morphine and Δ^9-tetrahydrocannabinol (Δ^9-THC). In general, the *Tetrahymena* were more

5. ACTION OF NARCOTIC AND HALLUCINOGENIC AGENTS

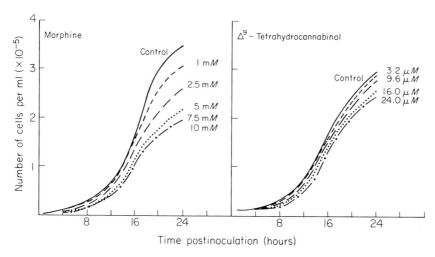

Fig. 3. The effects of morphine and Δ^9-tetrahydrocannabinol on the growth of log phase cultures of *Tetrahymena pyriformis*. (Data of McClean, 1972.)

sensitive to Δ^9-THC than to morphine. Log phase cultures exhibited reduced growth rate over a period of 24 hours when treated with 9.6 and 16 μM Δ^9-THC. When concentrations were increased above 16 μM, there was a further reduction in growth rate which was accompanied by extensive cytolysis. Morphine was less effective in reducing growth than Δ^9-THC when compared on a molar basis. A 1.0 mM concentration of morphine produced a significant reduction in the growth rate which was comparable to that found with 16 μM Δ^9-THC. At the highest concentration of morphine tested (10 mM) the growth rate was reduced by approximately 50% (Fig. 3). Both Δ^9-THC and morphine caused the cells to become ovoid or round. This was accompanied by a reduction of motility. However, in Δ^9-THC there was a distinctly observable recovery of cell shape and motility after 2–4 hours exposure to the drug; during the next 20 hours the cells continued to recover. No such transient effects were observed in the presence of morphine.

Since it has previously been demonstrated that amoebae develop a tolerance to morphine, studies were undertaken to establish whether or not *Tetrahymena* can develop tolerance to either morphine or Δ^9-THC. Cells were grown in low concentrations of morphine (2.5 mM) and after 4 days placed in a higher, more toxic concentration of the agent. From these studies it appeared that the cells did not develop tolerance to morphine as the growth kinetics of these morphine pretreated cells were similar to nonpretreated cells. A similar study with Δ^9-THC was not feasible. It was deter-

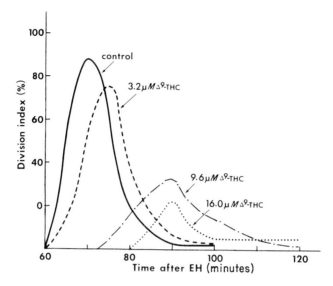

Fig. 4. The action of Δ⁹-tetrahydrocannabinol on heat-synchronized *Tetrahymena*. The development of division maxima and the determination of division delays in synchronized *Tetrahymena* continuously exposed to Δ⁹-THC commencing immediately after the heat-synchronizing treatment. Division delay was approximately 5 minutes in 3.2 μM Δ⁹-THC. Further delays were observed at higher concentrations of Δ⁹-THC. (Data of McClean, 1972.)

mined that the Δ⁹-THC lost its potency after 3–5 hours exposure to the cells. The loss of potency of the Δ⁹-THC may have a relationship to the observed recovery of cell shape and motility previously observed during the growth-kinetic studies.

The effects of drugs on synchronous cultures of *Tetrahymena* may be readily evaluated. The drug may alter the division schedule and/or block cytokinesis. Both of these effects can be readily determined and they are useful indices for analyzing drug action. The effects of drugs on the division schedule are dependent on the concentration and duration of exposure, as well as the stage during the cell cycle at which the agent is applied.

When cells were exposed to various concentrations of Δ⁹-THC, in which treatment was initiated, various division delays were recorded immediately after the end of the last heat treatment. In addition to the delay in division there was a reduction in division index (that is, the percentage of cells showing division furrows) which was determined to be a function of the drug concentration. The division delays and division indexes for concentrations of 3.2, 9.6, and 16 μM of Δ⁹-THC are shown in Fig. 4. Similarly, a comparison of the effects of Δ⁹-THC, morphine, levorphanol, and levallorphan on the division schedule at varying concentrations is illustrated in Fig. 5. It is evident from these studies that the cells are most sensitive to Δ^9-THC and least sensitive to morphine.

Shorter exposures of synchronized cells to pulses of the various agents (morphine, levorphanol, and Δ^9-THC) resulted in division delays and division indexes which were related to the duration of exposure and the concentration of the drug. It was also established that the cells were most sensitive to the drugs during the first half of the cell cycle.

C. MACROMOLECULAR SYNTHESIS

Although it is difficult to establish the specific molecular mechanism through which these drugs act, there is sufficient information to indicate that macromolecular synthesis is affected in cells treated with morphinans and Δ^9-THC. In cells exposed to Δ^9-THC, morphine, levorphanol, or levallorphan, there was a reduction in RNA, DNA, and protein synthesis and a slight stimulation in lipid synthesis.

Synchronized *Tetrahymena* exposed to 3.2 μM Δ^9-THC immediately

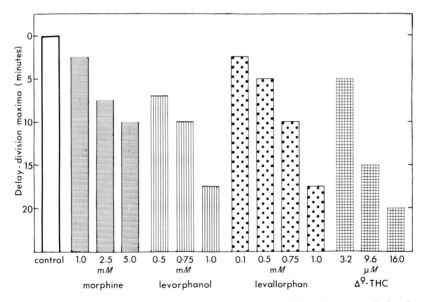

Fig. 5. The effects of various concentrations of morphine, levorphanol, levallorphan, and Δ^9-tetrahydrocannabinol on synchronized *Tetrahymena*. The drug treatment was initiated immediately after the heat-synchronizing treatment. The division delays were determined at each of the concentrations shown. (Data of McClean, 1972; Stephens and Zimmerman, 1972.)

following the last heat treatment displayed a marked reduction in the incorporation of ^{3}H-uridine into the acid-insoluble fraction. When the cells were incubated in ^{14}C-thymidine or ^{14}C-phenylalanine incorporation was only slightly reduced, whereas in ^{14}C-labeled sodium acetate incorporation was

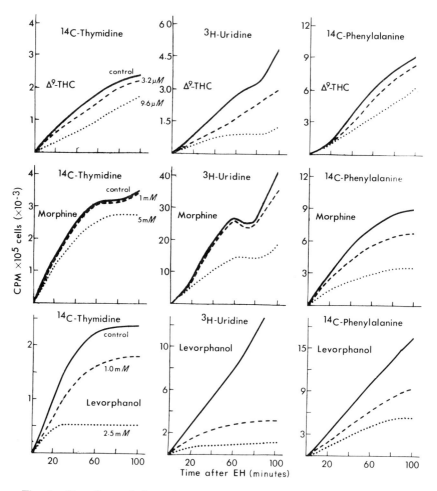

Fig. 6. The effects of Δ^9-tetrahydrocannabinol, morphine, and levorphanol on the incorporation of radioactive thymidine, uridine, and phenylalanine into DNA, RNA, and protein, respectively. The incorporation into controls are shown by the solid line. Synchronized *Tetrahymena* were added to the drug, and the appropriate isotope at 0 EH. The incorporation into the TCA-insoluble material was determined at various times thereafter. (Data of McClean, 1972.)

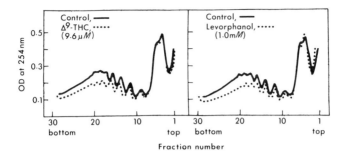

Fig. 7. The effects of Δ⁹-tetrahydrocannabinol (9.6 μM) and levorphanol (1.0 mM) on the polysomes of synchronized *Tetrahymena*. The sucrose density gradient profiles of the 10,000 g supernatant fraction from drug-treated *Tetrahymena* (.....) and control cells (———). The drug was added to the cells at 0 EH; at 60 minutes EH the cells were suspended in a buffer (0.05 M tris at pH 7.4 containing 5 mM magnesium) and lysed with 0.3% sodium deoxycholate. The 12,000 g supernatant was centrifuged at 27,000 rpm (2.5 hours) in a 15%–30% surcose gradient using a SW 25.2 rotor. The gradients were scanned at OD_{254} (OD-optical density) units with an ISCO Model D fractionator equipped with a 1.0 cm flow cell. (Data of McClean, 1972.)

slightly stimulated. Proportionally greater reduction in macromolecular synthesis was observed when Δ⁹-THC concentrations were increased to 9.6 μM. Cells treated with morphine showed a somewhat different incorporation profile. Cells incubated with 1.0 mM morphine displayed marked reduction in ¹⁴C-phenylalanine incorporation but only a moderate decrease in ³H-uridine incorporation and a negligible decrease in ¹⁴C-thymidine and ¹⁴C-labeled sodium acetate incorporation. At 5.0 mM morphine the reductions in ³H-uridine and ¹⁴C-phenylalanine incorporation were quite extensive (greater than 50%), whereas the reductions in ¹⁴C-thymidine and ¹⁴C-labeled sodium acetate incorporation were not significantly affected. Levorphanol and levallorphan, at concentrations in the range of 1–2 mM, showed a more extensive inhibitory activity than that of morphine as demonstrated by similar, labeled precursor incorporation studies (Fig. 6).

D. NATURE OF RNA INHIBITION

Recent reports have shown that in both prokaryotic and eukaryotic cells there is a reduction in the content of polyribosomes following treatment with narcotic agents. In the present study using sucrose density gradient analysis, it has been established that a significant reduction in the amount of recoverable polysomal material occurred when *Tetrahymena* cells were treated with 9.6 μM Δ⁹-THC during the first 60 minutes of the division schedule. Simi-

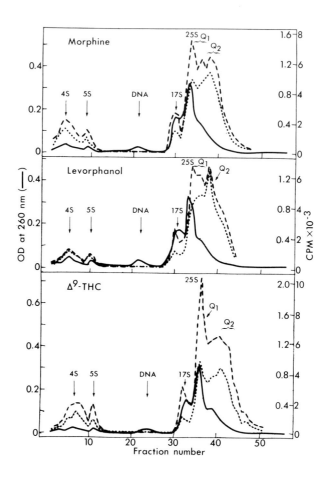

Fig. 8. MAK chromatographic profiles of cellular RNA from drug-treated *Tetrahymena*. Synchronized *Tetrahymena* were pulsed with ^{14}C-uridine (———) (2.0 μCi/ml) for a period of 15 minutes (from 35–50 minutes EH). Simultaneously, other cells were pulsed with ^{3}H-uridine (. . .) (20 μCi/ml) in the presence of the test concentration of drug (10 mM morphine, 2.5 mM levorphanol or 32 μM Δ^9-THC). After the 15-minute pulse period, both groups of cells were mixed and the RNA was extracted by the cold phenol-sodium dodecylsulfate procedure and separated by MAK column chromatography using a 0.2–1.2 M NaCl linear gradient. The morphine-treated cells display a reduction in both the Q_1 and Q_2 RNA. The levorphanol treated cells show a marked reduction in Q_1 RNA; however, the Q_2 RNA is not altered. The Δ^9-THC affects both the Q_1 and Q_2 RNA species. (Data of McClean, 1972).

lar reductions were determined for cells incubated with 1.0 mM levorphanol and 1.0 mM levallorphan. However, the quantity of polysomal material recovered from morphine-treated (5 mM) cells was not appreciably reduced as determined by the sucrose density gradient analysis (Fig. 7).

Reduction in the amount of polysomal material recovered suggests that the drugs (levorphanol and Δ^9-THC) affect the transcription of mRNA or the assembly of the polyribosomes. In order to further investigate the action of the drugs on cellular synthesis, cells were double labeled with ^{3}H-uridine and ^{14}C-labeled amino acids in the presence of one of the experimental drugs (Δ^9-THC, morphine, levorphanol, or levallorphan) for a 15-minute pulse late in the cell cycle. The polysomes from these cells were subjected to sucrose density gradient analysis. Each of the drugs caused a dramatic reduction of ^{3}H-uridine and ^{14}C-labeled amino acid incorporation in the polysome region of the sucrose density gradient profiles.

The previous studies have clearly shown that Δ^9-THC and the morphinans cause reduction in RNA and protein synthesis in synchronized *Tetrahymena*. In order to determine the nature of the RNA inhibition it is essential to investigate the kinetics of RNA labeling and to determine the effect of the drugs on the synthesis of the various species of RNA.

In selected experiments, cells were treated with either Δ^9-THC, morphine, levorphanol or levallorphan in the presence of ^{3}H-uridine at various stages during the cell cycle. Following a 15-minute pulse of ^{3}H-uridine and drug, the cells were mixed with non-drug-treated control cells, which were incubated at the same time in ^{14}C-uridine and the nucleic acids were extracted. In order to standardize the radioactivity between ^{3}H and ^{14}C, two additional groups of cells at the same stage during the cell cycle were separately treated with either ^{3}H-uridine or ^{14}C-uridine for 15 minutes. These two groups of cells were mixed and the nucleic acids were extracted and fractionated; the actual counts for the ^{14}C and ^{3}H were determined and the ratio of these counts allowed a valid comparison to be made between the ^{14}C and ^{3}H in the experimental series.

Methylated albumin keiselguhr (MAK) column chromatography is an especially useful technique for separating ribosomal RNA (25 S and 17 S RNA) and its precursor RNA from DNA-like RNA as well as 4 S and 5 S RNA. As shown in Fig. 8 there are significant reductions in the synthesis of ribosomal RNA precursor (which is designated Q_1; see, for example, Yoshikawa-Fukada, 1966) and DNA-like RNA (which is designated Q_2) following the treatment with 32.0 μM of Δ^9-THC for 15 minutes. Accompanying these effects, there are also reductions in the ribosomal RNA fractions (17 S and 25 S RNA). Similar treatment with 10 mM morphine results in profiles that indicate a reduction in the synthesis of Q_1 and Q_2

RNA but not as marked as that seen with the $^9\Delta$-THC. Levorphanol, at concentrations of 2.5 mM for 15 minutes, shows a reduction in the synthesis Q_1 RNA similar to that of morphine, but there is only a slight reduction in the synthesis of Q_2 fraction. Both morphine and levorphanol treatment cause depression of ribosomal RNA synthesis similar to Δ^9-THC.

VI. Concluding Remarks

From the previous discussion it is evident that the mechanisms of action of the narcotic and the hallucinogenic drugs on cells have not been resolved. Although the early work tended to support the hypothesis that the agents were acting through interference with the synthesis of RNA, which resulted in a general decrease in cellular biosynthesis, other theories have received strong support. One hypothesis which has been proposed relates the action of the morphinans to the cellular ATP system. This theory is based on the increased amounts of adenine nucleotides found in the surrounding medium following drug treatment.

The more recent studies on the morphine-related drugs indicate that there may indeed be another mechanism to explain the action of these drugs. Recent data has been accumulated that supports the hypothesis that the drug action is mediated through an alteration of the cell surface as well as on intracellular membrane systems. Thus the drugs may also affect the transport of material across the cell surface in addition to membrane-essential cellular biosynthesis.

The aforementioned theories have been used to explain the action of morphinans on cellular systems. However, the mechanism of action of the active marijuana ingredient, Δ^9-THC, is as yet unclear. From the present research it is evident that macromolecular synthesis is markedly reduced by Δ^9-THC, however, it would be presumptuous at this time to propose a mechanism.

It is apparent that the action of drugs on cellular systems has only recently become of scientific interest, and it is hoped that with additional research information will be available to assist in the clinical evaluation of these drugs.

Acknowledgment

The unpublished experiments from the authors' laboratory have been carried out with the support of research grants from the National Research Council of Canada. The authors are grateful for the Δ^9-tetrahydrocannabinol which was made available to them from the Food and Drug Directorate, Department of National Health and Welfare and for Levo-Dromoran (levorphanol tartrate) and Lorfan (levallorphan tartrate) from Hoffmann-La Roche Ltd., Montreal, Canada.

References

Binz, C. (1867). Über die Einwirkung des Chinin auf Protoplasma-Bewegungen. *Arch. Mikrosk. Anat.* **3**, 383.

Brdar, B., and Fromageot, P. (1970). Inhibition of viral RNA synthesis by levallorphan. *FEBS Lett.* **6**, 190–192.

Casy, A. F. (1971). The structure of narcotic analgesic drugs. *In* "Narcotic Drugs: Biochemical Pharmacology" (H. Clouet, ed.), pp. 1–16. Plenum, New York.

Chari-Bitron, A. (1971). Stabilization of rat erythrocyte membrane by Δ^1-tetrahydrocannabinol. *Life Sci.* **10**, Part II, 1273–1279.

Clouet, D. H. (1968). Biochemical responses to narcotic drugs in the nervous system and in other tissues. *Int. Rev. Neurobiol.* **11**, 99–128.

Clouet, D. H., ed. (1971a). "Narcotic Drugs: Biochemical Pharmacology." Plenum, New York.

Clouet, D. H. (1971b). Protein and nucleic acid metabolism. *In* "Narcotic Drugs: Biochemical Pharmacology" (H. Clouet, ed.), pp. 216–228. Plenum, New York.

Clouet, D. H., and Ratner, M. (1968). The effect of morphine administration on the incorporation of ^{14}C leucine into protein in cell-free systems from rat brain and liver. *J. Neurochem.* **15**, 17–23.

Cohen, G. M., Peterson, D. W., and Mannering, G. J. (1971). Interactions of Δ^9-tetrahydrocannabinol with the hepatic microsomal drug metabolizing system. *Life Sci.* **10**, Part I, 1207–1215.

Corssen, G., and Skora, I. A. (1964). Addiction reactions in cultured human cells. *J. Amer. Med. Ass.* **187**, 328–332.

Cotten, M. de V. (1971). Marihuana and its surrogates. *Pharmacol. Rev.* **23**, 263–380.

Cox, B. M., and Osman, O. H. (1970). Inhibition of the development of tolerance to morphine in rats by drugs which inhibit ribonucleic acid or protein synthesis. *Brit. J. Pharmacol.* **38**, 157–170.

Datta, R. K., and Antopol, W. (1971). Effects of morphine on mouse liver and brain ribonuclease and deoxyribonuclease activities. *Toxicol. Appl. Pharmacol.* **18**, 851–858.

Dewey, W. L., Harris, L. S., Howes, J. F., Kennedy, J. S., Granchelli, F. E., Pars, H. G., and Razdan, R. K. (1970). Pharmacology of some marijuana constitutents and two heterocyclic analogues. *Nature (London)* **226**, 1265–1267.

Gale, E. F. (1970a). Effects of diacetylmorphine and related morphinans on some biochemical activities of *Staphylococcus aureus*. *Mol. Pharmacol.* **6**, 128–133.

Gale, E. F. (1970b). Effect of morphine derivatives on lipid metabolism in *Staphylococcus aureus*. *Mol. Pharmacol.* **6**, 134–145.

Goodman, L. S., and Gilman, A., eds. (1970). "The Pharmacological Basis of Therapeutics," 4th ed. Macmillan, New York.

Greene, R., and Magasanik, B. (1967). The mode of action of levallorphan as an inhibitor of cell growth. *Mol. Pharmacol.* **3**, 453–472.

Heubner, W., Albrecht, M., Barocke, E., and Kewitz, H. (1952). Gewöhnungsversuche an fibroblastenkulturen. *J. Mt. Sinai Hosp., New York* **19**, 47–52.

Hollister, L. E. (1971). Marihuana in man: Three years later. *Science* **172**, 21-29.
Joyce, C. R. B., and Curry, S. H., eds. (1970). "The Botany and Chemistry of Cannabis." Churchill, London.
Krueger, H. N., Eddy, N. B., and Sumwalt, M. (1941). "The Pharmacology of the Opium Alkaloids." *Pub. Health Rep., Suppl.* **165**, Part I, 1-816.
Kuschinsky, K. (1971). Effect of morphine on protein synthesis in synaptosomes and mitochondria of mouse brain *in vivo*. *Naunyn-Schmiedebergs Arch. Pharmakol. Exp. Pathol.* **271**, 294-300.
Landau, J. V., Zimmerman, A. M., and Marsland, D. (1954). Temperature-pressure experiments on *Amoeba proteus*: Plasmagel structure in relation to form and movement. *J. Cell. Comp. Physiol.* **44**, 211-232.
LeDain, G., Campbell, I. L., Lehmann, H. E., Stein, J., and Bertrand, M. (1972). "Cannibis. A Report of the Commission of Inquiry into the Non-Medical Use of Drugs." Information Canada, Ottawa.
Löser, R., Boquet, P. L., and Röschenthaler, R. (1971). Inhibition of R-factor transfer by levallorphan. *Biochem. Biophys. Res. Commun.* **45**, 204-211.
McClean, D. K. (1972). Cell division and macromolecular synthesis in *Tetrahymena pyriformis*: The action of tetrahydrocannabinol, morphine, levorphanol and levallorphan. Ph. D. Thesis, University of Toronto, Toronto.
Mechoulam, R. (1970). Marihuana chemistry. *Science* **168**, 1159-1166.
Mechoulam, R., Shani, A., Edery, H., and Grunfeld, Y. (1970). Chemical basis of hashish activity. *Science* **169**, 611-612.
Neresheimer, E. R. (1903). Über die Höhe histologischer Differenzierung bei heterotrichen Ciliaten. *Arch. Protistenk.* **2**, 305.
Neu, R. L., Powers, H. O., King, S., and Gardner, L. I. (1970). Δ^8 and Δ^9-tetrahydrocannabinol: Effects on cultured human leucocytes. *J. Clin. Pharmacol. J. New Drugs* **10**, 228-230.
Neumeyer, J. L., and Shagoury, R. A. (1971). Chemistry and pharmacology of marijuana. *J. Pharm. Sci.* **60**, 1433-1457.
Noteboom, W. D., and Mueller, G. C. (1966). Inhibition of protein and RNA synthesis in HeLa cells by levallorphan and levorphanol. *Mol. Pharmacol.* **2**, 534-542.
Noteboom, W. D., and Mueller, G. C. (1969). Inhibition of cell growth and the synthesis of ribonucleic acid and protein in HeLa cells by morphinans and related compounds. *Mol. Pharmacol.* **5**, 38-45.
Painter, J. T., Pomerat, C. M., and Ezell, D. (1949). The effect of substances known to influence the activity of the nervous system on fiber outgrowth from living embryonic chick spinal cords. *Tex. Rep. Biol. Med.* **7**, 417-455.
Pick, E. P., and Wasicky, R. (1915). Über die Wirkung des Papaverins und Emetins auf protozoen. *Wien. Klin. Wochenschr.* **28**, 590.
Pomerat, C. M., and Leake, C. D. (1954). Short term cultures for drug assays: General considerations. *Ann. N. Y. Acad. Sci.* **58**, 1110-1128.
Raina, A., and Cohen, S. S. (1966). Polyamines and RNA synthesis in a polyauxotrophic strain of E. coli. *Proc. Nat. Acad. Sci. U. S.* **55**, 1587-1593.
Raina, A., Jansen, M., and Cohen, S. S. (1967). Polyamines and the accumulation of ribonucleic acid in some polyauxotrophic strains of *Escherichia coli*. *J. Bacteriol.* **94**, 1684-1696.
Röschenthaler, R., Devynck, M. A., Fromageot, P., and Simon, E. J. (1969). Inhibition of the synthesis of 5-S ribosomal RNA in *Escherichia coli* by levallorphan. *Biochim. Biophys. Acta* **182**, 481-490.

Rossman, T., Becker, F. F., and Vilcek, J. (1971). An investigation into the mechanism of cytotoxicity of levorphanol. *Mol. Pharmacol.* **7**, 480–483.
Ruffin, N. E., Reed, B. L., and Finnin, B. C. (1969). Effects of morphine withdrawal on cells in continuous culture. *Life Sci.* **8**, Part II, 671–675.
Sakiyama, A., Usui, S., and Miura, Y. (1969). Regulation of ribosomal RNA synthesis. *Advan. Enzyme Regul.* **7**, 207–218.
Sasaki, M. (1938). Studies on phenomena of morphine abstinence of cultures *in vitro* of fibroblasts and on curative effect of morphine and its derivatives on them. *Arch. Exp. Zellforsch. Besonders Gewebezücht.* **21**, 289–307.
Scherbaum, O., and Zeuthen, E. (1954). Induction of synchronous cell division in mass cultures of *Tetrahymena pyriformis*. *Exp. Cell Res.* **6**, 221–227.
Seevers, M. H., and Deneau, G. A. (1963). Physiological aspects of tolerance and physical dependence. *Physiol. Pharmacol.* **1**, 565–640.
Simon, E. J. (1963). Inhibition of synthesis of RNA in *E. coli* by the narcotic drug levorphanol. *Nature (London)* **198**, 794–795.
Simon, E. J. (1964). Inhibition of bacterial growth by drugs of the morphine series. *Science* **144**, 543–544.
Simon, E. J. (1971). Single cells. *In* "Narcotic Drugs: Biochemical Pharmacology" (D. H. Clouet, ed.), pp. 310–341. Plenum, New York.
Simon, E. J., and Rosenberg, P. (1970). Effects of narcotics on the giant axon of the squid. *J. Neurochem.* **17**, 881–887.
Simon, E. J., and Van Praag, D. (1964a). Inhibition of RNA synthesis in *Escherichia coli* by levorphanol. *Proc. Nat. Acad. Sci. U. S.* **51**, 877–883.
Simon, E. J., and Van Praag, D. (1964b). Selective inhibition of synthesis of ribosomal RNA in *Escherichia coli* by levorphanol. *Proc. Nat. Acad. Sci. U. S.* **51**, 1151–1158.
Simon, E. J., Cohen, S. S., and Raina, A. (1966). Polyamines and inhibition of RNA synthesis in *E. coli* by levorphanol. *Biochem. Biophys. Res. Commun.* **24**, 482–488.
Simon, E. J., Garwes, D. J., and Rand, J. (1970a). Reversible inhibition of RNA phage replication and macromolecular synthesis by levorphanol. *Biochem. Biophys. Res. Commun.* **40**, 1134–1141.
Simon, E. J., Schapira, L., and Wurster, N. (1970b). Effect of levorphanol on putrescine transport in *Escherichia coli*. *Mol. Pharmacol.* **6**, 577–587.
Singer, A. J., ed. (1972). Marijuana: chemistry, pharmacology and patterns of social use. *Ann. N.Y. Acad. Sci.* **191**, 1–269.
Stephens, R. E., and Zimmerman, A. M. (1973). Action of levallorphan: macromolecular synthesis and cell division. *Mol. Pharmacol.* (in press).
Tjio, J., Pahnke, W. N., and Kurland, A. A. (1969). LSD and chromosomes. *J. Amer. Med. Ass.* **210**, 849–856.
Wolstenholme, G. E. W., and Knight, J., eds. (1965). "Hashish: Its Chemistry and Pharmacology." Churchill, London.
Wurster, N., Elsbach, P., Rand, J., and Simon, E. J. (1971a). Effects of levorphanol on phospholipid metabolism and composition in *Escherichia coli*. *Biochim. Biophys. Acta* **248**, 282–292.
Wurster, N., Elsbach, P., Simon, E. J., Pettis, P., and Lebow, S. (1971b). The effects of the morphine analogue levorphanol on leukocytes. Metabolic effects at rest and during phagocytosis. *J. Clin. Invest.* **50**, 1091–1099.
Yoshikawa-Fukada, M. (1966). The nature of the two rapidly labelled ribonucleic acid components of animal cells in culture. *Biochim. Biophys. Acta* **123**, 91–101.

Zeuthen, E., ed. (1964). The temperature-induced division synchrony in *Tetrahymena*. *In* "Synchrony in Cell Division and Growth," pp. 99–158. Wiley (Interscience), New York.

Zimmerman, A. M. (1967). Sensitivity of *Amoeba proteus* to morphine and N-allylnormorphine. A pressure study. *J. Protozool.* **14**, 451–455.

Zimmerman, A. M., and Zimmerman, S. (1970). Biostructural, cytokinetic and biochemical aspects of hydrostatic pressure on protozoa. *In* "High Pressure Effects on Cellular Processes" (A. M. Zimmerman, ed.), pp. 179–210. Academic Press, New York.

Zucker-Franklin, D., Elsbach, P., and Simon, E. J. (1971). The effect of the morphine analog levorphanol on phagocytosing leukocytes. *Lab. Invest.* **25**, 415–421.

6

Effects of Drugs on Hepatic Cell Proliferation

J. W. GRISHAM

Abbreviations	95
I. Introduction	96
II. Kinetics of Hepatic Cell Proliferation	97
A. In the Maturing Liver	97
B. After Cell Loss	100
C. Hepatocellular Proliferation in Mice	108
D. Biochemistry of the Hepatic Cell Cycle	109
E. Cyclic Status of Normal (Nonproliferative) Hepatocytes	109
III. Regulation of Hepatic Cell Proliferation	110
IV. Drugs and Hepatic Cell Proliferation	111
A. Introduction	111
B. Classic Hepatotoxins	112
C. Hepatocarcinogens	113
D. Metabolic Inhibitors	117
E. Hormones in Pharmacologic Doses	122
F. Miscellaneous Drugs Apparently not Hepatotoxic	124
V. Concluding Remarks	126
References	127

Abbreviations

C and t_C	cell cycle and its length (in hours)
G_1 and t_{G_1}	pre-DNA synthesis phase and its length
S and t_S	DNA synthesis phase and its length
G_2 and t_{G_2}	post-DNA synthesis phase and its length
M and t_M	mitotic phase and its length
G_0	hypothetical proliferative rest phase

^{3}H-TdR tritium-labeled thymidine
I_S percent of cells in S or DNA synthetic index (after single pulse of ^{3}H-TdR)
I_M percent of cells in M or mitotic index
I_P percent of cells tagged after continuous or frequently repeated pulses of ^{3}H-TdR over protracted period of time

I. Introduction

This review surveys effects of drugs on proliferation of hepatocytes,* a subject that is of interest to both laboratory and clinical scientists who work with hepatic pathophysiology. A variety of drugs have marked and, apparently, specific effects on different phases of the cell cycle. Studies utilizing such agents have provided insight into the biochemistry of prereplicative (G_1) and premitotic (G_2) phases (see Baserga, 1968, for review). The liver is intimately involved in the metabolism of most drugs; agents that influence cell proliferation may have especially pronounced effects on formation of hepatocytes because they are concentrated by this tissue. Attention has previously been focused on drugs that are overtly toxic and/or carcinogenic to the liver, and hepatic proliferative responses to drugs that do not cause cell death have not been extensively studied. Hepatocarcinogens produce complex and progressive alterations in the regulation of hepatocellular proliferation long before neoplastic transformation is detectable. Continued study of this phenomenon may demonstrate that it is causally related to development of hepatic cancer and, thus, may provide insight into the mechanism of hepatocarcinogenesis. Further studies of the proliferative responses of nonhepatotoxic drugs are, perhaps, of equal importance; knowledge of the action of these drugs may provide insight into the mechanism of regulation of hepatocytic proliferation, insight not derivable from studies utilizing hepatotoxins and hepatocarcinogens. Furthermore, some of the nonhepatotoxic drugs that affect hepatocellular proliferation are widely used therapeutically, and it is likely that other commonly used drugs may have identical though as yet undetected effects. Similar drugs, when given to patients with liver diseases, may influence substantially the reparative response of this tissue. More data are needed to allow more intelligent use of drugs in patients with liver damage.

To provide a base for interpreting the effects of drugs on hepatic cell formation, data are reviewed on the kinetics of hepatic cell proliferation during

---

* The terms hepatic cell and hepatocyte are used synonymously to designate the parenchymal cell of the liver; as used here these terms do not include any of the other several types of cells contained in this tissue.

postnatal growth and following controlled production of hepatic deficit (partial resection of liver). The liver is apparently unique among adult mammalian viscera in its capacity to massively reinitiate cell division after long dormancy. Although the literature on hepatic regeneration is extensive and has been reviewed many times, proliferative kinetics of hepatocytes have received comparatively little coverage and this review concentrates on this facet. Awareness of characteristics and peculiarities of the hepatic cell cycle are necessary in order to understand and interpret the action of drugs on these reactions.

This review contains data that pertain almost exclusively to hepatocellular proliferation in the rat. Because some important studies have been performed with mice, a brief discussion of the most significant documented differences in cellular proliferation in rats and mice is included. Concentration on rodents is not the author's wish, but it is necessitated by the availability of published data. Studies on other species are greatly needed.

II. Kinetics of Hepatic Cell Proliferation

A. In the Maturing Liver

Following birth both I_S and I_M in hepatocytes fall rapidly. Two types of temporal curves for I_S and I_M have been described, either a steady decline from a high point on the day of birth (Post and Hoffman, 1964) or a biphasic pattern characterized by a rapid decline during the first few days after delivery, followed by a secondary rise at about 2–3 weeks of age (McKellar, 1949; Stöcker and Butter, 1968; Grisham, 1969). The reasons for these differences are unclear but may be related to the frequency of tissue sampling; if sampling is infrequent the postnatal peak may be missed. By 5–6 weeks of age, I_S and I_M have decreased to the low level characteristic of adult rats. Both I_S and I_M vary with the time of day at which ^{3}H-TdR is injected or tissue is sampled for mitotic counts (Jackson, 1959; Messier and Leblond, 1960; Echave Llanos et al., 1971). Although lighting and feeding schedules, especially the latter, have not always been well controlled, I_S appears to peak toward the end of the normal dark period (Echave Llanos et al., 1971). This opinion has received strong support from a study in which the normal photoperiod was inverted; maximal DNA synthesis still occurs toward the end of the synthetic dark period (Barbiroli and Potter, 1971). I_M peaks about 8 hours following the peak in I_S (Echave Llanos et al., 1971).

At birth, hepatocytes in M or S phase are scattered randomly along the entire length of hepatic cell plates between adjacent terminal branches of

TABLE I
Lengths of Phases of the Hepatic Cell Cycle in Intact Livers of Maturing Rats

Age or weight	Phase time (hours)					Reference
	t_C	t_{G_1}	t_S	t_{G_2} [a]	t_M	
1 day	13.75	4.95	7.0	1.5	0.3	Post and Hoffman, 1964
3 weeks	21.5	9.0	9.0	1.3	1.7	Post et al., 1963; Post and Hoffman, 1964
4–5 weeks	20–26 [b]	9–14	8–9	1–3	0.8–1.0	Grisham, 1969, 1971
8 weeks	—	—	8.2	>2.0	0.8	Edwards and Koch, 1964
8 weeks and older	(47.5) [c]	(28.0)	(16.0)	(1.8)	(1.7	Post and Hoffman, 1964
8 weeks and older	21.5	9.0	9.0	1.8	1.7	Post and Hoffman, 1965
300 gm	—	—	15–18	2.5	5.5	Stöcker and Heine, 1965

[a] In some studies data listed for t_{G_2} represent $t_{G_2} + \frac{1}{2} t_M$.
[b] These values represent ranges derived from study of different lots of rats.
[c] Data from which the phase times in parentheses were derived were reinterpreted to give the times listed in the next line.

6. DRUGS AND THE HEPATIC CELL CYCLE

portal and hepatic venules (McKellar, 1949; LeBouton and Marchand, 1970). Within the first 2 days after delivery there is a marked decline in I_S in hepatic cells adjacent to terminal hepatic venules, followed by a similar decline in hepatocytes located midway between terminal portal and hepatic venules. Since I_S in hepatocytes adjacent to terminal portal venules does not greatly change during the first 12 days after birth, at this time the majority (80%) of all labeled hepatic cells are located periportally (LeBouton and Marchand, 1970). Dynamics of change in location of S phase hepatocytes between 12 and 24 to 28 days of age is undocumented. However, at the latter age proliferating hepatocytes (as indicated by continuous infusion or repeated injection of ^{3}H-TdR for 20–24 hours) are said to be located randomly along hepatocellular plates (Bucher et al., 1964). Although it is commonly stated that tagged hepatocytes are located equally in all parts of hepatic cell plates in adult rats, the paucity of S phase cells in rats this age makes precise determination of maximal location impossible, even after repeated infusions of ^{3}H-TdR (Bucher et al., 1964). The cause of the postnatal shift in locus of dividing hepatocytes is uncertain but may reflect waves of proliferation moving along hepatic cell plates or result from birth-related changes in the characteristics of hepatic tissue perfusion.

The hepatic cell cycle apparently lengthens sometime during the first three postnatal weeks and thereafter t_C and its phases appear to be fairly constant (see Table I). In intact livers of adult rats, measurement of the cell cycle is unreliable by the method of fraction of labeled mitoses because of the low mitotic rate. This may explain the very long cycle described in some studies on adult rats (Post et al., 1963; Post and Hoffman, 1964; Stöcker and Heine, 1965). The results of one of these studies was reinterpreted by the investigators as being compatible with a shorter, poorly synchronized cycle (Post and Hoffman, 1965). In most studies, t_S, t_{G_2}, and t_M (measured by a variety of methods) have been found to be relatively constant in rats older than 3 weeks (Table I). Further study of the murine hepatic cell cycle is needed between 1 and 21 days of age and in adults.

The growth fraction (the proportion of hepatocytes that are involved in the replicative cycle) may be calculated from values for t_C, t_S, and I_S (Post and Hoffman, 1964) or by determining the fraction of cells labeled (I_P) after continuous infusion or frequently repeated injections of ^{3}H-TdR (Stöcker, 1966). Results obtained by each of the two methods are reasonably close, when comparable data are available. The growth fraction declines markedly as rats age; from a high of about 36% on the first day after birth it declines to 4%–6% in rats 6–8 weeks old (Post and Hoffman, 1964). I_P in rats this age is about 10% (repeated injections of ^{3}H-TdR during a period of over 60 hours) (Fabrikant, 1969).

At birth, murine hepatocytes uniformly contain a single diploid nucleus, but in adult rats there is marked heterogenetiy in number and ploidy of nuclei, cells with polyploid nuclei predominating (Alfert and Geschwind, 1958; Nadal and Zajdela, 1966; Nadal, 1970). Although the causation and functional implications of hepatic polyploidy are unknown, formation of higher ploid nuclei involves cell proliferation (Nadal and Zajdela, 1966; Nadal, 1970). Cytokinesis does not take place in some hepatocytes during mitosis, resulting in binuclear cells (Nadal, 1970); appearance of binuclear cells of one ploidy always precedes formation of nuclei of higher ploidy (Alfert and Geschwind, 1958; Nadal and Zajdela, 1966). Normally, the shift of hepatic cell nuclei to higher ploidy levels occurs by fusion of two mitotic spindles in binuclear cells passing through a replicative cycle (Nadal and Zajdela, 1966; Nadal, 1970). Development of polyploidy thus depends on the occurrence of a complete replicative cycle in binuclear cells. Although G_2 arrest (DNA synthesis not followed by mitosis) has been proposed as the cause of polyploid cells in normal mouse liver (Perry and Schwartz, 1967; Epstein, 1967), the evidence for a subpopulation of hepatocytes in G_2 in normal rodent liver is tenuous (Grisham, 1969). However, this mechanism may underlie development of polysomaty after exposure to some chemicals (see below). Hepatocytes containing nuclei of all ploidy classes proliferate during postnatal growth (Busanny-Caspari, 1962; Carriere, 1967), although diploid cells are said to be predominantly involved (Post and Hoffman, 1965). Polyploid cells must proliferate or else continued cell proliferation would result in their decrease.

Once formed, hepatocytes are long lived in normal circumstances, with a "life-span" or turnover time of 160 to over 500 days (MacDonald, 1961b; Stöcker and Altmann, 1964; Cameron, 1970).

B. After Cell Loss

Reduction in the number of hepatocytes in rat liver by partial hepatic resection results in rapid and marked increase in I_S. Characteristics of the proliferation of hepatocytes induced by partial hepatic resection vary with both the amount of liver removed and the age of the rat. Most studies have been made after a standard partial hepatectomy (removal of left lateral and median lobes which constitute 68%–70% of the hepatic mass and number of hepatocytes) in rats 2–5 months of age (body weight 120–250 gm). Results of these studies will be discussed first, after which variations in the hepatoecllular proliferative response with age of host and amount of liver resected will be considered.

I_S first increases above baseline levels between 15 and 18 hours after standard partial hepatectomy of 2–5 month-old rats (Grisham, 1962; Oehlert et al., 1962; Edwards and Koch, 1964; Fabrikant, 1968). Following this initiation, I_S reaches a peak at 20–25 hours after surgery. The maximal I_S is 250–350 times the I_S in nonoperated or sham-operated controls (Grisham, 1962; Bucher et al., 1964). After reaching the maximal level, I_S declines markedly during the next 24 hours but remains slightly elevated for as long as 2 weeks after hepatic resection. The curve for I_M follows that for I_S by 6–8 hours and is generally similar in shape though different in absolute magnitude (Cater et al., 1956; Grisham, 1962; Oehlert et al., 1962). Although these data describe the fraction of S and M phase hepatocytes in the residual liver, they do not reflect that the population of hepatocytes is increasing during this time. Correction for the increasing population size shows that the proliferation of hepatocytes, in terms of total number of cells involved, is not nearly as episodic as the curves for I_S appear to imply (Fabrikant, 1968). The entry of cells into S (expressed as fraction of the number of residual cells remaining after surgery) reaches a maximum rate of about 8%/hour at 20–22 hours and then continues at the rather steady rate of about 5%/hour for the next 40 hours (Fabrikant, 1968).

I_S and I_M vary diurnally in proliferating hepatic cells after partial hepatectomy, with peaks occurring in the morning hours (Jaffe, 1954; Günther et al., 1968; Klinge and Mathyl, 1969). There is disagreement as to whether or not the time of occurrence of I_S and I_M peaks can be shifted by altering the time of operation (and, thus, of tissue sampling). When the time of surgery or tissue sampling is either 10:00 A.M. or 12:00 P.M., the initial mitotic wave appears at a constant time (32 hours) after surgery, but the magnitude of the peak is greatest when sampling occurs during the early morning hours (6:00–10:00 A.M.) (Klinge and Mathyl, 1969). However, other studies have shown that the time of occurrence of peak I_S and I_M can be shifted by varying the time of surgery (Barbason and Lelièvre, 1970; Barbiroli and Potter, 1971). When rats are partially hepatectomized at either 10:00 A.M. or 12:00 P.M., peaks I_S and I_M occur at the expected times after surgery (i.e., 25–26 and 31–32 hours after operation, respectively). However, if the liver is resected at 4:00 P.M., both peaks are delayed by 8–10 hours; this delay occurs without alteration in the length of G_2 or S phases of the heaptic cell cycle (Barbason and Lelièvre, 1970). In a study in which light and dark photoperiods were inverted and feeding time was rigidly controlled, partial hepatectomy initiated DNA synthesis in what was interpreted as two distinct populations of hepatocytes (Barbiroli and Potter, 1971). One group of cells was stimulated to enter S phase at a constant time after surgery and independent of the time of day, while a second group entered S phase at the time of the diurnal maximum in intact liver (i.e., near the end of the dark photoperiod). Resection of the liver at different times in relation to light–dark photoperiods allow these two hepatocytic subpopulations to be discerned (Barbiroli and Potter, 1971). Clearly, more work is needed in this area, but these data point out the necessity of carefully controlling this variable when studies on the effects of exogenous agents on the heaptic cell cycle are to be made.

The localization within hepatocellular plates of hepatocytes in S or M phases varies with time after hepatic resection. At the time of maximal I_S, labeled hepatocytes are sharply localized to the third of the hepatic plate nearest to terminal portal venules (Grisham, 1962; Oehlert et al., 1962; Bucher et al., 1964; DeRecondo and Frayssinet, 1963; Fabrikant, 1967). With increasing time after partial hepatectomy, the location of S phase hepatic cells becomes more diffusely distributed along the distance between terminal portal and hepatic venules. M phase hepatocytes show the same pattern of localization occurring 6–8 hours later (Harkness, 1952; Grisham, 1962). The localization of S phase cells has been analyzed with more precision in a study in which the distance between adjacent portal and hepatic venules was divided into 10 to 12 equal-sized segments (Rabes and Tuczek, 1970). At 20 hours after partial hepatectomy, the site of maximal concentration of S phase cells was located just downstream from portal venules and the segment (one-tenth of the length) of the hepatic plate immediately adjacent to portal tracts contained about the same proportion of labeled cells as did segments in the centers of hepatic plates. At 27 hours, hepatocytes were maximally labeled and tagged cells were concentrated in the periportal four-tenths of hepatic plates. At 34 hours, the number of hepatic cells in S had declined markedly, but of these a maximal concentration occurred in the third of the acini adjacent to terminal hepatic veins, and this pattern was accentuated at 40 hours. At 48 hours, the number of S phase hepatocytes was low in all parts of hepatic plates and there was no focal concentration; at 56 hours, a second but lesser increase in S phase cells occurred, with maximal concentration in the third of the hepatic plates adjacent to portal venules.

These findings on locations of focal concentrations of S phase hepatocytes at different times after partial hepatectomy suggest that activation of cell proliferation occurs in a wave, which, after initiation at 15–18 hours in hepatocytes just downstream from terminal portal venules, progresses along the hepatic plates toward the terminal hepatic venules. A second wave of lesser magnitude begins periportally between 48 and 56 hours and moves toward adjacent hepatic venules. Since maximal I_S and concentration of labeled cells in periportal parenchyma occur at the same time, the majority of all new hepatocytes produced after partial hepatectomy are derived from residual cells located in the third of hepatic plates adjacent to terminal portal venules (Grisham, 1962).

6. DRUGS AND THE HEPATIC CELL CYCLE

After a standard partial hepatectomy, activation of increased cell proliferation occurs earlier in weanling (22–25 days old, about 45 gm body weight) rats than in young adults (about 4 months of age and 235 gm body weight) (Bucher et al., 1964; Klinge, 1968); similarly, proliferation is initiated earlier in young adult rats than it is in old animals (12–15 months of age and about 350 gm body weight). The times after partial hepatectomy when a significant increase in I_S is first apparent are as follows: weanlings, between 13 and 15 hours; young adults, between 15 and 18 hours; and old adults, between 18 and 23 hours (Bucher et al., 1964). There is also a difference in the shape of the temporal curves for hepatic DNA synthesis in rats of these three ages (Bucher et al., 1964). Weanlings demonstrate a rapidly accelerating sharp peak followed by a rapid decline and a second peak at about 33 hours. Distinct from this pattern, occurrence of S in hepatocytes of young adult and old rats begins and peaks at progressively later intervals; the declining phases never reach baseline during the interval studied but are skewed toward longer times. These patterns suggest that initiation of S phase is fairly well synchronized in weanlings (the beginning of the rise and the end of the decline are separated by only 10–12 hours, little more than the 8–9 hours needed to complete S in individual hepatocytes) and that synchrony deteriorates as rats age. Occurrence of the second peak in weanlings is compatible with this contention; this peak may represent a second proliferative cycle by some of the hepatocytes that divided earlier (see below). In rats of all ages, S phase hepatocytes are concentrated periportally during peak I_S (Bucher et al, 1964). This is also true of the second peak that occurs in weanlings.

Hepatocytic proliferation is related, in a somewhat complicated fashion, to both age of host and amount of liver resected (MacDonald et al., 1962; Bucher and Swaffield, 1964; Menyhárt and Szabó, 1968). In young adults (defined previously) resection of 9% or 34% of the liver results in DNA labeling of small magnitude, which continues for over 24 hours without occurrence of a distinct peak (Bucher and Swaffield, 1964), whereas removal of 43% to 68% causes initial sharp bursts of labeling followed by lower plateaus which persist. The time of onset of increased labeling, whether a burst or a plateau, does not vary with the amount of liver resected but is the same in all instances. In contrast to this pattern, in either weanling or old rats initial bursts of DNA labeling, followed by prolonged plateau phases, occur after all amounts of liver resected, even 9% (Bucher and Swaffield, 1964). In all instances, S phase cells are initially concentrated in parenchyma adjacent to portal venules. In weanlings, however, the initial periportal localization of S phase cells is less sharp, perhaps as a result of the relatively large number of randomly distributed S phase hepatocytes in rats of this age

(Bucher et al., 1964). These data were interpreted as indicating that labeling of DNA is directly related to the amount of liver removed, but when liver loss exceeds a certain critical amount (which depends on age), an additional rapidly occurring burst of DNA synthesis is produced.

In spite of the extensive changes in I_S and I_M, which follow partial hepatic resection, present evidence suggests that the length of C (t_C) or of its phases is altered very little if at all (Table II). The lengths of S, G_2, and M are remarkably constant in proliferating hepatocytes after partial hepatectomy and do not differ significantly from similar values determined for young growing rats (Table I). In most studies, two consecutive curves of labeled mitotic figures have not been found after partial hepatectomy of adult rats, and it has usually been impossible to measure G_1. Weanling rats show two consecutive peaks of replicating hepatocytes following partial hepatectomy, with maxima occurring at approximately 18 and 36 hours after surgery (Bucher et al., 1964). This pattern is compatible with fairly well synchronized cohorts of proliferating cells having a t_C of about 18 hours. Studies with rats slightly older than weanlings (4–5 weeks old, about 65 gm body weight) show two consecutive peaks of mitotic figures after a pulse of ^{3}H-TdR at 16 hours after partial hepatectomy, a time which falls early on the ascending curve of increasing I_S (Grisham, 1969). Results of further studies of this nature in different groups of partially hepatectomized rats and in control animals of the same age strengthen the opinion that the length of the hepatic cell cycle and its phases is probably not greatly affected by partial hepatectomy. Although the median hepatocytic t_C is somewhat shorter after partial hepatectomy than it is in intact livers of similarly aged rats, the range of variation of t_C and of its phases are considerable in both situations and the data overlap (Grisham, 1969, 1971).

In support of the validity of these measurements for t_C, one can estimate a maximal value for hepatocytic t_C after partial hepatectomy on the basis of the measured values of t_S, t_M, and t_{G_2}, and on the length of time between surgery and occurrence of the first detectable increase in I_S. The latter value varies, depending on the age of the host, from about 13 to about 18 hours. This value is most likely longer than the true G_1, since if cells are in G_0 or if they are arrested in G_1, it may take some time before they begin to proceed in an orderly and continuous fashion through G_1. With these limitations in mind, maximum t_C may be estimated to vary from 25 to 30 hours, depending on age of the animal. This calculated value is not greatly different from t_C determined from curves of labeled mitotic figures for normal and regenerating hepatocytes (Tables I and II).

It has been proposed, mainly on the basis of finding only one distinct cycle of labeled mitotic figures in regenerating hepatocytes, that these cells

TABLE II
LENGTHS OF PHASES OF THE HEPATIC CELL CYCLE IN REGENERATING LIVERS OF RATS

Age or weight	Phase time (hours)					Reference
	t_C	t_{G_1}	t_S	t_{G_2}[a]	t_M	
4–5 weeks	18–21[b]	6–12	7–9	1–3	0.8–1.0	Grisham, 1969, 1971
100–180 gm	—	—	—	—	0.8–0.9	Brues and Marble, 1937
6–8 weeks	—	—	8.0	3.0	~1.0	Fabrikant, 1968
8 weeks	—	—	8.2	>2.0	0.0	Edwards and Koch, 1964
200–300 gm	—	—	8.0	—	—	Looney, 1960
300 gm	—	—	—	—	~1.0	Cater et al., 1956
330 gm	—	—	7.2	2–5	1–1.5	Stöcker and Pfeifer, 1967

[a] In some studies data listed for t_{G_2} represent $t_{G_2} + \frac{1}{2} t_M$.
[b] These values represent ranges derived from study of different lots of rats.

pass through only one division cycle following partial hepatectomy (Fabrikant, 1968). Several lines of evidence suggest that this proposal is incorrect and indicate that a significant proportion of the residual hepatocytes replicate more than one time following partial hepatectomy. In order for 90% (the approximate growth fraction) of the hepatic cells remaining after standard partial hepatectomy to repopulate the regenerated liver, at least 50% of this residual population must replicate twice. Furthermore, occurrence of two peaks of DNA labeling in weanling rats (Bucher et al., 1964) and of two consecutive peaks of labeled mitotic figures when early replicating cells are tagged in young rats (Grisham, 1969) supports this contention. That at least some of the hepatocytes that replicate earliest following partial hepatectomy, proliferate a second time, has been demonstrated by tagging the same cell with thymidine labeled with different isotopes as it passes through S phase of two separate replicative cycles (Grisham, 1969). Almost 70% of hepatocytes proliferating at 38 hours after partial hepatectomy are progeny of cells that replicated 20 hours earlier (Grisham, 1969).

If many residual hepatocytes do replicate more than once after standard partial hepatectomy, why has only one cycle of labeled mitotic figures typically been found? Of considerable potential importance is the possibility of rapid loss of synchrony after initial tagging. This possibility is compatible with the observed variability in cycle phases. Of perhaps equal importance is the focal location in hepatic cell plates of residual hepatocytes initiating S phase. For instance, cells adjacent to terminal hepatic venules are just beginning to enter S phase at approximately the same time that hepatocytes adjacent to portal venules are entering a second round of DNA synthesis (Rabes and Tuczek, 1970). Thus, if mitoses are evaluated in hepatocytes in all parts of the hepatic cell plate, after a good first curve of labeled mitoses (all periportal) the second curve could be obscured by the unlabeled mitoses arising in perihepatic cells. To obviate this problem, labeled mitoses should be evaluated only in a circumscribed area of hepatic cell plates (the one-third nearest to terminal portal venules). Finally, another consideration may account for finding only one cycle of labeled mitotic figures. Occurrence of S phase in proliferating residual hepatocytes is highly episodic. Hepatocytes that enter S earliest are more likely to undergo a second replicative cycle than are cells that initiate S at a later time after hepatic resection. It follows, then, that in order to demonstrate two cycles of labeled mitotic figures hepatocytes must be initially labeled considerably before the peak of I_S, i.e., before 16 or 18 hours after partial hepatectomy.

Hepatocytes in the inner half of the lobule or acinus are apparently not activated to begin S as rapidly as are those hepatic cells near portal venules. The data discussed previously (Rabes and Tuczek, 1970) suggest that a wave of activation proceeds from portal toward central venules. However, an alternative explanation is that cells in all parts of hepatic plates are activated simultaneously and that the difference in time of occurrence of S reflects variations in t_{G_1}. This would mean that t_{G_1} and, in turn, t_C would be dependent on the location of the cell in the hepatic plate. Evidence from studies with heterotopic partial hepatic autografts indicate that this possibility is unlikely. Lobes of dog liver may be surgically removed from the main hepatic mass and grafted into the neck where they are perfused by the carotid artery and drained by the jugular vein (Sigel et al., 1968). Artery and vein can be connected so that the grafts are perfused through either portal or hepatic veins, i.e., blood flow either orthograde or retrograde. Resection of a major fraction of the liver remaining in the abdomen causes proliferation of hepatocytes in both residual liver and in hepatic graft. S phase cells are initially localized periportally in residual livers, but their position in grafts depends on the direction of blood flow; with orthograde flow they are periportal but with retrograde flow they are perihepatic (Sigel et al., 1968). This result indicates that all hepatocytes are capable of responding to hepatocellular deficiency and that the cells responding soon after partial hepatectomy are those that first contact the perfusing blood. Variation in the time of onset of S phase is, therefore, probably related to the removal of some critical factor from blood.

Following partial hepatectomy, new hepatic cells are not derived primarily from stem cells or from a subpopulation of hepatocytes, which replicates more than two consecutive times (Klinman and Erslev, 1963). Continuous infusion (over 60 hours) of ^{3}H-TdR into partially hepatectomized rats results in tagging of almost all hepatic cells remaining after partial hepatectomy ($I_P = 85\%-95\%$) (Stöcker, 1966; Fabrikant, 1969).

Two types of apparent movement of labeled cells along hepatic plates have been described: fast (occurring within 48–72 hours; Grisham, 1962) and slow (occurring between 5 and 91 days; Scherer and Friedrich-Freksa, 1970). If hepatocytes are flash labeled with ^{3}H-TdR at the time of maximal S and the animals are then killed at progressively later intervals, the initially sharp periportal localization of labeled cells gradually becomes more diffuse (Grisham, 1962). This spreading results from mitotic division of labeled cells and is correlated in time with a progressive diminution in average grain count per cell (Grisham, 1962). Some labeled cells may be forced from their initial periportal location by population pressure resulting from localized formation of new cells. It is probable, however, that most of the rapid migration described in these earlier studies is only apparent, and that it results from tagging of late replicating cells located in hepatic plates near terminal hepatic venules (Rabes and Tuczek, 1970) by reused ^{3}H-TdR (Bryant, 1962, 1963).

Slow movement of labeled hepatocytes from portal toward hepatic regions has recently been described (Scherer and Friedrich-Freksa, 1970). When hepatocytes were tagged with ^{3}H-TdR at 21 and 27 hours after standard partial hepatectomy, 5 days after surgery localization of radioactivity (as indicated by grain count) over hepatocytes was maximal in parenchyma adjacent to portal venules. In contrast, 90 days later maximal radioactivity was concentrated over hepatocytes in the halves of hepatic plates near hepatic venules (Scherer and Friedrich-Freksa, 1970). The investigators proposed that this pattern reflected the normal migration of cells during their life span; new cells, which were formed periportally, migrated toward hepatic veins, apparently there to die. Consistent with this finding, they calculated (from data on t_M) a life span of 90 to 135 days for hepatocytes. However, the t_M used by them (20–30 minutes) is about half of that ascertained by most investigators (Tables I and II). Based on I_S (McDonald, 1961b) or on long-term continuous labeling (Cameron, 1970), the life span or turnover time of hepatocytes has been estimated to be as long as 400 days. This would be incompatible with their hypothesis. It is possible that this slow movement of cells is also more apparent than real, reflecting remodeling of parenchymal units (acini) due to splitting of terminal hepatic venules in a manner similar to that which occurs in the actively growing liver postnatally (McKellar, 1949). However, the exact mechanism underlying redistribution of newly formed hepatocytes in relation to portal and hepatic venules remains undetermined.

Polyploid cells increase in number after partial hepatectomy, apparently as a consequence of the stimulation of the replicative cycle in binucleate cells (Nadal and Zajdela, 1966). Once stimulated to enter S phase, all residual hepatocytes appear to complete mitosis; there is apparently no report of a G_2 population in regenerated rat liver.

C. Hepatocellular Proliferation in Mice

Many studies of drug effects on hepatic cellular proliferation have been performed in mice. The kinetics of parenchymal cell formation in livers of mice and rats are quantitatively similar, but differ qualitatively in significant ways. The major distinction between cell formation in the livers of these two species is the considerably longer time necessary for activation of hepatic cell proliferation in mice following partial hepatic resection or other perturbation. Both hepatocellular DNA synthesis and mitosis begin about 12 hours later in mice than in rats [data on the kinetics of hepatocellular proliferation in mice are detailed by Edwards and Klein (1961) Kinosita and Sisken (1962), Bade *et al.* (1966), and Chernozemski and Warwick (1970)]. Cycling of hepatocytes through S is as slow in adult mice as it is

6. DRUGS AND THE HEPATIC CELL CYCLE 109

in adult rats (Edwards and Klein (1961). Livers of adult mice normally contain hepatocytes that are more highly polyploid than are similar cells in rat liver (Epstein, 1967). The lengths of the hepatic cell cycle and its phases in mice has apparently not been documented. A subpopulation of hepatocytes arrested in G_2 may exist in normal mouse liver (Perry and Schwartz, 1967) but the evidence is not strong (Grisham, 1969).

D. Biochemistry of the Hepatic Cell Cycle

Recent reviews present a detailed consideration of contemporary insight into the biochemistry of the hepatic cell cycle (Baserga, 1968; Bucher and Malt, 1971). Insight into the biochemistry of G_1 and G_2 phases of the cell cycle is imperfect, but some appreciation of these chemical events is necessary in order to understand the effects of drugs on cell proliferation.

Precise analysis of the mammalian cell cycle is currently limited by the paucity of cellular markers (biochemical reactions or products of such reactions) that unequivocally indicate the position of the cell within the cycle. The only readily demonstrable markers presently available are DNA synthesis and mitosis; G_1 and G_2 phases are distinguishable because they are delimited by S and M phases and, as yet, no specific events have been discovered that unambiguously characterize G_1 or G_2. However, it is now obvious that many biochemical events occurring in both G_1 and G_2 are essential for progression of the cell into and through S or M phase, respectively; e.g., synthesis of several types of macromolecules during G_1 are necessary for entry into S phase (Baserga, 1968). These essential biochemical reactions are so obscured by other reactions of similar chemical nature, but unrelated to cell proliferation, that they can be discerned only by indirect means, such as by using metabolic inhibitors. Some of these events are probably involved in the regulation of cell proliferation (see below). Furthermore, certain drugs affect specifically some essential events in G_1 or G_2 periods, their apparent effect on DNA synthesis or mitosis being only indirect. Extensive study of the diverse effects of various drugs may be expected to have an important role in the further elucidation of cycle-determining biochemical events.

E. Cyclic Status of Normal (Nonproliferative) Hepatocytes

Proliferation kinetics in normal adult murine hepatocytes are unusual in that few cells continuously cycle through S and M phases; yet these cells can be readily stimulated to rapidly cycle through these phases. To account for these characteristics, a G_0 phase has been hypothesized (see Baserga, 1968). While in this phase the cells are considered to be outside the replicative cycle in a type of holding pattern from which they can be recalled by appropriate stimuli. This theory is hypothetically attractive, but no means has been found to distinguish cells in either G_1 or G_0 phases. Furthermore, no compelling reasons for existence of G_0 phase has yet been advanced, and it is possible that infrequent cycling through S and M are the result of blocked or slowed passage through G_1 phase.

III. Regulation of Hepatic Cell Proliferation

The mechanism by which proliferation of hepatocytes is regulated is obscure. Substantial evidence indicates that control of hepatic cell formation is effected by substances that are disseminated throughout the organism by the blood stream. The reader is referred to recent reviews of the data leading to this opinion (Bucher and Malt, 1971; Sigel, 1969; Grisham, 1969). Although detailed knowledge of the extracellular events that regulate hepatocytic proliferation is still lacking, the investigator who studies the effect of chemicals on hepatic cellular proliferation must be aware of the possibilities, since certain drugs may affect formation of hepatocytes by interfering with this extracellular feedback mechanism. The following hypothesis attempts to explain the results of recent studies designed to elucidate the mechanism of extracellular regulation of hepatocyte formation. It is included here to introduce this presently obscure aspect of hepatocellular proliferation.

Control of hepatocyte formation is postulated to depend on the ability of these cells to excrete rapidly into or remove efficiently from blood substances, which interact with this specific type of cell during the G_1 period of the cycle, and thereby influence its progression through the cycle (to inhibit cell birth if formed by hepatocytes or to stimulate proliferation if inactivated by them). Thus, the extent of hepatic cell formation is thought to depend on the concentration of regulatory substances in the cellular microenvironment and the humoral level is proposed to be proportional to hepatic cell number. Rapidity of proliferative response to cell loss is assumed to depend on the rapid turnover of humoral effectors. The hypothesis can be satisfied by either inhibitory or stimulatory substances acting alone or by a balanced combination of both types of effectors. To provide stable negative feedback, inhibitory influences would necessarily be formed by liver and inactivated or removed by extrahepatic tissues, whereas stimulatory substances would have to be formed by tissues outside the liver and inactivated by this organ. Since the characteristics of hepatic blood flow and hepatic tissue perfusion influence intimately the ability of hepatocytes to release into or remove from blood substances which are, respectively, synthesized or catabolized by these cells, this hypothesis is consistent with the periportal localization of newly activated proliferating cells and also explains the localization of dividing cells in reversed-flow grafts. Further studies of certain drugs that block hepatocellular proliferation may provide data helpful in elucidating this problem.

IV. Drugs and Hepatic Cell Proliferation

A. INTRODUCTION

1. *Drug Metabolism by Residual Liver*

Hepatic metabolism of drugs cannot be discussed in depth in this review. However, the effect of partial hepatic resection on metabolism of drugs merits fuller comment, since this represents a potentially serious pitfall when the effects of drugs are compared in animals with intact and partly resected livers. Not only does partial hepatectomy result in a reduction in the number of cells in the liver, but the ability of the residual tissue to metabolize a variety of drugs is also depressed on a unit or cell basis (Von Der Decken and Hultin, 1960; Fouts et al., 1961; Henderson and Kersten, 1970). The most marked depressions of drug metabolizing enzymes appear to occur during the period of most rapid cellular proliferation, although there is considerable variation in both timing and degree of impairment of different enzymes (Von Der Decken and Hultin, 1960; Fouts et al., 1961; Henderson and Kersten, 1970). Metabolism of some drugs may not be depressed at all (Henderson and Kersten, 1970), whereas metabolism of others may be maximally depressed as late as 4 to 7 days after surgery (Fouts et al., 1961), a time when a large fraction of the resected cells have already been replaced. The mechanism of this postsurgery depression in metabolism of drugs is undocumented. Residual liver is still capable of responding vigorously to induction of drug-metabolizing enzymes, leading to increase in even those enzymes that are depressed (Seidman et al., 1967; Gram et al., 1968; Chiesara et al., 1970; Henderson and Kersten, 1970). The effect of combined reduction in the number of cells and in the capacity of residual cells to metabolize drugs is clear, however; the same dose of drug given to normal or partially hepatectomized animals may have quite different quantitative effects in each situation. Drugs that depend on hepatic metabolism for their elimination from the body will have a much prolonged and heightened effect, whereas drugs that depend on hepatic activation for their pharmacologic action may be less effective in the partially hepatectomized animal, particularly if they can be eliminated from the body by extrahepatic pathways.

If the latter situation does not obtain, their action may be delayed until the liver's capacity to metabolize them has been regained. This situation makes it difficult to design adequate controls.

2. *General Effects of Drugs on Hepatic Cell Proliferation*

Drugs influencing hepatic cell proliferation may either stimulate or inhibit cell formation, or they may produce both effects at different doses or at different times after exposure. Induction of cell proliferation may occur as a compensatory response to hepatocellular necrosis. In this situation, hepatocellular proliferation is a secondary response that results only because of the liver's ability to reinitiate cell proliferation after creation of a deficit of hepatocytes. This response is, in some characteristics, analogous to the proliferative response after partial hepatic resection. A large number of drugs, including hepatocarcinogens, cause hepatocellular necrosis and stimulate hepatic cell formation secondarily. It is possible, but undocumented, that compensatory proliferation could occur without preceding cellular necrosis, if the drug blocked some essential hepatocellular function or otherwise interfered with the extracellular feedback loop that appears to be involved in regulation of hepatocellular formation.

Noncompensatory increases in hepatic cell proliferation may be produced potentially by direct effects of drugs so that the growth fraction is increased or transit through the replicative cycle is shortened (primarily by shortening G_1). No drugs are known to increase hepatocellular proliferation by such a mechanism. Apparent, but not true, stimulation of cell proliferation may occur after transiently inhibited cells escape from blockade. This type of response appears to be common and probably underlies most reports of stimulation of hepatocellular proliferation by drugs. It is caused by a temporary partial synchronization of dividing cells and can occur whenever there is any degree of proliferation going on at the time the drug is given; the magnitude of the rebound peak varies directly with the magnitude of cell proliferation at the time reversible inhibition is produced.

Inhibition of hepatic cell proliferation is common and is effected by drugs that specifically block occurrence of biochemical events during particular stages of the cell cycle. These events include the synthesis of RNA during G_1 and G_2, the synthesis of proteins during all phases of the cell cycle, the synthesis of DNA during S, and the assembly of spindle proteins and function of the mitotic apparatus during M. Inhibition can also be effected by general protoplasmic poisons that are effective at any stage of the cell cycle.

B. CLASSIC HEPATOTOXINS

A large number of toxic compounds cause hepatocellular necrosis and compensatory hepatocellular proliferation. The effects of carbon tetrachloride, the archtype hepatotoxin, will be discussed as representative of this large group.

Carbon tetrachloride causes necrosis of hepatocytes located at the hepatic venous end of hepatocellular plates. Administration to rats of a single dose leads to an increase in I_S of hepatocytes sometime between 24 and 30 hours later (Leevy et al., 1959). Peak I_S occurs at about 48 hours following which there is a gradual decline to basal level at 96 hours after treatment. Proliferating hepatocytes are concentrated in the portal half of hepatocellular plates (Melvin, 1968). When carbon tetrachloride is given to partially hepatectomized rats as a single dose 3 hours after surgery, the typical posthepatectomy burst of cell proliferation is delayed by 8 hours, the timing of hepatic cell formation conforming closely to that occurring after treatment with carbon tetrachloride alone (Stöcker and Boecker, 1970). The magnitude of cell proliferation that occurs following carbon tetrachloride treatment plus partial hepatectomy appears compatible with an additive effect of both procedures in producing hepatocellular deficiency (Stöcker and Boecker, 1970). The longer time taken for activation of cell proliferation by carbon tetrachloride (as compared to partial hepatectomy) may reflect the extra time needed to produce hepatocellular necrosis. However, the fact that carbon tetrachloride also delays occurrence of hepatic cell proliferation in partially hepatectomized animals suggests that this chemical may transiently inhibit hepatocellular proliferation. This subject needs further study.

C. HEPATOCARCINOGENS

As a group, hepatocarcinogens have complex effects on proliferation of hepatocytes; these effects are often difficult to separate and analyze precisely. Most hepatocarcinogens are markedly toxic to the hepatocyte, and some of their ability to stimulate cell proliferation probably is secondary to the cell death they cause. Furthermore, most hepatocarcinogens inhibit strongly certain phases of the hepatic cell cycle, with the result that chronic administration produces a complex mixture of both stimulation and inhibition of proliferation. Some carcinogens may also have pervasive effects on the extracellular regulation of hepatic cell proliferation, resulting in atypical responses to partial hepatectomy.

Feeding rats a diet containing dimethylaminoazobenzene (DAB) results in increase in I_S and I_M of hepatocytes (MacDonald, 1961a; Daoust, 1962; Maini and Stich, 1962; Daoust and Molnar, 1964). Increase in I_S shows cyclic elevations during 180 days of feeding this carcinogen (MacDonald, 1961a). Although the endogenous rate of hepatocytic proliferation increases, hepatocytes of treated animals fail to respond to partial hepatectomy by increasing their rate of mitosis (Maini and Stich, 1962).

This altered mitotic response to partial hepatectomy is seen as early as 3 weeks after instituting the diet; normal mitotic response to partial hepatectomy is regained within 3 weeks after removing DAB from the diet (Maini and Stich, 1962). Posthepatectomy DNA synthesis in DAB-fed rats is not as strongly inhibited as is mitosis, leading to an accumulation of hepatocytes blocked apparently in G_2 (Banerjee, 1965). The mechanism of this G_2 block is not known.

Three morphologically different populations of hepatocytes, each with a characteristic rate of mitotic proliferation, have been noted during feeding of DAB (Daoust and Molnar, 1964). Hypobasophilic cells, located around terminal hepatic venules and thought to be degenerating, have a low rate of cell proliferation (Daoust and Molnar, 1964). Foci of basophilic hepatocytes, which gradually develop adjacent to portal tracts, show a considerably elevated rate of cellular proliferation; these cells are considered to be regenerating in response to degeneration of hypobasophilic cells (Daoust and Molnar, 1964). Hyperbasophilic cells develop from basophilic cells and form nodular aggregates; proliferation of cells in these nodules is elevated markedly and they are considered to give origin to hepatic neoplasms (Daoust and Molnar, 1964). Hepatocellular mitoses in animals fed DAB exhibit a high proportion of abnormal figures (Maini and Stich, 1961).

Giving rats with intact livers a single injection (either intravenous or intraperitoneal) of dimethylbenzanthracene (DMBA) causes an increase in DNA labeling beginning between the third and fifth days after treatment and peaking on the seventh day (Juhn and Prodi, 1965). If there is active proliferation in livers of treated rats (young animals) this is inhibited within the first 6 hours after treatment and before augmentation occurs (Marquardt and Philips, 1970). This result takes place without morphologic evidence of hepatocellular necrosis (Juhn and Prodi, 1965). DMBA depresses the normal posthepatectomy rise in DNA synthesis and mitosis (Juhn and Prodi, 1965; Marquardt and Philips, 1970). DNA labeling in regenerating hepatocytes is depressed as early as 2 hours after exposure to DMBA, whereas mitotic inhibition does not begin until 6 hours after treatment (Marquardt and Philips, 1970). When DMBA is given to rats 24 hours after partial hepatectomy, the mitotic peak is delayed greatly and does not become maximal until 72 hours (Marquardt and Philips, 1970). The mechanism of action of DMBA on the hepatic cell cycle is not clear. Rapid depression of DNA synthesis when animals are treated at 24 hours after partial hepatectomy suggests that DMBA acts directly on S phase. Delayed effect on mitosis could mean that this result is secondary to a primary action occurring at some earlier time, perhaps during S phase. However, studies in partially hepatectomized mice suggest that S phase is insensitive to the action of DMBA and that the block occurs somewhat earlier, possibly at the point of transition between G_1 and S (Raick and Ritchie, 1969).

Feeding rats a diet containing acetylaminofluorene (AAF) leads to an increased rate of I_S and I_M as early as 2 weeks after beginning treatment (Becker and Klein, 1971; Peraino et al., 1971). Proliferation of hepatocytes is disorganized and delayed following partial hepatectomy of rats fed AAF for 3 or more weeks (Laws, 1959; Becker and Klein, 1971). Regeneration occurs in nodules of tissue by rapid proliferation of a small number of hepatocytes (Laws, 1959; Becker and Klein, 1971). Proliferation of hepatocytes in nonnodular areas of parenchyma appears to be inhibited (Becker and Klein, 1971).

A single dose (subcutaneous or intraperitoneal) of thioacetamide (TA) causes appreciable elevation in DNA labeling and mitosis in hepatocytes of intact rat liver (Reddy et al., 1969). Initiation of elevated DNA synthesis begins between 18 and 24 hours after treatment and peaks at about 36 hours; mitosis follows DNA synthesis by about 6–8 hours, and during a 30-hour interval (24–54 hours after treatment) over 54% of hepatocytes enter M phase (Reddy et al., 1969). Hepatocytes in S and M are augmented comparably by TA treatment, leading to the suggestion that accumulation of polyploid nuclei does not result from G_2 arrest (Stöcker and Altmann, 1964). Treatment with TA accelerates markedly the turnover time of hepatocytes (Stöcker and Altmann, 1964). Proliferating cells are located predominantly in the periportal half of hepatocellular plates (Reddy et al., 1969). Treatment of rats with TA for 7 days either before or after partial hepatectomy does not interfere with posthepatectomy mitosis in hepatocytes (Kleinfeld and von Haam, 1959). However, a single large dose of TA given at different times after partial hepatectomy causes striking effects on cell proliferation, which are related to the phase of the cell cycle in which exposure occurs (Mironescu and Ciovîrnache, 1971). When TA is given during the first 14 hours of the prereplicative period (early G_1), I_S is increased twofold at 19–24 hours after surgery. The investigators suggest that TA shortens the G_1 period (Mironescu and Ciovîrnache, 1971). When given between 16 and 24 hours after partial hepatectomy (S phase), TA causes marked mitotic abnormalities that consist predominantly of bridges and chromosomal breaks (Mironescu, 1969; Mironescu and Ciovîrnache, 1971). TA causes no abnormality in proliferating hepatocytes when given at 26 hours after surgery (M phase) (Mironescu, 1969). These results were interpreted as indicating that S phase of proliferating hepatocytes was most sensitive to the effects of TA.

Feeding rats diethylnitrosamine (DEN) causes a progressive increase in hepatocytic I_S prior to the development of hepatomas at 140–160 days after beginning treatment (Côté et al., 1962; Rubin et al., 1964; Mohr and Speetzen, 1967; Rabes et al., 1970). It was suggested that increased DNA synthesis represents a secondary response to cellular degeneration and necrosis (Côté et al., 1962); hepatocytes of rats treated with DEN have a shorter than normal life span (Scherer and Friedrich-Freksa, 1970). Hepatocarcinomas appear to develop from nodular foci of rapidly proliferating hepatocytes located in the vicinity of terminal hepatic veins (Côté et al., 1962). Characteristic morphologic changes divide the hepatocytic population into subgroups during treatment with DEN; these subpopulations have differing rates of endogenous cellular proliferation (Rubin et al., 1964; Rabes et al., 1970) and differing responsiveness to partial hepatectomy (Rabes et al., 1970). Adjacent to hepatic veins, sharp foci containing cells with vacuolated cytoplasm (Rubin et al., 1964) with decreased glycogen and glucose-6-phosphatase (Rabes et al., 1970) develop initially (about 40 days after beginning treatment). Proliferation of cells in these foci is not increased within the first several weeks after treatment (Rubin et al., 1964; Rabes et al., 1970). Subsequently, cells in perihepatic foci develop the abnormal ability to respond to partial hepatectomy with proliferation equivalent to that of periportal hepatic cells (Rabes et al., 1970). The growth fraction of hepatocytes in these foci reaches 90% after partial hepatectomy, although normally less than 20% of hepatic cells in this location proliferate after surgery (Rabes et al., 1970). Following development of responsiveness of these hepatocytes to partial hepatectomy their endogenous rate of proliferation begins to increase and the foci become nodules of hyperplastic cells (Rubin et al., 1964; Rabes et al., 1970). Later, after hepatocytes in these hyperplastic foci have reached a very high rate of proliferation, they again become unresponsive to partial hepatectomy (Mohr and Speetzen, 1967; Mohr et al., 1967; Rabes et al., 1970). Hypothetically, hepatocarcinomas arise by direct transformation of these hyperplastic foci (Côté et al., 1962; Rubin et al., 1964; Rabes et al., 1970).

Pyrrolizidine alkaloids (lasiocarpine, lasiocarpine N-oxide, and heliotrine) cause marked inhibition of mitosis in hepatocytes of regenerating livers (Downing and Peterson, 1968). Within 2 months after giving lasiocarpine, hepatocytic karyomegaly is marked; the I_S of such cells is increased although no mitotic figures are found (Svoboda et al., 1971). Lasiocarpine N-oxide treatment for 3 months greatly modifies the proliferative response of hepatocytes to partial hepatectomy. Hepatocytic I_S is not greatly affected in regenerating hepatocytes of lasiocarpine-treated rats, whereas I_M is almost totally suppressed (Peterson, 1965). These studies suggest that a major effect of lasiocarpine is blockage of passage through the G_2 phase. Highly polyploid hepatocytes in livers of lasiocarpine-treated rats, therefore, probably arise from successive cycles of DNA synthesis in cells incapable of mitosis. The mechanism of this chronic action of lasiocarpine is unknown; this drug also inhibits DNA synthesis in regenerating hepatocytes when given at the time of partial hepatectomy (Svoboda et al., 1971).

Aflatoxin B inhibits I_S and I_M in proliferating hepatocytes of weanling rats (Rogers and Newberne, 1967) or of adult rats following partial hepatectomy (DeRecondo et al., 1966; Rogers et al., 1971). When aflatoxin B is given to rats at the time of partial hepatectomy or 14 hours later, DNA synthesis and mitosis are completely blocked at 24 hours after surgery (Rogers et al., 1971). When aflatoxin B is given at 30 hours after partial hepatectomy, DNA synthesis is depressed by about 70% and mitosis is lowered by about 60% at 48 hours after surgery (Rogers et al., 1971). When given at 24–36 hours (S phase) after partial hepatectomy, aflatoxin B acts rapidly to effect maximal inhibition of DNA synthesis within 2 hours (DeRecondo et al., 1966). Reduction in the rate of DNA labeling is effected by decreasing the number of cells involved, as well as by slowing the rate of synthesis in each cell (DeRecondo et al., 1966). These results suggest that aflatoxin B is effective during either G_1 or S phase of the hepatic cell cycle and that is probably acts by inhibiting the primer activity of DNA (DeRecondo et al., 1966).

Development of hepatocarcinomas in response to some chemical carcinogens is intimately related to the cycle phase represented by the majority of the hepatocytes at the time treatment is given (Chernozemski and Warwick, 1970; Craddock, 1971). Although related to the topic of this discussion, this subject is beyond the scope of this review. The reader is referred to the two articles quoted above and to a recent discussion of the subject (Warwick, 1971).

D. METABOLIC INHIBITORS

The large group of drugs categorized as metabolic inhibitors (including agents currently used for chemotherapy of neoplastic diseases) have been extensively studied for their effect on hepatic cellular proliferation. In many instances, liver regenerating after partial hepatectomy has been used as a convenient *in vivo* system in which to analyze the effects of these agents on the proliferation of nonneoplastic cells. Studies with certain metabolic inhibitors have led to much of our present understanding of the biochemistry of G_1 and G_2 phases of the hepatic cell cycle. (See also Hoffman and Post, Chapter 9, this volume.)

Actinomycin D blocks initiation of DNA labeling and mitosis that follow partial hepatectomy in rats (Fujioka et al., 1963; Schwartz et al., 1965b). It produces maximal inhibition of DNA synthesis only if given during the first 12–16 hours after surgery. Tiny doses of this drug, when given at the time of partial hepatectomy and at 2 and 4 hours later, delay the onset of S phase by 4 hours (Fujioka *et al.*, 1963). Inhibition of DNA synthesis is apparently secondary to inhibition of RNA synthesis and the subsequent blockage of formation of special enzymes necessary for production of DNA (Fujioka *et al.*, 1963; Giudice and Novelli, 1963). Reversal of the suppression of RNA and DNA synthesis is rapid, and both RNA and DNA synthesis resume following cessation of treatment (Fujioka *et al.*, 1963; Schwartz *et al.*, 1965b). When given to rats 24 hours after partial hepatectomy, actinomycin D rapidly inhibits mitosis (Schwartz *et al.*, 1965b). Mitosis is inhibited as quickly as 1 hour after treatment and inhibition is maximal after 16 hours (Schwartz *et al.*, 1965b). These results suggest that initiation of mitosis is blocked by an effect on some process occurring late during S or G_2. An influence of actinomycin D on G_2 phase of dividing hepatocytes is supported by a study in which proliferation was induced in mouse liver with carbon tetrachloride (Melvin, 1967). Although in this instance actinomycin D given at a time corresponding to late S and G_2 did not prevent entry into M, it did slow passage appreciably through this phase (Melvin, 1967). These data suggest that actinomycin D blocks critical biochemical reactions in both G_1 and G_2 phases of the cell cycle. It is toxic to proliferating hepatocytes (Schwartz *et al.*, 1965b) but has no apparent toxic morphologic effect on nonproliferating hepatic cells. Toxicity to the regenerating liver may be related to the fact that actinomycin D is markedly concentrated in the liver and that hepatic metabolism of this drug is slowed greatly after partial hepatectomy (Schwartz *et al.*, 1965b). No apparent hepatic damage occurs after small doses that are, nevertheless, capable of reversibly delaying passage through the cell cycle (Fujioka *et al.*, 1963).

Nogalomycin appears to influence hepatic cell proliferation in a manner similar to actinomycin D; both RNA and DNA synthesis are inhibited by treatment during the early prereplicative period (Gray *et al.*, 1966). Inhibition of both RNA and DNA synthesis is dose dependent (Gray *et al.*, 1966). Daunomycin, however, may exert an influence that differs from that of the other two drugs, since DNA synthesis is inhibited when treatment is given late during the prereplicative period, whereas RNA synthesis is much less inhibited (Theologides *et al.*, 1968).

Chlortetracycline also effects proliferation of hepatocytes in a manner clearly different from the mode of action of actinomycin D. When given at the time of partial hepatectomy, chlortetracycline almost completely depresses DNA labeling 22 hours later (Hurwitz and Carter, 1969). The inhibition is reversible and both thymidine kinase activity and the level of DNA labeling recover by 30 hours after surgery (Hurwitz and Carter, 1969). The mechanism of action of chlortetracycline is unknown, although it appears to directly prevent protein synthesis (translation) rather than to block mRNA synthesis (transcription).

Several well-documented inhibitors of protein synthesis markedly affect hepatocellular proliferation. Cyclohexamide delays DNA labeling in regenerating hepatocytes when given during late G_1 (14 hours after surgery) or during S (22 hours after surgery), and it also delays M when given at almost any time during the cycle, but especially during late S or early G_2 (24 hours after surgery) (Verbin et al., 1969). Cyclohexamide rapidly suppresses protein synthesis in regenerating hepatocytes when given during S phase and inhibition of DNA synthesis quickly follows (Brown et al., 1970). Sparsomycin and pactomycin have effects similar to cyclohexamide on protein and DNA synthesis in S phase hepatocytes (Brown et al., 1970). The effect of all three of these drugs on protein and DNA synthesis is dose dependent (Brown et al., 1970). Blockage of the various phases of the cell cycle by inhibitors of protein synthesis is postulated to indicate that synthesis of a protein that is necessary for initiation or continuation of the phase in question has been prevented (Verbin et al., 1969; Brown et al., 1970).

Puromycin, when given in multiple doses between 16 and 21 hours after partial hepatectomy or between 32 and 37 hours after surgery, markedly inhibits DNA labeling occurring at 24 or 40 hours, respectively, after the operation (Gottlieb et al., 1964). This depression of S is associated with marked inhibition of incorporation of leucine into protein but with lesser depressions of nuclear and cytoplasmic RNA labeling (Gottlieb et al., 1964). Puromycin may affect as many phase-specific events as does cyclohexamide, but data are presently incomplete. One such event, the posthepatectomy rise in DNA polymerase, has been studied in some detail (Giudice et al., 1964). The degree of inhibition of the activity of DNA polymerase is related to the time in the prereplicative period when puromycin is given. Maximal depression is effected by treatment between 13 and 19 hours after surgery; inhibition is less marked if the drug is given either earlier (3 to 9 hours after surgery) or slightly later (18–23 hours after surgery), and treatment between 20 and 26 hours causes no depression of DNA polymerase (Giudice et al., 1964). These results may indicate the part of the G_1 period during which DNA polymerase is being synthesized. However, puromycin is also effective in suppressing DNA synthesis in regenerating liver 40 hours after surgery, a time when DNA polymerase has already been synthesized (Gottlieb et al., 1964); at this time some other event is apparently blocked.

Several other drugs, which inhibit protein synthesis, have also been studied for their effect on DNA synthesis when given to rats shortly after partial hepatectomy. Ethionine, p-fluorophenylalanine, β-thienylalanine, and chloramphenicol all markedly depress DNA labeling 20–22 hours after partial hepatectomy when the drugs are given during the early prereplicative period (2–10 hours after surgery) (Schneider et al., 1960; Fujioka et al., 1963). Effects of most drugs are dose dependent (Schneider et al., 1960) and rapidly reversible (Fujioka et al., 1963). Effects of these inhibitors of protein synthesis on other phases of the hepatic cell cycle have not been documented.

Hydroxyurea markedly and rapidly depresses DNA synthesis in proliferating hepatocytes when given at 24 hours after surgery (Schwartz et al., 1965a; Yarbro et al., 1965; Hill and Gordon, 1968). Inhibition of DNA synthesis is dose dependent and depression is rapidly reversed (Schwartz et al., 1965a; Baugnet-Mahieu et al., 1971). Daily administration of hydroxyurea completely suppresses the normal posthepatectomy proliferative response in hepatocytes (Lea et al., 1970). Hydroxyurea does not affect incorporation of precursors into RNA or most protein (Schwartz et al., 1965a; Yarbro et al., 1965), although glycine incorporation into histones is said to be inhibited markedly (Yarbro et al., 1965). The major mechanism of action of hydroxyurea probably involves inhibition of ribonucleotide reductases (Schwartz et al., 1965a). Although hydroxyurea is quite toxic to intestinal epithelium and to certain other continuously replicating cells, there is no evidence that it causes necrosis of hepatocytes whether they are cycling or not (Schwartz et al., 1965a).

Suppressing *de novo* biosynthesis of purines, 6-mercaptopurine abolishes completely the normal posthepatectomy hepatocytic proliferative response when given in daily doses beginning immediately following surgery (Lea et al., 1970). Azathioprine, which is converted to 6-mercaptopurine in the body, similarly suppresses hepatic DNA synthesis when given daily after partial hepatectomy (Gonzalez et al., 1970). This effect is reversible and within 36 hours after the last dose of azathioprine a rebound increase in DNA labeling occurs and formation of DNA remains elevated for the next 4 days (Van Vroonhoven and Malt, 1971). Incorporation of precursors into RNA and protein is not affected by treatment with this drug (Gonzalez et al., 1970). 5-Fluorouracil blocks mitosis following partial hepatectomy (Paschkis et al., 1959). This effect on mitosis is probably indirect since the primary action of this drug is blockage of DNA synthesis by inhibition of thymidylate synthetase (Hartmann and Heidelberger, 1961).

Methotrexate depresses labeling of DNA which follows partial hepatectomy, only when rats are treated with the drug several days before surgery, as well as following operation (Barton and Laird, 1957). However, the effect of methotrexate on proliferation of regenerating hepatocytes appears to

6. DRUGS AND THE HEPATIC CELL CYCLE

be highly variable (Meschan and Castells, 1964). Similarly, the effect of this drug on folate reductase, its primary site of action in the regenerating hepatocyte, is inconstant (Ngu *et al.*, 1964; Brown *et al.*, 1965). The reason for this variability is unknown.

Asparaginase blocks the initial increase in I_S and I_M that follows partial hepatectomy of rats (Becker and Broome, 1967; Becker *et al.*, 1970). This drug acts by reducing enzymatically the extracellular level of asparagine; proliferation is inhibited in cells that have a low capacity to synthesize this amino acid (Becker and Broome, 1967). Depression of I_S is elicited only when asparaginase is given during the first 6 hours after partial hepatectomy (Becker *et al.*, 1970). Diminished availability of asparagine may depress the formation of essential proteins made during the early prereplicative period. The inhibitory effect of asparaginase on hepatic cell proliferation is self-limited; recovery of I_S and I_M occurs after a delay of about 12 hours, even when the drug is given continuously (Becker and Broome, 1967). This may reflect induction of increased asparagine synthetase activity in residual hepatocytes (Becker *et al.*, 1970).

Colchicine and podophyllin block the passage of hepatocytes through M phase by preventing formation of the mitotic spindle (Brues, 1936; Brues and Cohen, 1936; Brues and Jackson, 1937; Seidlová-Mašínová *et al.*, 1957; Kleinfeld and Sisken, 1966). Colchicine does not stimulate entry of cells into M phase; the accumulation of cells blocked in M can be accounted for by the normal rate of entry (Brues, 1936). After blockage of over 98% of M phase hepatic cells, recovery from a single dose of colchicine begins about 6 hours after treatment, but complete recovery takes 24–48 hours more (Brues and Jackson, 1937; Kleinfeld and Sisken, 1966). Occurrence of grossly abnormal mitotic figures is associated with recuperation from the effects of colchicine or podophyllin (Brues and Jackson, 1937; Seidlová-Mašínová *et al.*, 1957; Kleinfeld and Sisken, 1966). Vinblastine, which has a pronounced colchicine-like effect, also markedly depresses DNA labeling in regenerating hepatocytes when rats are treated during the first 9 hours after partial hepatectomy (Luyckx and Van Lancker, 1966). Although the mechanism of this action of vinblastine is not clear, it may result from depression of the priming ability of DNA; labeling of nuclear RNA is depressed by treatment with this drug but much less than is labeling of DNA (Luyckx and Van Lancker, 1966). For a further discussion of the mechanism of action of the vinca alkaloids see Taylor, Chapter 2, this volume.

Mustards powerfully inhibit I_M after partial hepatectomy (Marshak, 1946; Landing *et al.*, 1949) without influencing DNA labeling (Marshak, 1946). Inhibition of mitosis is ineffective when nitrogen mustard is given at the time of surgery; when rats are treated at 24 or 36 hours after surgery depression of mitosis is marked but recovery occurs after a short delay (Landing et al., 1949). It was concluded that nitrogen mustard slows both the rate of entry into M as well as the rate of passage through this phase (Landing et al., 1949).

Sodium fluoroacetate inhibits markedly the mitotic response following partial hepatectomy if given early during the prereplicative period (Weinbren and Fitschen, 1959). This chemical is ineffective when given at 24 hours after surgery (Weinbren and Fitschen, 1959).

E. Hormones in Pharmacologic Doses

Growth hormone enhances hepatic DNA synthesis and mitosis when given to rats with intact adrenal and pituitary glands. Treatment of rats with intact livers with growth hormone for 4 days results in a small increase in I_S, which peaks on the third day after starting treatment, following which hepatocellular proliferation spontaneously declines (Nettesheim and Oehlert, 1962). A single dose of growth hormone given 4.5 hours before partial hepatectomy shortens the length of the prereplicative period following surgery (Moolten et al., 1970). Similarly, treatment of partially hepatectomized rats with 3 doses of growth hormone (at 2 and 8 hours after surgery and 5–6 hours before killing at various times between 22 and 30 hours after surgery) causes earlier onset and a two- to threefold elevation of posthepatectomy DNA synthesis and mitosis (Cater et al., 1957). The mechanism of this effect is not clear, but several forms of nonspecific stress applied shortly before partial hepatectomy potentiate the proliferative response to the operation (Simek et al., 1968; Moolten et al., 1970). Hypophysectomy delays the onset of posthepatectomy hepatocellular proliferation by about 12 hours (Hemingway and Cater, 1958; Rabes et al., 1965; Rabes and Brändle, 1969). Although onset of both I_S and I_M are delayed, the magnitude of the delayed response is not appreciably affected (Rabes and Brändle, 1969). Growth hormone treatment of hypophysectomized animals is ineffective in reversing this delay, but certain combinations of growth hormone and cortisone are said to be effective (Hemingway and Cater, 1958).

Pharmacologic doses of adrenal cortical hormones (cortisone, corticosterone, and hydrocortisone) all depress the proliferative response of rat liver to partial hepatectomy (Roberts et al., 1952; Perez-Tamayo et al., 1953; Horváth and Kovács, 1956; Guzek, 1964, 1968; Lahtiharju and Teir, 1964; Schwartz et al., 1965b; Davis and Hyde, 1966; Sakuma and Terayama, 1967; Raab and Webb, 1969; Moolten et al., 1970; Rizzo et al., 1971). Similarly, cortisone inhibits DNA synthesis in intact livers of young growing rats (Henderson et al., 1971). Cortisone is cleared quickly and depression of hepatocellular proliferation is rapidly reversed (Henderson et al., 1971; Rizzo et al., 1971); most studies have used repeated doses. Posthepatectomy DNA synthesis is not inhibited by single doses of cortisone if the drug is given prior to 12 hours after partial hepatectomy (Sakuma and Terayama, 1967; Raab and Webb, 1969). Single doses of cortisone are maximally effective in inhibiting DNA synthesis and mitosis when given between 17 and 19 hours after partial hepatic resection (Raab and Webb, 1969); inhibition of DNA synthesis is dose dependent (Raab and Webb, 1969). Following cortisone treatment of partially hepatectomized rats at 19 hours after surgery, DNA labeling is delayed by about 11 hours. However, after the cortisone-induced inhibition is released, the subsequent peak of DNA synthesis is higher and narrower than the normal, suggesting that hepatocytic proliferation has been synchronized partially by the transient inhibition (Rizzo et al., 1971). Continuous treatment of rats with desoxycorticosterone acetate (subcutaneous pellets) for 6 months prior to partial hepatectomy results in a delayed but augmented miotic response to this operation (Symeonidis et al., 1955).

The mechanism of action of cortisone on the hepatic cell cycle is not clear. Although DNA synthesis and mitosis are inhibited, synthesis of total hepatocellular RNA and protein are augmented (Roberts et al., 1952; Raab and Webb, 1969; Rizzo et al., 1971). The time of maximal effectiveness of single doses given after partial hepatectomy suggests that cortisone acts directly on the S phase to inhibit DNA synthesis (Raab and Webb, 1969). This opinion is supported by observations that cortisone treatment decreases not only the number of nuclei that incorporate ^{3}H-TdR but also the intensity of labeling of tagged nuclei (Guzek, 1964, 1968). However, DNA polymerase is said to be inhibited by cortisone, suggesting that this hormone may also act during G_1 (Henderson et al., 1971).

Adrenalectomy produces no significant effect on hepatocytic proliferation in intact liver or following partial hepatectomy (Moolten et al., 1970). Treatment of rats with ACTH only partly reproduces the action of cortisone on hepatic cell proliferation (Šimek et al., 1968; Davis and Hyde, 1966; Guzek, 1964, 1968; Moolten et al., 1970).

Adrenalin given hourly after partial hepatectomy causes pronounced depression of DNA labeling. This drug is most effective when given during the first 15 hours after surgery (Sakuma and Terayama, 1967). Adrenal medullectomy does not influence the proliferative response to partial hepatectomy (Weinbren et al., 1967).

Triiodothyronine treatment of euthyroid or hypothyroid rats results in an approximately threefold increase in I_S and I_M in hepatocytes of intact livers (Lee et al., 1968). Thyroidectomy causes decreased hepatocellular proliferation which is rapidly reversed by treatment with triiodothyronine (Lee et al., 1968). Effects of triiodothyronine or thyroidectomy on hepatic cell proliferation after partial hepatectomy have apparently not been documented.

Infusion of 20% glucose and crystalline insulin into nondiabetic rats hourly after partial hepatectomy causes depression of both I_S and I_M at 34 hours after surgery (Šimek et al., 1967a). Treatment of nondiabetic rats with protamine–zinc insulin potentiates DNA labeling at 24 and 44 hours after partial hepatectomy of rats fed ad libitum (Šimek et al., 1969). In rats fasted for 4 days before surgery and fed ad libitum thereafter, protamine-zinc insulin treatment inhibits DNA labeling at 24 hours after surgery but increases it at 44 hours (Šimek et al., 1969). These reports are difficult to evaluate because of the limited number of time points studied. However, insulin clearly induces hepatic cell proliferation in livers of severely alloxan diabetic rats (Younger et al., 1966). When severely diabetic rats are treated with one dose of crystalline insulin followed by protamine-zinc insulin every 24 hours, DNA synthesis and I_S increase after a lag period of about 24 hours following the first dose and reach a peak between 48 and 72 hours (Younger et al., 1966). DNA polymerase and total hepatic DNA also increase. The mechanism of action of insulin on hepatocytic proliferation in diabetic rats is unknown. Diabetes does not affect hepatocellular proliferative response to partial hepatectomy (Younger et al., 1966). Cell proliferation may be a side effect of the general metabolic stimulation caused by insulin.

F. MISCELLANEOUS DRUGS APPARENTLY NOT HEPATOTOXIC

Phenobarbital given repeatedly either in the diet or parenterally increases I_S (Schlicht et al., 1968; Peraino et al., 1971) and I_M (Argyris and Magnus, 1968; Schulte-Hermann et al., 1968) in hepatocytes of intact livers of young rats. Elevations of I_S and I_M are modest (two- to threefold) and they are either only maintained transiently (Argyris and Magnus, 1968) or vary in magnitude for several days (Schlicht et al., 1968; Schulte-Hermann et al., 1968). Pretreatment of rats with phenobarbital produces a fivefold increase in the magnitude of the rise in I_M that typically follows partial hepatectomy, without affecting the timing of the mitotic peak (Japundžić et al., 1967). However, other investigators claim that phenobarbital suppresses I_S and I_M in hepatic cells in livers of growing rats or in livers in which cell proliferation is stimulated by treatment with isoproterenol (Barka and Popper, 1967). Furthermore, one dose of phenobarbital given immediately after partial hepatectomy is said to delay the normal increases in DNA labeling and mitosis (Schindler and Burki, 1971). The reason for these divergent

6. DRUGS AND THE HEPATIC CELL CYCLE

observations is unknown. The data demonstrating inhibition of proliferation suggest that the reported elevations may be due to escape from an inhibitory blockage of the hepatic cell cycle.

Methylcholanthrene causes a slight increase in DNA labeling and mitosis in hepatocytes of intact livers of young rats (Barka and Popper, 1967; Argyris and Layman, 1969). A single dose of α-benzene hexachloride causes a tenfold increase in I_S and I_M in hepatocytes of young rats, with a peak occurring on the second day after treatment (Schlicht et al., 1968; Schulte-Hermann et al., 1968). Nikethamide produces a variable but often marked increase in I_M of hepatocytes when fed to mice of various ages (Wilson and Leduc, 1954). Nicotinic acid, structurally related to nikethamide, is reported to markedly depress the posthepatectomy mitotic peak 30 hours later when the drug is given at the time of surgery or 12 hours afterwards (Šimek et al., 1967b). Allylisopropylacetamide in repeated doses causes a transient increase in DNA labeling, maximal on the second day after starting treatment (Barka and Popper, 1967).

All the drugs discussed in the preceding two paragraphs cause enlargement of the liver, hypertrophy of the smooth endoplasmic reticulum, and induction of a variety of drug metabolizing enzymes (Barka and Popper, 1967). It is possible that these cellular alterations relate to the effects that these drugs have on cell proliferation. The mechanism by which drugs of this group affect hepatocellular formation is not apparent, but the possibility that increases in I_S and I_M represent rebound phenomena following a transient block of the cell cycle seems probable. Drugs representative of this interesting group merit further study as to their effects on proliferation of hepatic cells.

Isoproterenol produces transient hepatocellular proliferation without causing hypertrophy of hepatocytes or increase in liver weight (Barka and Popper, 1967). A single dose of isoproterenol causes a threefold increase in hepatic DNA labeling within 24 hours (Barka and Popper, 1967). Repeated treatments with this drug produce a three- to eightfold increase in hepatic DNA labeling (Barka, 1965; Barka and Popper, 1967; Chang et al., 1969). DNA labeling is maximally increased 24 hours after beginning treatment, and the increased cell proliferation is self-limiting, declining to basal levels even when treatment is continued (Barka and Popper, 1967). Female rats are considerably more sensitive to the action of isoproterenol than are male rats (Chang et al., 1969). This drug causes a considerably greater increase in proliferation of salivary glandular epithelial cells than it does of hepatocytes (Barka, 1965). The mechanism of action of isoproterenol, a relatively simple chemical, is unknown although many studies have been directed toward its elucidation (Baserga, 1968).

Heparin is said to stimulate DNA labeling and I_M in hepatocytes of adult rats (Zimmerman and Celozzi, 1961). Heparitin sulfate and chondroitin sulfate B also stimulate hepatic cell proliferation although they are less potent than heparin (Zimmerman and Celozzi, 1961). Heparin is also said to depress posthepatectomy I_M at 50 hours when given at 24 hours after surgery (Räsänen et al., 1966). However, no effect on I_S in either intact (sham hepatectomy) or partly resected livers of rats occurs when a single large dose of heparin is given 6 hours after surgery (Grisham et al., 1966). These divergent results may reflect differences in timing of exposures or variations in different sources of heparin.

Serotonin in a single dose causes a three- to fivefold increase in I_S 24 hours after treatment of young rats (MacDonald et al., 1959). Sensitivity to serotonin appears to be age dependent since rats weighing more than 115 gm are not affected significantly (MacDonald et al., 1959). This drug is also said to potentiate I_M in regenerating liver (Pukhalskaya, 1964).

Excessive hepatic storage of iron (produced in rats by parenteral treatment with iron dextran for 16 weeks) is associated with an augmented mitotic response to partial hepatectomy; both I_S (Volini et al., 1965) and I_M (Brain, 1964) are elevated when compared to the response in rats with normal hepatic iron stores. Folic acid causes a modest (twofold) potentiation of I_M at 27.5 hours after partial hepatectomy (Härkönen and Kiviranta, 1958).

Interferon inducers (Poly I-C, Newcastle disease virus, and Statolon) inhibit markedly the mitotic response of mouse hepatocytes to partial hepatectomy when given at 24 hours after surgery (Jahiel et al., 1971). Inhibition of I_M by Poly I-C is dose related; Poly I is ineffective. Interferon, rather than the inducer molecules, is postulated to be the direct cause of mitotic inhibition (Jahiel et al., 1971). Since treatment with these drugs is effective when they are given 24 hours before the mitotic peak normally occurs, it is postulated that DNA synthesis is directly inhibited (Jahiel et al., 1971).

V. Concluding Remarks

This survey discloses that knowledge of the effects of drugs on proliferation of hepatocytes is fragmentary. Although many drugs have been studied, investigations have frequently been incomplete and the results have often provided only tantalizing bits of information, insufficient to support firm conclusions. Future studies must be designed to take into account the unusual characteristics of drug metabolism and cell proliferation in the liver.

Acknowledgments

Supported by grants from the United States Public Health Service (AM-07568) and the John A. Hartford Foundation, Inc. The author acknowledges the assistance of Mrs. Linda Fulmer and Miss Mary Stenstrom.

References

Alfert, M., and Geschwind, I. I. (1958). The development of polysomaty in rat liver. *Exp. Cell Res.* **15**, 230–232.
Argyris, T. S., and Layman, D. L. (1969). Liver growth associated with induction of demethylase activity after injection of 3-methylcholanthrene in immature rats. *Cancer Res.* **29**, 549–553.
Argyris, T. S., and Magnus, D. R. (1968). The stimulation of liver growth and demethylase activity following phenobarbital treatment. *Develop. Biol.* **17**, 187–201.
Bade, E. G., Sadnik, I. L., Pilgrim, C., and Maurer, W. (1966). Autoradiographic study of DNA-synthesis in the regenerating liver of the mouse. *Exp. Cell Res.* **44**, 676–678.
Banerjee, M. R. (1965). Mitotic blockage at G_2 after partial hepatectomy during 4-dimethylaminoazobenzene hepatocarcinogenesis. *J. Nat. Cancer Inst.* **35**, 585–589.
Barbason, H., and Lelièvre, P. (1970). Influence du rythme de l'activité circadienne sur les différentes phases du premier cycle cellulaire suivant une hépatectomie partielle. *C. R. Acad. Sci.* **271**, 1798–1801.
Barbiroli, B., and Potter, V. R. (1971). DNA synthesis and interaction between controlled feeding schedules and partial hepatectomy in rats. *Science* **172**, 738–741.
Barka, T. (1965). Induced cell proliferation. The effect of isoproterenol. *Exp. Cell Res.* **37**, 662–679.
Barka, T., and Popper, H. (1967). Liver enlargement and drug toxicity. *Medicine (Baltimore)* **46**, 103–117.
Barton, A. D., and Laird, A. K. (1957). Effects of amethopterin on nucleic acid metabolism in mitotic and non-mitotic growth. *J. Biol. Chem.* **227**, 795–803.
Baserga, R. (1968). Biochemistry of the cell cycle. A review. *Cell Tissue Kinet.* **1**, 167–191.
Baugnet-Mahieu, L., Goutier, R., and Baes, C. (1971). Differential response of mitochondrial and nuclear DNA synthesis to hydroxyurea in normal and regenerating rat liver. *Biochem. Pharmacol.* **20**, 141–149.
Becker, F. F. and Broome, J. D. (1967). L-asparaginase: Inhibition of early mitosis in regenerating rat liver. *Science* **156**, 1602–1603.
Becker, F. F., and Klein, K. M. (1971). The effect of L-asparaginase on mitotic activity during N-2-fluorenylacetamide hepatocarcinogenesis: Subpopulations of nodular cells. *Cancer Res.* **31**, 169–173.
Becker, F. F., Baserga, R., and Broome, J. D. (1970). Effect of L-asparaginase on DNA synthesis in regenerating liver and in other dividing tissues. *Cancer Res.* **30**, 133–137.
Brain, M. C. (1964). Iron loading and the liver: The effect on regeneration after partial hepatectomy. *Gut* **5**, 374–378.

Brown, R. F., Umeda, T., Takai, S.-I., and Lieberman, I. (1970). Effect of inhibitors of protein synthesis on DNA formation in liver. *Biochim. Biophys. Acta* **209**, 49–53.

Brown, S. S., Neal, G. E., and Williams, D. C. (1965). Lack of effect of methotrexate on hepatic regeneration. *Nature (London)* **206**, 1007–1009.

Brues, A. M. (1936). The effect of colchicine on regenerating liver. *J. Physiol. (London)* **86**, 63P-64P.

Brues, A. M., and Cohen, A. (1936). Effects of colchicine and related substances on cell division. *Biochem. J.* **30**, 1363–1368.

Brues, A. M., and Jackson, E. B. (1937). Nuclear abnormalities resulting from inhibition of mitosis by colchicine and other substances. *Amer. J. Cancer* **30**, 504–511.

Brues, A. M., and Marble, B. B. (1937). An analysis of mitosis in liver restoration. *J. Exp. Med.* **65**, 15–28.

Bryant, B. J. (1962). Reutilization of leukocyte DNA by cells of regenerating liver. *Exp. Cell Res.* **27**, 70–79.

Bryant, B. J. (1963). Reutilization of lymphocyte DNA by cells of intestinal crypts and regenerating liver. *J. Cell Biol.* **18**, 515–523.

Bucher, N. L. R., and Malt, R. A. (1971). "Regeneration of Liver and Kidney." Little, Brown, Boston, Massachusetts.

Bucher, N. L. R., and Swaffield, M. N. (1964). The rate of incorporation of labeled thymidine into the deoxyribonucleic acid of regenerating rat liver in relation to the amount of liver excised. *Cancer Res.* **24**, 1611–1625.

Bucher, N. L. R., Swaffield, M. N., and DiTroia, J. F. (1964). The influence of age upon the incorporation of thymidine-2-C^{14} into the DNA of regenerating rat liver. *Cancer Res.* **24**, 509–512.

Busanny-Caspari, W. (1962). Autoradiographische Untersuchungen mit H^3-Thymidin über die DNS-Syntheses in Leberzellen verschiedener Ploidiestufen. *Frankfurt. Z. Pathol.* **72**, 123 134.

Cameron, I. L. (1970). Cell renewal in the organs and tissues of the nongrowing adult mouse. *Tex. Rep. Biol. Med.* **28**, 203–248.

Carriere, R. (1967). Polyploid cell reproduction in normal adult rat liver. *Exp. Cell Res.* **46**, 533–540.

Cater, D. B., Holmes, B. E., and Mee, L. K. (1956). Cell division and nucleic acid synthesis in the regenerating liver of the rat. *Acta Radiol.* **46**, 655–667.

Cater, D. B., Holmes, B. E., and Mee, L. K. (1957). The effect of growth hormone upon cell division and nucleic acid synthesis in the regenerating liver of the rat. *Biochem. J.* **66**, 482–486.

Chang, L. O., Morris, H. P., and Looney, W. B. (1969). The modification of labeled cytidine and thymidine incorporation into mitochondrial and nuclear DNA in normal liver, hepatoma 3924A and its host liver by isoproterenol. *Brit. J. Cancer* **23**, 868–874.

Chernozemski, I. N., and Warwick, G. P. (1970). Liver regeneration and induction of hepatomas in B6AF$_1$ mice by urethan. *Cancer Res.* **30**, 2685–2690.

Chiesara, E., Conti, F., and Meldolesi, J. (1970). Influence of partial hepatectomy on the induction of liver microsomal drug-metabolizing enzymes produced by phenobarbital. *Lab. Invest.* **22**, 329–338.

Côté, J., Oehlert, W., and Büchner, F. (1962). Autoradiographische Untersuchungen zur DNS-Syntheses während der experimentellen Kanzerisierung der Leberparenchymzelle der Ratte durch Diäthylnitrosamin. *Beitr. Pathol. Anat. Allg. Pathol.* **127**, 450–473.

Craddock, V. M. (1971). Liver carcinomas induced in rats by single administration of dimethylnitrosamine after partial hepatectomy. *J. Nat. Cancer Inst.* **47**, 899-907.

Daoust, R. (1962). The mitotic activity in rat liver during DAB carcinogenesis. *Cancer Res.* **22**, 743–747.

Daoust, R., and Molnar, F. (1964). Cellular populations and mitotic activity in rat liver parenchyma during azo dye carcinogenesis. *Cancer Res.* **24**, 1898–1909.

Davis, J. C., and Hyde, T. A. (1966). The effect of corticosteroids and altered adrenal function on liver regeneration following chemical necrosis and partial hepatectomy. *Cancer Res.* **26**, 217–220.

DeRecondo, A. M., and Frayssinet, C. (1963). Etude autohistoradiographique après injection de thymidine tritiée des cellules synthétisant de l'ADN dans le foie de rat en hypertrophie compensatrice. *J. Physiol. (Paris)* **55**, 242–243.

DeRecondo, A. M., Frayssinet, C., Lafarge, C., and LeBreton, E. (1966). Action de l'aflatoxine sur le métabolisme du DNA au cours de l'hypertrophie compensatrice du foie après hepatectomie partielle. *B˙ ˙chim. Biophys. Acta* **119**, 322–330.

Downing, D. T., and Peterson, J. E. (1968). Quantitative assessment of the persistent antimitotic effect of certain hepatotoxic pyrrolizidine alkaloids on rat liver. *Aust. J. Exp. Biol. Med. Sci.* **46**, 493–502.

Echave Llanos, J. M., Aliosso, M. D., Souto, M., Balduzzi, R., and Surur, J. M. (1971). Circadian variations of DNA synthesis, mitotic activity, and cell size of hepatocyte population in young immature male mouse growing liver. *Virchows Arch., B* **8**, 309–317.

Edwards, J. L., and Klein, R. E. (1961). Cell renewal in adult mouse tissues. *Amer. J. Pathol.* **38**, 437–453.

Edwards, J. L., and Koch, A. (1964). Parenchymal and littoral cell proliferation during liver regeneration. *Lab. Invest.* **13**, 32–43.

Epstein, C. (1967). Cell size, nuclear content, and the development of polyploidy in the mammalian liver. *Proc. Nat. Acad. Sci. U. S.* **57**, 327–334.

Fabrikant, J. I. (1967). The spatial distribution of parenchymal cell proliferation during regeneration of the liver. *Johns Hopkins Med. J.* **120**, 137–147.

Fabrikant, J. I. (1968). The kinetics of cellular proliferation in regenerating liver. *J. Cell Biol.* **36**, 551–565.

Fabrikant, J. I. (1969). Size of proliferating pools in regenerating liver. *Exp. Cell Res.* **55**, 277–279.

Fouts, J. R., Dixon, R. L., and Shultice, R. W. (1961). The metabolism of drugs by regenerating liver. *Biochem. Pharmacol.* **7**, 265–270.

Fujioka, M., Koga, M., and Lieberman, I. (1963). Metabolism of ribonucleic acid after partial hepatectomy. *J. Biol. Chem.* **238**, 3401–3406.

Giudice, G., and Novelli, G. D. (1963). Effect of actinomycin D on the synthesis of DNA polymerase in hepatectomized rats. *Biochem. Biophys. Res. Commun.* **12**, 383–387.

Giudice, G., Kenney, F. T., and Novelli, G. D. (1964). Effect of puromycin on deoxyribonucleic acid synthesis by regenerating rat liver. *Biochim. Biophys. Acta* **87**, 171–173.

Gonzalez, E. M., Krejczy, K., and Malt, R. A. (1970). Modification of nucleic acid synthesis in regenerating liver by azathioprine. *Surgery* **68**, 254–259.

Gottlieb, L. I., Fausto, N., and Van Lancker, J. L. (1964). Molecular mechanism of liver regeneration. The effect of puromycin on deoxyribonucleic acid synthesis. *J. Biol. Chem.* **239**, 555–559.

Gram, E., Guarino, A. M., Greene, F. E., Gigon, P. L., and Gillette, J. R. (1968). Effect of partial hepatectomy on the responsiveness of microsomal enzymes and cytochrome P-450 to phenobarbital or 3-methylcholanthrene. *Biochem. Pharmacol.* **17**, 1769–1778.

Gray, G. D., Camiener, G. W., and Bhuyan, B. K. (1966). Nogalamycin effects in rat liver: Inhibition of tryptophan pyrrolase induction and nucleic acid biosynthesis. *Cancer Res.* **26**, 2419–2424.

Grisham, J. W. (1962). A morphologic study of deoxyribonucleic acid synthesis and cell proliferation in regenerating rat liver: Autoradiography with thymidine-^{3}H. *Cancer Res.* **22**, 842–849.

Grisham, J. W. (1969). Cellular proliferation in the liver. *In* "Normal and Malignant Cell Growth" (R. J. M. Fry, M. L. Griem, and W. H. Kirsten, eds.), pp. 28–43. Springer-Verlag, Berlin and New York.

Grisham, J. W. (1971). Unpublished observations.

Grisham, J. W., Leong, G. F., Albright, M. L., and Emerson, J. D. (1966). Effect of exchange transfusion on labeling of nuclei with thymidine-^{3}H and on mitosis in hepatocytes of normal and regenerating rat liver. *Cancer Res.* **26**, 1476–1485.

Günther, G., Hübner, K., and Paul, A. (1968). Mitose-Rhythmen der Leber nach Teilhepatektomie. *Virchows Arch., B* **1**, 69–79.

Guzek, J. W. (1964). Effect of adrenocorticotrophic hormone and cortisone on the uptake of tritiated thymidine by regenerating liver tissue in the white rat. *Nature (London)* **201**, 930–931.

Guzek, J. W. (1968). Effect of corticotrophin and cortisone on the incorporation of ^{3}H-thymidine into the deoxyribonucleic acids (DNA) of the regenerating liver in the white rat. *Acta Endocrinol. (Copenhagen)* **59**, 10–22.

Harkness, R. D. (1952). The spatial distribution of dividing cells in the liver of the rat after partial hepatectomy. *J. Physiol. (London)* **116**, 373–379.

Härkönen, M., and Kiviranta, A. (1958). Effects of folic acid on cell division. Studies in normal and regenerating organs and in Yoshida sarcoma in rats. *Ann. Med. Exp. Biol. Fenn.* **36**, 213–221.

Hartmann, K.-U., and Heidelberger, C. (1961). Studies on fluorinated pyrimidines. XIII. Inhibition of thymidylate synthetase. *J. Biol. Chem.* **236**, 3006–3013.

Hemingway, J. T., and Cater, D. B. (1958). Effects of pituitary hormones and cortisone upon liver regeneration in the hypophysectomized rat. *Nature (London)* **181**, 1065–1066.

Henderson, I. C., Fischel, R. E., and Loeb, J. N. (1971). Suppression of liver DNA synthesis by cortisone. *Endocrinology* **88**, 1471–1476.

Henderson, P. T., and Kersten, K. J. (1970). Metabolism of drugs during rat liver regeneration. *Biochem. Pharmacol.* **19**, 2343–2351.

Hill, R. B., Jr., and Gordon, J. A. (1968). Effect of hydroxyurea analogues in regenerating rat liver. *Exp. Mol. Pathol.* **9**, 71–76.

Horváth, E., and Kovács, K. (1956). Beiträge zur Rolle der Nebenniere in der Regeneration der Leber. *Z. Gesamte Exp. Med.* **127**, 236–240.

Hurwitz, A., and Carter, C. E. (1969). Effect of chlortetracycline on regenerating liver. *Mol. Pharmacol.* **5**, 350–357.

Jackson, B. (1959). Time-associated variations of mitotic activity in livers of young rats. *Anat. Rec.* **134**, 365–377.

Jaffe, J. J. (1954). Diurnal mitotic periodicity in regenerating rat liver. *Anat. Rec.* **120**, 935–954.

Jahiel, R. I., Taylor, D., Rainford, N., Hirschberg, S. E., and Kroman, R. (1971). Inducers of interferon inhibit the mitotic response of liver cells to partial hepatectomy. *Proc. Nat. Acad. Sci. U.S.* **68**, 740–742.

Japundzic, M., Knezevic, B., Djordjevic-Camba, V., and Japundzic, I. (1967). The influence of phenobarbital-NA on the mitotic activity of parenchymal liver cells during rat liver regeneration. *Exp. Cell Res.* **48**, 163–167.

Juhn, S. K., and Prodi, G. (1965). The effect of 7,12-dimethylbenz (α) anthracene on the incorporation of thymidine-H^3 into deoxyribonucleic acid in normal and regenerating liver. *Experientia* **21**, 473–474.

Kinosita, R., and Sisken, J. E. (1962). Mitosis and incorporation of tritiated thymidine in liver lobules after partial hepatectomy. *In* "Biological Interactions in Normal and Neoplastic Growth" (M. Brennen and W. Simpson, eds.), 37–41. Little, Brown, Boston, Massachusetts.

Kleinfeld, R. G., and Sisken, J. E. (1966). Morphological and kinetic aspects of mitotic arrest by and recovery from colcemid. *J. Cell Biol.* **31**, 369–379.

Kleinfeld, R. G., and von Haam, E. (1959). The effect of thioacetamide on rat liver regeneration. I. Cytological studies. *Cancer Res.* **19**, 769–778.

Klinge, W. (1968). Altersabhängige Beeinträchtigung der Zellvermehrung in der regenerierenden Rattenleber. *Virchows Arch., B* **1**, 342–345.

Klinge, W., and Mathyl, J. (1969). Tageszeitliche Mitose-Rhythmen in der teilektomierten Rattenleber. *Virchows Arch., B* **2**, 154–162.

Klinman, N. R., and Erslev, A. J. (1963). Cellular response to partial hepatectomy. *Proc. Soc. Exp. Biol. Med.* **112**, 338–340.

Lahtiharju, A., and Teir, H. (1964). Inhibition of DNA synthesis in regenerating liver and of mitosis in liver cell culture following a single corticosteroid administration. *Ann. Med. Exp. Biol. Fenn.* **42**, 136–138.

Landing, B. H., Seed, J. C., and Banfield, W. G. (1949). The effects of a nitrogen mustard [Tris (2-chloroethyl) amine] on regenerating rat liver. *Cancer* **2**, 1067–1074.

Laws, J. O. (1959). Tissue regeneration and tumor development. *Brit. J. Cancer* **13**, 669–674.

Lea, M. A., Sasovetz, D., Musella, A., and Morris, H. P. (1970). Effects of hydroxyurea and 6-mercaptopurine on growth and some aspects of carbohydrate metabolism in regenerating and neoplastic liver. *Cancer Res.* **30**, 1994–1999.

LeBouton, A. V., and Marchand, R. (1970). Changes in the distribution of thymidine-^{3}H labeled cells in the growing liver acinus of neonatal rats. *Develop. Biol.* **23**, 524–533.

Lee, K. L., Sun, S. C., and Miller, O. N. (1968). Stimulation of incorporation by triiodothyronine of thymidine-methyl-^{3}H into hepatic DNA of the rat. *Arch. Biochem. Biophys.* **125**, 751–757.

Leevy, C. M., Hollister, R. M., Schmid, R., MacDonald, R. A., and Davidson, C. S. (1959). Liver regeneration in experimental carbon tetrachloride intoxication. *Proc. Soc. Exp. Biol. Med.* **102**, 672–675.

Looney, W. B. (1960). The replication of deoxyribonucleic acid in hepatocytes. *Proc. Nat. Acad. Sci. U.S.* **46**, 690–698.

Luyckx, A., and Van Lancker, J. L. (1966). Metabolic effects of vinblastine. II. The effect of vinblastine on deoxyribonucleic acid and ribonucleic acid synthesis of regenerating liver. *Lab. Invest.* **15**, 1301–1303.

MacDonald, R. A. (1961a). Experimental carcinoma of the liver. "Regeneration" of liver cells in premalignant stages. *Amer. J. Pathol.* **39**, 209–220.

MacDonald, R. A. (1961b). "Lifespan" of liver cells. Autoradiographic study using tritiated thymidine in normal, cirrhotic and partially hepatectomized rats. *Arch. Intern. Med.* **107**, 335–343.

MacDonald, R. A., Schmid, R., Hakala, T. R., and Mallory, G. K. (1959). Effect of serotonin upon liver cells of young rats. *Proc. Soc. Exp. Biol. Med.* **101**, 83–86.

MacDonald, R. A., Rogers, A. E., and Pechet, G. (1962). Regeneration of the liver. Relation of regenerative response to size of partial hepatectomy. *Lab. Invest.* **11**, 544–548.

McKellar, M. (1949). The postnatal growth and mitotic activity of the liver of the albino rat. *Amer. J. Anat.* **85**, 263–295.

Maini, M. M., and Stich, H. F. (1961). Chromosomes of tumor cells. II. Effects of various liver carcinogens on mitosis of hepatic cells. *J. Nat. Cancer Inst.* **26**, 1413–1424.

Maini, M. M., and Stich, H. F. (1962). Chromosomes of tumor cells. III. Unresponsiveness of precancerous hepatic tissues and hepatomas to a mitotic stimulus. *J. Nat. Cancer Inst.* **28**, 753–762.

Marquardt, H., and Philips, F. S. (1970). The effects of 7,12-dimethylbenz (α) anthracene on the synthesis of nucleic acids in rapidly dividing hepatic cells in rats. *Cancer Res.* **30**, 2000–2006.

Marshak, A. (1946). Effect of mustard gas on mitosis and P^{32} uptake in regenerating liver. *Proc. Soc. Exp. Biol. Med.* **63**, 118–120.

Melvin, J. B. (1967). The effect of actinomycin D on mitosis in regenerating mouse liver. *Exp. Cell Res.* **45**, 559–569.

Melvin, J. B. (1968). The localization of mitotic figures in regenerating mouse liver. *Anat. Rec.* **160**, 607–618.

Menyhárt, J., and Szabó, K. (1968). Dissimilar character of proliferative responses elicited by partial hepatectomy of different extent in the rat. *Acta Physiol.* **34**, 161–174.

Meschan, I., and Castells, J. (1964). The utilization of methotrexate, aminopterin, spleen homogenate, thymus homogenate, and DNA as test agents for suppression of DNA synthesis in regenerating rat liver. *Radiology* **83**, 520–527.

Messier, B., and Leblond, C. P. (1960). Cell proliferation and migration as revealed by radioautography after injection of thymidine-H^3 into male rats and mice. *Amer. J. Anat.* **106**, 247–265.

Mironescu, S. (1969). Mitotic abnormalities in proliferating hepatocytes induced by thioacetamide at certain periods after partial hepatectomy. *Exp. Cell Res.* **55**, 435–437.

Mironescu, S., and Ciovîrnache, M. (1971). Mitotic, chromosomal, and nucleolar alterations induced by thioacetamide in relation to the mitotic cycle after partial hepatectomy. *J. Nat. Cancer Inst.* **46**, 49–61.

Mohr, U., and Speetzen, R. (1967). Zur Regeneration der DÄNA-geschädigten Leber nach Teilhepatektomie. *Naturwissenschaften* **54**, 321.

Mohr, U., Speetzen, R., and Wrba, H. (1967). Zur Proliferationsaktivität der durch Diäthylnitrosamin geschädigten Leber nach Teilhepatektomie. *Naturwissenshaften* **54**, 566.

Moolten, F. L., Oakman, N. J., and Bucher, N. L. R. (1970). Accelerated response of hepatic DNA synthesis to partial hepatectomy in rats pretreated with growth hormone or surgical stress. *Cancer Res.* **30**, 2353–2357.

Nadal, C. (1970). Polyploïdie hépatique du rat. Mole de formation des cellules binucléées. *J. Microsc. (Paris)* **9**, 611–618.

Nadal, C., and Zajdela, F. (1966). Polyploidie somatique dans le foie de rat. I. Le rôle des cellules binucléées dans la genèse des cellules polyploïdes. *Exp. Cell Res.* **42**, 99–116.

Nettesheim, P., and Oehlert, W. (1962). Die Wirkung des Wachstumshormons auf die parenchymatösen Organe der ausgewachsenen weissen Maus unter besonderer Berücksichtigung der Leber. *Beitr. Pathol. Anat.* **127**, 193–212.

Ngu, V. A., Roberts, D., and Hall, T. C. (1964). Studies on folic reductase. I. Levels in regenerating rat liver and the effect of methotrexate administration. *Cancer Res.* **24**, 989–993.

Oehlert, W., Hämmerling, W., and Büchner, F. (1962). Der zeitliche Ablauf und das Ausmass der Desoxyribonukleinsäure-Synthese in der regenerierenden Leber der Ratte nach Teilhepatektomie. *Beitr. Pathol. Anat.* **126**, 91–112.

Paschkis, K. E., Bartuska, D., Zagerman, J., Goddard, J. W., and Cantarow, A. (1959). Effect of 5-fluorouracil on noncancerous tissue growth. *Cancer Res.* **19**, 1196–1203.

Peraino, C., Fry, R. J. M., and Staffeldt, E. (1971). Reduction and enhancement by phenobarbital of hepatocarcinogenesis induced in the rat by 2-acetylaminofluorene. *Cancer Res.* **31**, 1506–1512.

Perez-Tamayo, R., Murphy, W. R., and Ihnen, M. (1953). Effect of cortisone and partial starvation on liver regeneration. *AMA Arch. Pathol.* **56**, 629–636.

Perry, L. D., and Schwartz, F. J. (1967). Evidence for a subpopulation of cells with an extended G_2 period in normal adult mouse liver. *Exp. Cell Res.* **48**, 155–157.

Peterson, J. E. (1965). Effects of the pyrrolizidine alkaloid, lasiocarpine N-oxide, on nuclear and cell division in the liver of rats. *J. Pathol. Bacteriol.* **89**, 153–171.

Post, J., and Hoffman, J. (1964). Changes in the replication times and patterns of the liver cell during the life of the rat. *Exp. Cell Res.* **36**, 111–123.

Post, J., and Hoffman, J. (1965). Further studies on the replication of rat liver cells *in vivo*. *Exp. Cell Res.* **40**, 333–339.

Post, J., Huang, C.-Y., and Hoffman, J. (1963). The replication time and pattern of the liver cell in the growing rat. *J. Cell Biol.* **18**, 1–12.

Pukhalskaya, E. C. (1964). Mechanism of antimitotic and antitumor action of 5-hydroxytryptamine (serotonin). *Acta Unio Int. Contra Cancrum* **20**, 131–134.

Raab, K. H., and Webb, T. E. (1969). Inhibition of DNA synthesis in regenerating rat liver by hydrocortisone. *Experientia* **25**, 1240–1242.

Rabes, H. M., and Brändle, H. (1969). Synthesis of RNA, protein, and DNA in the liver of normal and hypophysectomized rats after partial hepatectomy. *Cancer Res.* **29**, 817–822.

Rabes, H., and Tuczek, H. V. (1970). Quantitative autoradiographische Untersuchung zur Heterogenität der Leberzellproliferation nach partieller Hepatektomie. *Virchows Arch., B* **6**, 302–312.

Rabes, H., Wrba, H., and Brändle, H. (1965). Synthesis of deoxyribonucleic acid in the liver of hypophysectomized rats after partial hepatectomy. *Proc. Soc. Exp. Biol. Med.* **120**, 244–246.

Rabes, H., Hartenstein, R., and Scholze, P. (1970). Specific stages of cellular response to homeostatic control during diethylnitrosamine-induced liver carcinogenesis. *Experientia* **26**, 1356–1359.

Raick, A. N., and Ritchie, A. C. (1969). Differential inhibitory effect on mouse liver DNA synthesis by 9,12-dimethylbenzanthracene in relation to the mitotic cycle after partial hepatectomy. *Fed. Proc., Fed. Amer. Soc. Exp. Biol.* **28**, 365.

Räsänen, R. T., Cederberg, A., and Taskinen, E. (1966). The mitotic count in the gastrointestinal epithelium and regenerating liver of heparinized rats. *Gastroenterology* **50**, 41-44.

Reddy, J., Chiga, M., and Svoboda, D. (1969). Initiation of the division cycle of rat hepatocytes following a single injection of thioacetamide. *Lab. Invest.* **20**, 405-411.

Rizzo, A. J., Heilpern, P., and Webb, T. E. (1971). Temporal changes in DNA and RNA synthesis in the regenerating liver of hydrocortisone-treated rats. *Cancer Res.* **31**, 876-881.

Roberts, K. B., Florey, H. W., and Joklik, W. K. (1952). The influence of cortisone on cell division. *Quart. J. Exp. Physiol.* **37**, 239-257.

Rogers, A. E., and Newberne, P. M. (1967). The effects of aflatoxin B_1 and dimethylsulfoxide on thymidine-^{3}H uptake and mitosis in rat liver. *Cancer Res.* **27**, 855-864.

Rogers, A. E., Kula, N. S., and Newberne, P. M. (1971). Absence of an effect of partial hepatectomy on aflatoxin B_1 carcinogenesis. *Cancer Res.* **31**, 491-495.

Rubin, E., Masuko, K., Goldfarb, S., and Zak, F. G. (1964). Role of cell proliferation in hepatic carcinogenesis. *Proc. Soc. Exp. Biol. Med.* **115**, 381-384.

Sakuma, K., and Terayama, H. (1967). Effects of adrenal hormones upon DNA synthesis in regenerating rat liver and tumors. *J. Biochem. (Tokyo)* **61**, 504-511.

Scherer, E., and Friedrich-Freksa, H. (1970). Zue zentralvene gerichtete Wanderung von Leberzellen der Ratte nach partiellen Hepatektomie und nach Verabfolgung von Diäthylnitrosamin. *Z. Naturforsch. B* **25**, 637-642.

Schindler, R., and Burki, K. (1971). Effects of phenobarbital on rat liver regeneration. *Experientia* **27**, 730.

Schlicht, I., Koransky, W., Magour, S., and Schulte-Hermann, R. (1968). Grösse und DNS-Synthese der Leber unter dem Einfluss körperfremder Stoffe. *Naunyn-Schmiedebergs Arch. Pharmakol. Exp. Pathol.* **261**, 26-41.

Schneider, J. H., Cassir, R., and Chordikian, F. (1960). Inhibition of incorporation of thymidine into deoxyribonucleic acid by amino acid antigonists *in vivo*. *J. Biol. Chem.* **235**, 1437-1440.

Schulte-Hermann, R., Thom, R., Schlicht, I., and Koransky, W. (1968). Zahl und Plodiegrad der Zellkerne der Leber unter dem Einfluss körperfremder Stoffe. Analysen mit Hilfe eines elektronischen Partikelzählgerates. *Naunyn-Schmiedebergs Arch. Pharmakol. Exp. Pathol.* **261**, 42-58.

Schwartz, H. S., Garofalo, M., Sternberg, S. S. and Philips, F. S. (1965a). Hydroxyurea: Inhibition of deoxyribonucleic acid synthesis in regenerating liver of rats. *Cancer Res.* **25**, 1867-1870.

Schwartz, H. S., Sodergren, J. E., Garofalo, M., and Sternberg, S. S. (1965b). Actinomycin D effects on nuclei acid and protein metabolism in intact and regenerating liver of rats. *Cancer Res.* **25**, 307-317.

Seidlová-Mašinová, V., Malinsky, J., and Santavy, F. (1957). The biological effects of some podophyllin compounds and their dependence on chemical structure. *J. Nat. Cancer Inst.* **18**, 359-371.

Seidman, I., Teebor, G. W., and Becker, F. F. (1967). Hormonal and substrate induction of tryptophane pyrrolase in regenerating liver. *Cancer Res.* **27**, 1620-1625.

6. DRUGS AND THE HEPATIC CELL CYCLE

Sigel, B. (1969). The extracellular regulation of liver regeneration. *J. Surg. Res.* **9**, 387–394.

Sigel, B., Baldia, L. B., Brightman, S. A., Dunn, M. R., and Price, R. I. M. (1968). Effect of blood flow reversal in liver autotransplants upon the site of hepatocyte regeneration. *J. Clin. Invest.* **47**, 1231–1237.

Šimek, J., Chmelař, V., Mělka, J., Pazderka, J., and Charvát, Z. (1967a). Influence of protracted infusion of glucose and insulin on the composition and regeneration activity of liver after partial hepatectomy in rats. *Nature (London)* **213**, 910–911.

Šimek, J., Husáková, A., Lejsek, Kanta, J., and Pospíšil, M. (1967b). The effect of nicotinic acid on the development of liver regenerating activity in rats with partial hepatectomy. *Naturwissenschaften* **54**, 251–252.

Šimek, J., Erbenová, Z., Deml, F., and Dvořáčková, I. (1968). Liver regeneration after partial hepatectomy in rats exposed before the operation to the stress stimulus. *Experientia* **24**, 1166–1167.

Šimek, J., Husáková, A., Deml, F., and Dvořáčková, I. (1969). Hepatic DNA synthesis after partial hepatectomy in rats treated with protamin-Zn-insulin under different nutritional conditions. *Experientia* **25**, 791–792.

Stöcker, E. (1966). Der Proliferationsmodus in Niere und Leber. *Verh. Deut. Ges. Pathol.* **50**, 53–74.

Stöcker, E., and Altmann, H.-W. (1964). Die DNS-Synthese in Leberparenchymzellen und Gallengangsepithelien von normalen und mit Thioacetamid behandelten Ratten. *Naturwissenschaften* **51**, 15.

Stöcker, E., and Boecker, W. (1970). Zur Leberregeneration der Ratte nach Teilhepatektomie während akuter CCl_4-Intoxikation. *Experientia* **26**, 763–765.

Stöcker, E., and Butter, D. (1968). Die DNS-Synthese in Leber und Niere junger Ratten. Autoradiographische Untersuchungen mit ^{3}H-Thymidin. *Experientia* **24**, 704–705.

Stöcker, E., and Heine, W. D. (1965). Über die Proliferation von Nieren-und Leberepithel unter normalen und pathologischen Bedingungen. Autoradiographische Untersuchungen mit ^{3}H-Thymidin an der Ratte. *Beitr. Pathol. Anat.* **131**, 410–434.

Stöcker, E., and Pfeifer, U. (1967). Autoradiographische untersuchungen mit ^{3}H-Thymidin an der regenerierenden rattenleber. *Z. Zellforsch. Mikrosk. Anat.* **79**, 374–388.

Svoboda, D., Reddy, J., and Bunyaratvej, S. (1971). Hepatic megalocytosis in chronic lasiocarpine poisoning. *Amer. J. Pathol.* **65**, 399–409.

Symeonidis, A., Mulay, A. S., and Trams, E. G. (1955). Effect of prolonged pretreatment with desoxycorticosterone on the liver of hepatectomized rats. *Endocrinology* **57**, 550–558.

Theologides, A., Yarbro, J. W., and Kennedy, B. J. (1968). Daunomycin inhibition of DNA and RNA synthesis. *Cancer* **21**, 16–21.

Van Vroonhoven, T. J., and Malt, R. A. (1971). Rebound hyperplasia of regenerating liver after cessation of azathioprine. *Surg. Forum* **22**, 339–340.

Verbin, R. S., Sullivan, R. J., and Farber, E. (1969). The effects of cycloheximide on the cell cycle of the regenerating rat liver. *Lab. Invest.* **21**, 179–182.

Volini, F., Orfei, E., Baserga, R., Madera-Orsini, F., Minick, T., and Kent, G. (1965). Effect of iron loading upon liver cell regeneration. *Fed. Proc., Fed. Amer. Soc. Exp. Biol.* **24**, 167.

Von Der Decken, A., and Hultin, T. (1960). The enzymatic composition of rat liver ribosomes during liver regeneration. *Exp. Cell. Res.* **19**, 591–604.

Warwick, G. P. (1971). Effect of the cell cycle on carcinogenesis. *Fed. Proc., Fed. Amer. Soc. Exp. Biol.* **30**, 1760–1765.

Weinbren, K., and Fitschen, W. (1959). The influence of sodium fluoroacetate on regeneration of the rat's liver. *Brit. J. Exp. Pathol.* **40**, 107–112.

Weinbren, K., Bezmalinovic, Z., and Daniller, A. I. (1967). The effect of catecholamines on the delay of the restorative response after subtotal hepatectomy. *Brit. J. Exp. Pathol.* **48**, 305–308.

Wilson, J. W., and Leduc, E. H. (1954). The effect of coramine on mitotic activity and growth in the liver of the mouse. *Growth* **14**, 31–48.

Yarbro, J. W., Niehaus, W. G., and Barnum, C. P. (1965). Effect of hydroxyurea on regenerating rat liver. *Biochem. Biophys. Res. Commun.* **19**, 592–597.

Younger, L. R., King, J., and Steiner, D. F. (1966). Hepatic proliferative response to insulin in severe alloxan diabetes. *Cancer Res.* **26**, 1408–1414.

Zimmerman, A. M., and Celozzi, E. (1961). Stimulation by heparin of parenchymal liver cell proliferation in normal adult rats. *Nature (London)* **191**, 1014–1015.

7

Effects of Mitogens on the Mitotic Cycle: A Biochemical Evaluation of Lymphocyte Activation

HERBERT L. COOPER

I.	Introduction	138
II.	Background: Peripheral Blood Lymphocytes	138
III.	Nature of Mitogenic Substances	139
	A. "Nonspecific" Substances of Plant or Bacterial Origin	140
	B. Specific Antigens	141
	C. Mixed Leukocyte Reaction	141
	D. Small Molecule Stimulants	142
	E. Antisera to Cell Surface Components	142
	F. Blastogenic or Mitogenic Factors	143
IV.	General Characteristics of Mitogen-Induced Lymphocyte Growth	145
	A. Morphological Changes	145
	B. Biological Activity	147
V.	Biochemical Events in Lymphocyte Transformation	148
	A. Overview	148
	B. Initial Interaction of Mitogens with the Lymphocyte Surface	149
	C. Biochemical Alterations Related to the Cell Membrane	150
	D. Activation of Inactive Transport Sites and Enzymes	153
	E. Early Nuclear Changes	155
	F. Energy Metabolism	156
	G. RNA Metabolism	156
	H. Protein Synthesis	166
	I. DNA Synthesis	175
VI.	Discussion and Conclusions	177
	A. Sequence of Events in Lymphocyte Activation	178
	B. Restoration of the Resting State	180
	References	181

I. Introduction

Our knowledge of the events of the mitotic cycle derives largely from investigations restricted to rapidly and continuously dividing cells. This approach has been fruitful, and we owe a vast increase in our understanding of the details of cellular and molecular biology of animal cells to studies with HeLa cells, mouse L fibroblasts, and similar cell lines adapted for continuous growth *in vitro*. However, our interest in the dynamic events of the mitotic cycle should not divert our attention from the great bulk of normal animal cells that spend most of their existence either arrested at some stage in that cycle or, possibly, outside it entirely. It is of considerable importance to examine experimental systems in which one can ask questions about the mechanism regulating the entry of cells into active growth from a previous condition of physiological nongrowth.

Several experimental systems are available for the study of those events that relate to the stimulation of growth in physiologically resting cells. This review will consider the small lymphocyte and its mitotic response to various agents. This system has the advantage that the shift from nongrowth to growth is induced by a well-defined chemical substance administered *in vitro* whose initial interaction with the cell may be studied. The term "mitogen" has been applied to any substance capable of eliciting the *in vitro* lymphocyte growth response.

II. Background: Peripheral Blood Lymphocytes

The population of small lymphocytes that circulates in the peripheral blood of mammals (and of all other species above and including the hagfish) appears to be composed largely of long-lived cells (Everett and Tyler, 1970; Papermaster *et al.,* 1964). These cells may circulate for months or years without replicating their DNA or undergoing mitosis (Fitzgerald, 1964). During this time the cells remain small, with condensed nucleus and scanty cytoplasm. However, when confronted with an immunological stimulus, the lymphocyte shifts from this prolonged resting state to one of rapid enlargment, culminating in DNA synthesis and mitosis (Gowans and MacGregor, 1965). This blastogenic response, or transformation, may take different forms, depending on the origin of the lymphocyte. It is currently thought that the circulating small lymphocyte population consists of two major subtypes (Everett and Tyler, 1970).

The numerically predominant type (about 80%–90% of human small peripheral blood lymphocytes) is the so-called thymus-derived or T cell.

7. EFFECTS OF MITOGENS ON THE MITOTIC CYCLE

These cells have entered the circulation following, apparently, an initial period of residence in the thymus. Such cells are thought to be largely associated with the organism's ability to mount a cell-mediated type of immunological response, of the sort exemplified by delayed tuberculin hypersensitivity or the rejection of a skin or organ homograft. The second cell type is thought to enter the blood directly from the bone marrow. This cell is called the B cell. These cells traverse the blood stream on their way to lymphoid tissues, where they take up residence. There they respond to antigenic stimuli by differentiating into antibody-secreting cells. The T and B cells are morphologically indistinguishable by microscopy. However, it has been shown that B cells have large numbers of immunoglobulin-like components on their surface, whereas T cells are practically devoid of such determinants (Raff, 1970; Kincade et al., 1971; Biberfeld et al., 1971). The interaction of these cell types and their complicated mode of behavior in providing an effective immune response are outside the area to be covered in this chapter. What interests us here is the ability of most of these cells of both types to shift from a nongrowing condition to a growing one in response to a defined stimulus. Fortunately, this sequence may be studied *in vitro* under cell culture conditions.

Small resting lymphocytes may be obtained in high purity from the peripheral blood (Rabinowitz, 1964; Cooper, 1972a). When cultured under ordinary conditions, the cells remain in this nongrowing state for up to 10 days and then gradually die. At any time, this resting condition may be ended by the addition of one of a number of mitogenic agents (Oppenheim, 1968). A series of changes ensues rapidly, with progressive cellular enlargement culminating in DNA synthesis and mitosis. It is significant that lymphocytes from subjects who have impaired *in vivo* cell-mediated immunity also show reduced mitogenic responses *in vitro* (reviewed by Oppenheim, 1968). This gives us good reason to equate the *in vitro* response with the behavior of lymphocytes *in vivo* and thus to have some confidence that information obtained from a study of this system will have biological relevance. It should be noted that the techniques commonly used to purify lymphocytes from the peripheral blood may tend to eliminate B cells, so that the activities studied in cell cultures are largely those of the T cell (Plotz and Talal, 1967; Shortman et al., 1971; Rosenthal et al., 1972).

III. Nature of Mitogenic Substances

Materials that stimulate lymphocyte growth fall into a number of broad categories.

A. "Nonspecific" Substances of Plant or Bacterial Origin

Nonspecific materials have been so called because they evoke a response in the cells of virtually all normal donors, without evidence of any prior immunological sensitization. However, it cannot be ruled out that materials resembling such agents are universally present in the environment, thereby sensitizing all subjects. These agents act on the cells of newborns (Hirschhorn et al., 1964; Lindahl-Kiessling and Böök, 1964; Leikin et al., 1970), which would require sensitization *in utero*, a phenomenon that is not unknown. In any case, several of the nonspecific agents are known to interact with specific carbohydrate components of the lymphocyte surface. The term "non-specific" is therefore a misnomer but will be retained as a convenience to distinguish these agents from other types of mitogens.

1. *Phytohemagglutinin*

The first-discovered and most widely used mitogen of this sort is the phytohemagglutinin (PHA), isolated from the kidney bean (Rigas and Osgood, 1955; Nowell, 1960; Weber et al., 1967; Takahashi et al., 1967). PHA is a glycoprotein that strongly agglutinates erythrocytes (Rigas and Osgood, 1955; Takahashi et al., 1967). It may depend for its activity on the small amount of carbohydrate associated with it (Goldberg et al., 1969). PHA may be fractionated into erythroagglutinating and nonerythroagglutinating components, both of which are mitogenic (Weber et al., 1967). The cell surface receptor for erythroagglutinating PHA has been studied and will be discussed below. Most of the studies to be described in this section were performed with PHA as a mitogen because of its wide availability and fairly reproducible prompt, and potent effect. From 50% to 80% of the lymphocytes in a small lymphocyte preparation will respond to PHA, with a characteristic sequence of morphological and biochemical phenomena. There appears to be a requirement for divalent cations for the PHA molecule to remain active, since ethylenediaminetetraacetic acid (EDTA) will inhibit PHA activity. The inhibition affects the mitogen rather than the lymphocyte (Alford, 1970; Kay, 1971a).

2. *Concanavalin A*

Concanavalin A (jack bean agglutinin; Con A) is another erythroagglutinating protein of plant origin which is a powerful lymphocyte mitogen (Powell and Leon, 1970; Novogrodsky and Katchalski, 1971a). This material is a pure protein, and has a strong affinity for a variety of sugars possessing the D-arabinopyranoside configuration at carbons 3, 4, and 6 (Goldstein et al., 1965). Presumably, it acts through binding to carbohydrate-containing sites on the lymphocyte surface as an initial step.

3. Pokeweed Mitogen

Pokeweed mitogen is a glycoprotein derived from the roots of the pokeweed *(Phytolacca americana)* (Farnes *et al.*, 1964; Börjeson *et al.*, 1966). Like the preceding substances, it acts quickly on a large proportion of the lymphocyte population.

Microscopic studies show that a significant portion of the cells stimulated by pokeweed mitogen come to resemble plasma cells (Chessin *et al.*, 1966; Douglas *et al.*, 1967). This point will be considered further in Sections IV,A and V,H,6. It has been reported that agents blocking the H blood group antigen on cell surfaces will inhibit the action of pokeweed mitogen (Landy and Chessin, 1969), suggesting that the mitogen may bind to sites on the lymphocyte membrane which carry that configuration.

4. Bacterial Mitogens

Nonspecific mitogens of bacterial origin are staphylococcal filtrate (Ling and Husband, 1964; Ling *et al.*, 1965) and streptolysin S (K. Hirschhorn *et al.*, 1964). Less is known about these substances than about the plant mitogens.

B. Specific Antigens

Substances to which a subject has been previously sensitized immunologically will stimulate growth in cultures of his lymphocytes, while nonsensitized donors will not respond (Pearmain *et al.*, 1963; Elves *et al.*, 1963a; Coulson and Chalmers, 1967). The proportion of cells responding is smaller, and the peak of mitogenic activity occurs later (5 *vs.* 3 days) and at a lower level than with the nonspecific mitogens. The response of lymphocytes to specific antigens appears to require the presence of a small number of phagocytic cells (monocytes or macrophages) thought to facilitate the interaction between lymphocyte and antigen so as to evoke the mitogenic response (Oppenheim *et al.*, 1968).

A similar macrophage dependence for nonspecific mitogens has been suggested, but in this case, while the presence of macrophages may enhance the response somewhat, there is not an absolute requirement (Oppenheim *et al.*, 1968; Levis and Robbins, 1970).

C. Mixed Leukocyte Reaction

When lymphocytes from two nonidentical donors are mixed in the presence of a small number of macrophages, a mitogenic response occurs (Bain *et al.*, 1964; Bain and Lowenstein, 1964). The reaction is related presumably to genetic differences in cell surface components, primarily the histocompatibility antigens. The response has formed the basis of one approach to the matching of tissues between potential donors and homograft recipients. The degree of mitotic activity stimulated is thought to reflect the intensity of the rejection response which the donor will mount against the grafted tissue or organ on the basis of differences in histocompatibility antigen type (Bach and Hirschhorn, 1964; Rubin *et al.*, 1964). The mixed leukocyte response is slower in developing *in vitro* (peak: 8–9 days) than the response to nonspecific mitogens. The relationship of the mixed leukocyte response to the response caused by nonspecific or specific mitogens is unclear. While prior sensitization is not required, the mixed leukocyte response is clearly immunologically specific.

D. Small Molecule Stimulants

These mitogens, because they are small well-defined chemicals, offer the potentiality of a clearer understanding of their mode of interaction with the lymphocyte. However, insufficient work has been done to date to achieve this goal. Mercuric (Pauly *et al.*, 1969) and zinc salts (Kirchner and Rühl, 1970; Rühl *et al.*, 1971) have been shown to be mitogenic, although their mode of action is unknown. Sodium periodate has recently been shown to be an effective mitogen (Novogrodsky and Katchalski, 1971b). The well-known ability of this chemical to oxidize carbohydrates may form the basis of its initial effect on the cell.

These substances are mitogenic in very minute concentrations and are toxic at higher ones. This is true also of the nonspecific mitogens, particularly PHA, although the toxic potential may vary among different preparations of the mitogen.

E. Antisera to Cell Surface Components

A rabbit antiserum prepared against whole human white blood cells has been shown to cause a modest mitogenic response when added to human lymphocyte cultures (Gräsbeck *et al.*, 1963; Holt *et al.*, 1966; Claman and Brunstetter, 1968), and sera against particular immunoglobulins (Sell and Gell, 1965; Adinolfi *et al.*, 1967; Oppenheim *et al.*, 1969) have also been shown to be mitogenic when used under conditions where cell injury by the antisera is minimized. In the case of antiimmunoglobulin sera, the cells stimulated are most likely to be the B cells, since these possess immunoglobulin-like molecules on their surfaces (Raff, 1970; Kincade *et al.*, 1971; Biberfeld *et al.*, 1971).

The mitogenic action of agents such as antiimmunoglobulin and antilymphocyte sera, which are directed against cell surface components, supports the widely held notion that an initial step in stimulating the resting lymphocyte to enter the mitotic cycle is a binding step at the cell surface that alters the configuration of some preexisting surface component (Greaves and Bauminger, 1972).

F. BLASTOGENIC OR MITOGENIC FACTORS

During the mitogenic response of lymphocytes to many of the materials noted above, substances are elaborated into the culture medium which will induce other untreated lymphocytes of the same or a different donor to enter the mitotic cycle (Kasakura and Lowenstein, 1965; Gordon and MacLean, 1965; Maini et al., 1969; Wolstencroft, 1971). Production of this factor is induced by both specific and nonspecific mitogens (Wolstencroft, 1971). It is thought to be a protein, appearing with the serum albumin fraction in various purification procedures (Wolstencroft, 1971). However, a large proportion of the activity present in cell-free culture supernatants is sedimentable at 100,000 g (Kasakura and Lowenstein, 1965). This suggests either heterogeneity of such factors or aggregation of the mitogenic factor with itself or with other materials in the culture fluids.

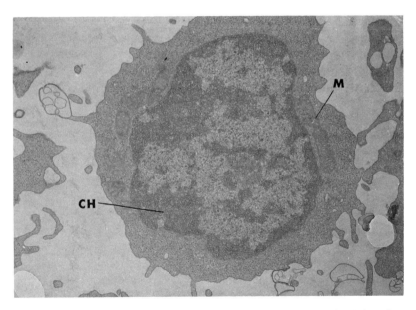

Fig. 1. Electron micrograph of normal human peripheral blood lymphocyte (resting). Note scanty cytoplasm, Nuclear chromatin is largely condensed (CH). There are many single ribosomes in the cytoplasm, a few mitochondria are present (M), and several small vesicles are evident. Magnification: 12,500. (Courtesy of Dr. Steven D. Douglas.)

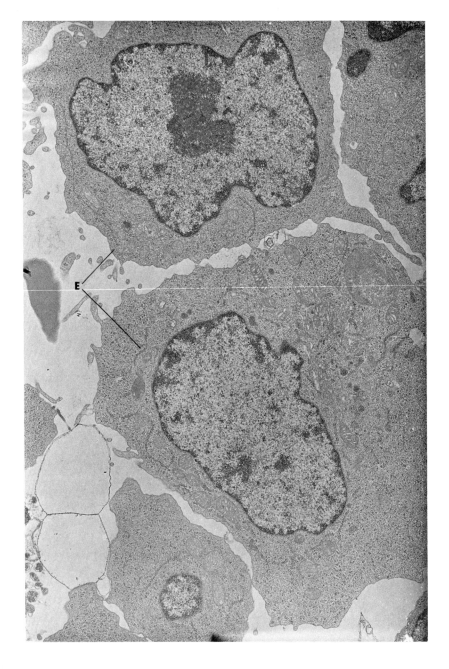

IV. General Characteristics of Mitogen-Induced Lymphocyte Growth

A. Morphological Changes

The resting lymphocyte is a small cell (9–12 μm diameter in smears; 5–8 μm in electron micrographs) with a densely staining nucleus and a scanty rim of cytoplasm. Electron microscopic studies reveal a nucleus with peripherally condensed chromatin (heterochromatin) and small nucleolus. The cytoplasm contains relatively sparse elements, including ribosomes, mitochondria, and infrequent strands of endoplasmic reticulum. The Golgi apparatus is poorly developed. The ribosomes are found primarily as free ribosomes, although polysomes are also present. Occasional lysosomelike electron-dense bodies are seen. This picture does not change during several days in culture if no mitogen is added (Fig. 1).

Following the addition of PHA, progressive cellular enlargement occurs. An early change is a reduction in the condensation of nuclear chromatin, which progresses over a period of 48–72 hours until virtually no heterochromatin is visable. Simultaneously, the nucleolus becomes progressively more prominent. The result is a much altered and enlarged nucleus which is observable well before the onset of DNA synthesis; even more striking enlargement of the cytoplasm occurs. Over a period of 48–72 hours, the cell may increase its diameter three- to four-fold. The cytoplasm now contains abundant ribosomes, with a tendency to a greater degree of organization into polysomes. The Golgi apparatus becomes prominent and vacuoles and lysosomelike bodies more frequent. Mitochondria are numerous and have been reported to show swelling and alteration of the cristae (reviewed by Douglas, 1972) (Fig. 2).

In PHA-stimulated cultures, endoplasmic reticulum is not usually prominent. However, after 72 hours of stimulation with pokeweed mitogen, 20%–30% of the enlarged cells show a marked development of the rough endoplasmic reticulum (Douglas *et al.*, 1967) (Fig. 3). Some of these cells, although enlarged, have condensed nuclear chromatin. These unusual cells possess many of the features of antibody-synthesizing plasma cells (Fawcett, 1966), and the resemblance has suggested to some that pokeweed mitogen induces differentiation of a portion of the lymphocyte population into such cells (Barker *et al.*, 1969; Douglas and Fudenberg, 1969).

Fig. 2. Electron micrograph of cells from a lymphocyte culture incubated for 72 hours with phytohemagglutinin (PHA blasts). Compare with Fig. 1. Note largely euchromatic nuclear chromatin, much enlarged cytoplasm. Numerous ribosomal aggregates (polysomes) are now evident. Golgi zone is enlarged and occasional strands of nondilated, rough-surfaced endoplasmic reticulum (E) are present. Magnification: 9000 (Courtesy of Dr. Steven D. Douglas).

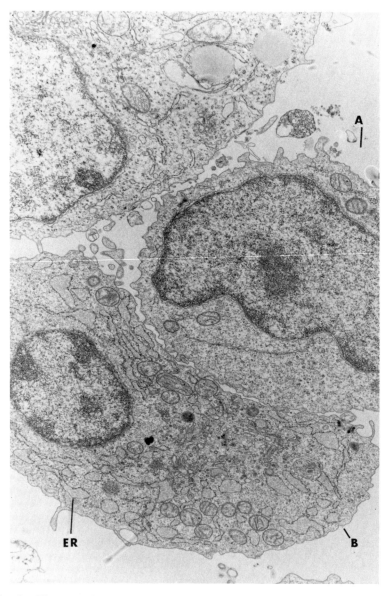

Fig. 3. Electron micrograph of cells from a lymphocyte culture incubated for 7 days with Pokeweed mitogen. Compare with Fig. 2. The cell at A is a blast cell similar to those seen in PHA cultures. The cell at B is a plasmacytoid cell typical of cultures exposed to pokeweed mitogen. Note well-developed dilated endoplasmic reticulum (ER). Magnification: 10,200. (Courtesy of Dr. Steven D. Douglas.)

7. EFFECTS OF MITOGENS ON THE MITOTIC CYCLE

In PHA or pokeweed-mitogen-stimulated cultures, mitotic figures are first seen about 40 hours after addition of the mitogen. Mitotic activity reaches a peak at about 72 hours (MacKinney *et al.,* 1962; Bender and Prescott, 1962). During this wave of mitotic activity, some cells may divide several times, whereas others divide only once (Bender and Prescott, 1962; Sasaki and Norman, 1966). After 7–10 days, if proper cell culture conditions are maintained, the cells may be restimulated with the same or another mitogen (Ling and Holt, 1967). However, secondary stimulation with various pairs of mitogens of different classes has not been extensively studied. In some cases it appears that primary stimulation with one mitogen produces a secondary population which fails to respond to a different mitogen (Möller, 1970; Ginsburg *et al.,* 1971).

Although a large proportion of the cell population present after 72 hours of PHA stimulation have undergone one or more cell divisions, actual cell numbers at this time seldom increase more than 50% (Stewart and Breitner, 1969). Measurements of total DNA content of such cultures give similar results (Handmaker *et al.,* 1969). About 70% of cells in culture at this time appear enlarged (blasts) and half the enlarged cells can be shown by autoradiography to be synthesizing DNA (Handmaker *et al.,* 1969). However, many of these cells may have already entered their second cycle of growth (Sasaki and Norman, 1966).

B. Biological Activity

Numerous studies have been undertaken to determine whether or not growth stimulation *in vitro* may be correlated with the various parameters of immunological activity *in vivo*. Conflicting reports have appeared about the production of specific antibodies by mitogen-stimulated cultures (reviewed by Ling, 1968). However, several other factors have been reported to be elaborated into the medium of mitogen-stimulated lymphocytes. These have been defined functionally, for the most part, with modest progress to date in their purification and identification (Lawrence and Landy, 1969; Bloom and Glade, 1971).

Some of the factors most widely studied are migration-inhibition factor (Bennett and Bloom, 1968; Lamelin, 1971), chemotactic factor (Ward, 1970), lymphotoxin (Kolb and Granger, 1968), and mitogenic factor (Kasakura and Lowenstein, 1965; Gordon and MacLean, 1965; Maini *et al.,* 1969; Wolstencroft, 1971). These activities are detected in lymphocyte culture media shortly after addition of mitogens and are measured by a functional assay. In addition, unexpectedly, the well-known antiviral substance, interferon, is also elaborated by mitogen-stimulated lymphocytes (Wheelock, 1965; Friedman and Cooper, 1967). Finally, stimulated lymphocytes

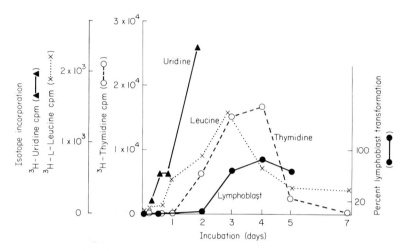

Fig. 4. Changes in macromolecular synthesis during PHA stimulation. Lymphocytes were incubated for various times with PHA, then labeled for 1 hour with ³H-uridine (RNA), ³H-leucine (protein), or ³H-thymidine (DNA). (From Loeb et al., 1970. With permission.)

may exert a direct immunogically specific cytotoxic action on other cells with which they come in contact (Holm and Perlmann, 1965). These various activities *in vitro* have been thought to represent components of the lymphocyte's mechanism for participation in the cell-mediated immune response (Bloom and Glade, 1971).

V. Biochemical Events in Lymphocyte Transformation

A. Overview

Addition of PHA to resting lymphocytes produces an increased level of activity in virtually every process that has been measured. An increased rate of incorporation of nucleosides into RNA is detectable almost immediately, which is at least partly due to increased synthesis of RNA (Cooper and Rubin, 1965; Pogo et al., 1966). An overall increase of amino acid incorporation into proteins is generally detectable after 2–4 hours (Hirschhorn et al., 1963; Kay, 1968a), although it is likely that stimulation of synthesis of specific proteins occurs more quickly.

Among the resting lymphocytes, a very small number of cells is found which spontaneously engage in DNA synthesis (perhaps one in 10^4) (MacKinney et al., 1962; McIntyre and Ebaugh, 1962; Cooper and Rubin, 1965). Following stimulation, significant elevation of DNA synthesis does

7. EFFECTS OF MITOGENS ON THE MITOTIC CYCLE   149

not begin until about 20 hours (MacKinney et al., 1962; McIntyre and Ebaugh, 1962; Loeb et al., 1970). After that time, the rate of DNA synthesis in cultures increases to a peak at 48–72 hours. Thereafter it declines, although a second peak may be detected. There is little synchrony of DNA synthesis in such cultures, although this can be partially achieved by the use of agents that block entry into S phase, such as methotrexate (Tormey and Mueller, 1965). By 9–10 days, DNA synthesis has fallen to near control levels (Fig. 4) (Ling and Holt, 1967; Cooper, 1969b).

B. INITIAL INTERACTION OF MITOGENS WITH THE LYMPHOCYTE SURFACE

1. *Nature of the Mitogen Binding Site*

A number of lines of evidence suggest that the primary event in stimulating lymphocyte growth is a combination between the mitogen and a specific receptor on the cell surface (Wigzell, 1970; Möller, 1970; Powell and Leon, 1970; Greaves and Bauminger, 1972). The presence of the sugar, N-acetyl galactosamine, will inhibit competitively PHA stimulation of lymphocytes (Borberg et al., 1968). Since such sugars are frequent components of cell surface antigens (Watkins, 1966; Lloyd and Kabat, 1968), it is possible that PHA binds to a cell surface component in a reaction analogous to that between an antigen and specific antibody. The surface receptor for PHA is inactivated temporarily by trypsin treatment, indicating that the combining site contains a protein component (Lindahl-Kiessling and Peterson, 1969). Most specifically, mild trypsinization releases a glycopeptide from the cell membranes of erythrocytes that binds specifically to erythroagglutinating PHA. Such binding prevents the mitogen from attaching to lymphocytes and thereby inhibits growth stimulation (S. Kornfeld and Kornfeld, 1969). Presumably, erythroagglutinating PHA binds to a site on the lymphocyte similar to that present on the erythrocyte. The structure of the carbohydrate moiety of this receptor has been determined and has been shown to contain sialic acid and galactose as branched terminal residues. The sialic-acid-terminated branch also contains a galactose in the penultimate position. The specificity of the receptor resides in the galactose moiety, since neuraminidase does not destroy its activity. However, the sialic acid appears to protect the galactose, since β-galactosidase alone is ineffective, whereas a combination of neuramindase and β-galactosidase will inactivate the receptor (R. Kornfeld and Kornfeld, 1970).

Concanavalin A activity may also be inhibited by the presence of various sugars, particularly methyl α-D-mannoside, which binds strongly to the mitogen (Goldstein et al., 1965; Powell and Leon, 1970; Novogrodsky and Katchlski, 1971a). Presumably, similar carbohydrates are present on specific surface sites of the lymphocyte. As noted previously, pokeweed mitogen may bind to cell surface sites resembling the H blood group antigen (Landy and Chessin, 1969).

Since competitive inhibitors show little cross reactivity among different mitogens (Novogrodsky and Katchalski, 1971a), and since combinations of mitogens may act synergistically (Möller, 1970), it is probable that different sites are bound by each of the mitogens. Nevertheless, binding of any of these various sites seems to set in motion a similar series of events in the cells involved.

2. *Characteristics of Mitogen–Lymphocyte Interaction*

The binding of PHA to lymphocyte surfaces is temperature dependent but is apparently not energy requiring. When incubated with lymphocytes at low temperatures, PHA can be effectively washed off cells by saline, preventing subsequent stimulation of DNA synthesis when the cells are returned to 37°C (Kay, 1971b; Lindahl-Kiessling and Mattson, 1971). However, incubation with PHA at 37°C in the presence of inhibitors of respiration and glycolysis (2,4-dinitrophenol, iodoacetate, sodium fluoride) does not prevent binding, since washing the cells will not reverse the mitogenic effect. Steps subsequent to PHA binding are affected by these inhibitors, since no growth occurs if inhibitors of glycolysis are not removed (Lindahl-Kiessling and Mattson, 1971). Interestingly, inhibition of oxidative phosphorylation need not be reversed to permit growth to proceed (Polgar *et al.*, 1968a) (see Section V, F).

PHA can be shown to have metabolic effects on RNA, phospholipid, and histone metabolism very rapidly after its addition to cultures (see Section V, C). Nevertheless, specific antisera to PHA, if given up to 1 hour after PHA treatment, will completely reverse the stimulation of DNA synthesis which ordinarily begins in these cells 24–48 hours later. Complete irreversibility does not develop for at least 6 hours (Kay, 1970). Similar results are obtained by repeated washing of treated cells with serum-containing medium (Mendelsohn *et al.*, 1971). A competitive inhibitor is apparently present in serum that effectively binds to PHA and can reverse its association with the lymphocyte membrane (Kay, 1971b). It appears that PHA remains accessible to antisera and competitive inhibitors even after provoking some initial biochemical responses, and that some extended action is required to permit the cellular response to become directed irreversibly toward DNA synthesis.

C. Biochemical Alterations Related to the Cell Membrane

1. *Phospholipid Metabolism*

A metabolic effect of PHA stimulation on the lymphocyte membrane may be detected as early as 10–30 minutes after addition of the mitogen.

Phospholipid components of the cell membrane, which are in a constant state of turnover in the resting cell, show an increase in their turnover rate soon after PHA treatment. Net increases of membrane phospholipids can be measured within several hours. However, the synthesis of the various types of phospholipids is not increased uniformly. An ordered sequence of stimulation occurs among the different classes, indicating a well-controlled process. The increased phospholipid turnover is independent of both RNA and protein synthesis, suggesting that the enzymes necessary for enhanced synthesis are already present in the resting cell (Fisher and Mueller, 1968, 1969; Kay, 1968b; Lucas et al., 1971).

2. *Glycoprotein Metabolism*

In similar fashion, the rate of synthesis of cell membrane glycoproteins was found to rise within 2 hours of PHA addition. This rise was shown by an increased rate of incorporation of radioactive glucosamine (Hayden et al., 1970). Such incorporation takes place on newly synthesized proteins and is thus inhibited by agents that interfere with protein synthesis (Hayden et al., 1970). It is not known, therefore, whether the response requires the synthesis of new enzymes in order to proceed.

3. *Pinocytosis*

The acceleration of turnover of membrane phospholipids and synthesis of glycoproteins correlates with an early functional alteration in cell membrane activity. Shortly after PHA treatment, an increase in pinocytosis is detectable (R. Hirschhorn et al., 1968). This process involves inward budding of portions of the cell membrane, and the changes in phospholipid and glycoprotein metabolism may reflect an increased rate of replacement of portions of cell membrane which have been internalized as pinocytotic vesicles.

4. *Transport Phenomena*

A major function of the cell membrane is the transport of materials into and out of the cell. A number of studies have demonstrated an early increase, following PHA stimulation, in the rates of active and facilitated transport of various substances across the lymphocyte membrane.

The active transport of potassium into lymphocytes is rapidly increased following PHA stimulation. If the Na-K-ATPase enzyme system involved in this transport is blocked by treatment with ouabain, the early alteration in transport is prevented, and the subsequent onset of DNA synthesis does not occur. The effect of ouabain is antagonized by increased potassium concentrations and is reversible on removal of the drug (Quastel and Kaplan, 1970, 1971). Thus, it seems that the inhibition of PHA action is related specifically to reduction of potassium influx rather than to some generalized toxicity. It is not known how growth stimulation may be related to increased potassium influx.

Analysis of the kinetics of potassium transport supports the conclusion that the enhanced movement of the ion across the cell membrane is due to an increase in the number of available transport sites rather than to an alteration in the activity of individual sites. This effect is not dependent on either protein or RNA synthesis, since neither cycloheximide nor actinomycin D inhibits the response (Quastel et al., 1970). One interpretation of this data is that inactive transport sites are present on the resting lymphocyte membrane and are activated by the binding of PHA to the cell (Quastel et al., 1970, 1972).

Other materials have been shown to undergo increased transport into lymphocytes following PHA treatment. The uptake of α-aminoisobutyric acid (Mendelsohn et al., 1971) and 3-O-methylglucose (Peters and Hausen, 1971b), nonmetabolizable analogs representing amino acids and glucose, respectively, show a rapid increase in uptake after PHA treatment. The same was true of uridine uptake (Peters and Hausen, 1971a). In the case of uridine, it appears that the step following transport, phosphorylation by uridine kinase, is also activated by PHA treatment (Lucas 1967; Hausen and Stein, 1968; Kay and Handmaker, 1970). Each of these substances behaved similarly to potassium in that the effect was apparently mediated by a PHA-induced increase in the number of transport sites or protein molecules rather than an altered activity of individual sites. All the changes in transport and in the activation of uridine kinase were independent of RNA and protein synthesis, again suggesting activation of previously nonfunctioning sites and proteins.

All the foregoing transport reactions are thought to occur through the mediation of a transport protein to which the entering material binds transiently. The material is then translocated to the internal portion of the transport site, where it is released (Pardee, 1968). The data summarized above indicate that the cell membrane of the resting lymphocyte contains a great many inactive transport complexes, each specific for one of a host of materials. The attachment of a mitogen to its own binding site activates all of these complexes, apparently in nonspecific fashion. The simplest hypothesis to explain these findings is that in binding to its own attachment sites on the cell surface, the mitogen causes a conformational change in a portion of the cell membrane. This change in configuration may bring previously hidden transport sites to the surface (Quastel et al., 1970, 1972).

5. How Many Mitogen Binding Sites?

Since many transport sites of different specificities are activated, it seems probable that a number of regions of the cell membrane must be subject to the steric activating process. This implies that more than one mitogen binding site per cell is probably involved in activation. Presumably, some optimum number of occupied binding sites exists for each cell, providing some latitude for partial and supraoptimal activation of single cells.

Studies involving various doses and combinations of mitogens have led to the conclusion that a critical number of mitogen binding sites must be occupied, regardless of the nature of the mitogen, in order for lymphocyte growth to ensue (Sell, 1967; Forster et al., 1969; Möller, 1970). It has also been observed, however, that simple binding of mitogen to the cell surface is not sufficient to induce lymphocyte growth. Inactive preparations of PHA will bind to cells but may fail to stimulate growth (Cooper and Rubin, 1965; Monjardino and MacGillivray, 1970). Thymus cells, which do not normally undergo growth in response to PHA, will bind the mitogen. They may grow, however, in response to another mitogen (Strobo et al., 1972).

D. Activation of Inactive Transport Sites and Enzymes

1. *Change in Configuration*

Inactive membrane transport sites may be activated by simple exposure or juxtaposition following a configurational change caused by mitogen binding (Quastel et al., 1970, 1972). This will be discussed in Section VI, A, 3, d.

More complicated behavior must be postulated for the mode of action of PHA in activating the enzyme systems related to phospholipid metabolism, pinocytosis, and nucleoside phosphorylation. These may be secondary phenomena, occurring in response to an increased input of previously limited extracellular materials, e.g., cofactors or sources of energy. Alternatively, PHA may have a more specific action in activating certain enzymes. This possibility is suggested by the reported changes of cyclic AMP levels and adenyl cyclase activity associated with PHA stimulation.

2. *Cyclic AMP*

As is now well known, cyclic adenosine $3',5'$-monophosphate (cAMP) plays an important role in regulating many biochemical processes within the cell (Robison et al., 1971). Its mode of action and end effect may differ, depending on the reaction involved. The level of cAMP within the cell depends on its rate of production by the enzyme, adenyl cyclase, and its rate of degradation by cAMP phosphodiesterase. Adenyl cyclase is a cell-membrane-associated enzyme that may exist in an inactive state. One group has reported that the adenyl cyclase activity of lymphocytes is elevated immediately following PHA treatment, and their data suggest that PHA may act directly to activate preexisting enzyme (Smith et al., 1971a). A rapid initial rise in cellular cAMP levels was reported within the first minutes of addition of PHA, which was followed by a fall of below baseline. Other groups have not detected these changes (Novogrodsky and Katchalski, 1970; McDonald, 1971), and it is known that, except for very minute concentrations, cAMP (or its dibutyryl analog) are inhibitory to PHA stimulation (Novogrodsky and Katchalski, 1970; R. Hirschhorn et al., 1970; Quastel et al., 1970; Henney and Lichtenstein, 1971; Smith et al., 1971b).

Theophylline or caffeine, prostaglandin E and epinephrine-like drugs such as isoproterenol will raise cellular cAMP levels through action at various points in cAMP metabolism (Butcher, 1968). These compounds also have generally been found to inhibit PHA-induced lymphocyte growth (Kay and Handmaker, 1970; Smith et al., 1971b). The small lymphocytes of chronic lymphatic leukemia are apparently more sensitive than normal lymphocytes to such inhibition (Johnson and Abell, 1970). It is of interest that these inhibitors and cAMP itself are most antagonistic to the mitogenic action of PHA if given during the first hour of exposure to PHA. Subsequently, they become less inhibitory (Johnson and Abell, 1970; Smith et al., 1971b). This suggests the existence of an early, essential step which is inhibited by elevated cAMP levels.

The situation remains to be clarified, but it is tempting to speculate that the transitory rise in cAMP is involved in the initial activation of repressed systems, while the later fall is essential to processes where cAMP is inhibitory normally. The detrimental effect of added cAMP would then be explained by a failure to lower internal cAMP concentrations to the levels necessary for growth-related reactions to proceed. Since virtually all studies demonstrating the inhibitory effect of cAMP have used eventual DNA synthesis as an endpoint, it is not known whether or not added cAMP can enhance those very early processes that might require elevated cAMP levels for activation.

A means by which enzymes may be activated by cAMP may be seen in the mode of activation of the protein-phosphorylating enzyme, protein kinase. In heart (Erlichman et al., 1971) and in skeletal muscle (Reimann et al., 1971), it appears that the inactive form of protein kinase is combined with an inhibitory protein that has binding affinity for cAMP. Apparently due to a steric change caused by binding cAMP, the inhibitor is dissociated from the enzyme which thereupon becomes functional.

The finding that phosphorylation of nuclear proteins is an early event following PHA treatment suggests that activation of protein kinase may be involved in lymphocyte stimulation by mitogens (Kleinsmith et al., 1966). The importance of protein kinase in regulating activities in animal cell systems is illustrated by the control of glycogenolysis, currently the best understood system in which cAMP regulates an animal metabolic activity (reviewed by Pastan and Perlman, 1971). In this system, cAMP acts reversibly in a dual fashion, functioning simultaneously in both an inhibitory and a stimulatory capacity. Elevation of cAMP concentration both activates the enzyme that breaks down glycogen (phosphorylase a) and inhibits the enzyme that synthesizes glycogen (glycogen synthetase I). Cyclic AMP performs both these functions through activation of protein kinase. The activated kinase phosphorylates the enzymes noted above, activating one and inactivating the other. The effect of cAMP in most animal cell systems appears to be mediated by protein kinase activity. In bacterial systems, another mode of cAMP action has been elucidated in which cAMP binds to a special protein; this cAMP-protein complex is then active in the initiation of the transcription of DNA by RNA polymerase (Pastan and Perlman, 1971).

3. *Proteolysis*

Another potential means of activating repressed proteins is suggested by the finding that, following PHA stimulation, there is a release of various acid hydrolases from the lysosomes of the treated lymphocytes (R. Hirschhorn *et al.,* 1968). These released enzymes may activate proteins through partial proteolysis by cleaving a segment of the protein responsible for preventing activity. The activation of fibrinogen and of various digestive enzymes are well-known examples of such proteolytic activation. In this connection it has been shown that ϵ-aminocaproic acid and other inhibitors of proteolysis will inhibit the response of lymphocytes to PHA (R. Hirschhorn *et al.,* 1971).

E. EARLY NUCLEAR CHANGES

A number of nuclear alterations have been reported to follow rapidly upon the addition of PHA to lymphocytes. Acetylation of nuclear histones (Pogo *et al.,* 1966; Mukherjee and Cohen, 1969; Ono *et al.,* 1969; Cross and Ord, 1970; Desai and Foley, 1970), phosphorylation of nuclear proteins (Kleinsmith *et al.,* 1966), and increased acridine orange binding (Killander and Rigler, 1969; Rigler and Killander, 1969) have been found to occur almost immediately after PHA treatment. This is followed by microscopically visible nuclear changes that have been interpreted as a loosening of the coils of heterochromatin. It has been suggested that the acetylation of histones and phosphorylation of nuclear proteins may reduce the binding of these inhibitory proteins to the DNA, permitting the released portions of the genome to function as templates for RNA synthesis (Pogo *et al.,* 1966). Increased acridine orange binding would reflect the greater accessibility of the activated chromatin (Killander and Rigler, 1969; Rigler and Killander, 1969). These responses do not require concomitant RNA or protein synthesis, suggesting activation or alteration of preexisting proteins.

Despite these generally repeatable early findings, lymphocyte stimulation is almost completely reversible at this stage by detachment of bound PHA (Kay, 1970). This suggests that subsequent changes are required to take advantage of the proposed genomic activation in order to permit blastogenesis to proceed.

One group has reported, however, that with specially purified mitogenically active PHA, no early increase in histone acetylation occurred. Conversely, with mitogenically inactive preparations of PHA they obtained rapid histone acetylation (Monjardino and MacGillivray, 1970). Thus, the relevance of this alteration to lymphocyte activation remains unproven. Similarly, increased acridine orange binding was also found in polymorphonuclear leukocytes after PHA treatment, although no stimulation of growth or of RNA synthesis occurs in these cells (Killander and Rigler, 1969; Rigler and Killander, 1969), provoking some uncertainty as to the biological significance of that phenomenon.

F. Energy Metabolism

During the response to PHA, lymphocytes elaborate large amounts of acid into the culture medium. This is due to the high production of lactate, which, in turn, relates to an apparent dependence of the growing lymphocyte on the glycolytic pathway as a primary means of glucose use (E. H. Cooper et al., 1963; MacHaffie and Wang, 1967; Pachman, 1967; Polgar et al., 1968a; Hedeskov, 1968). Inhibitors of glycolysis (iodoacetate, 2-deoxyglucose) are strongly inhibitory to progress through the mitotic cycle. However, 2,4-dinitrophenol, which inhibits oxidative phosphorylation, is much less inhibitory to growth (Polgar et al., 1968a; Lindahl-Kiessling and Mattson, 1971). Enhancement of the pentose phosphate pathway has also been noted (MacHaffie and Wang, 1967). Puromycin has been reported to inhibit the elevation of glycolysis in PHA-stimulated lymphocytes (Hedeskov, 1968). This suggests that increased activity of energy-producing pathways occurs secondarily, in response to more directly stimulated energy-requiring steps involving protein synthesis. Alternatively, stimulation of synthesis of hexokinase may be an early direct consequence of growth stimulation, which is essential for enhanced glycolysis, and which is inhibited by puromycin (Hedeskov, 1968).

G. RNA Metabolism

The various classes of RNA play an essential role in the transmission of genetic information, stored in DNA, to the protein-synthesizing machinery

7. EFFECTS OF MITOGENS ON THE MITOTIC CYCLE

of the cytoplasm. It is probably safe to assume that the regulation of lymphocyte growth is controlled, either directly or indirectly, by momentarily active portions of the cell's genome. Therefore, an examination of RNA metabolism is essential to an understanding of the mechanism by which mitogens induce resting lymphocytes to enter the mitotic cycle.

1. Difficulties of Quantitation

Initial studies in this area were concerned with changes in the overall rate of RNA synthesis in resting and stimulated lymphocytes. Measurements of total RNA content showed that a 50%–100% increase in total cellular RNA content occurred in the first 24–48 hours after PHA stimulation (Kay, 1966; Forsdyke, 1967, Cooper, 1969d). Similarly, measurements of the rate of incorporation of ^{3}H-uridine into RNA showed a rapid increase, detected within the first hour after addition of PHA (Cooper and Rubin, 1965; Pogo et al., 1966). An interpretation of these seemingly simple observations, however, is quite difficult, owing to the following complications.

Measurement of the overall rate of RNA synthesis by incorporation into RNA of ^{3}H-uridine or other radioactive nucleosides is confused by the fact that the figure obtained is a composite of the different rates of synthesis and degradation of several classes of RNA. The major factors contributing to this measurement are a rapidly turning over, small pool of heterogeneous nuclear RNA, and a large pool of relatively slowly synthesized stable ribosomal RNA (see Section G,2). The radioactivity accumulated in total RNA after any period of labeling will depend markedly on the degree to which the small labile pool is saturated with radioactive molecules (Fig. 5A). Total radioactivity will accumulate briskly, essentially at the rapid rate of synthesis of heterogeneous nuclear RNA, early in the labeling period. Once this pool is saturated, labeled heterogeneous RNA is degraded as rapidly as it is formed, and further net accumulation of radioactivity in that pool stops. Label then accumulates at the rate of synthesis of the slowly synthesized stable pool of ribosomal RNA. Thus, values for total labeling rates will differ depending on whether short or long pulses are employed. If the rates of synthesis of the two pools are affected differentially by PHA, then a comparison of resting and growing cells is questionable at any pulse length if a single time point is used, as is commonly the case. An additional serious complication is that the specific activity of the nucleoside triphosphate pools, from which new RNA molecules obtain nucleotides, is changing constantly during the labeling period (Fig. 5C). This is due to the relatively slow rate at which labeled nucleosides from the cell culture medium are transported into the cell and phosphorylated (Cooper, 1972b). As a result, the rate of entry of radioactivity into RNA will be less during short labeling periods than during long ones despite a constant true rate of RNA synthesis.

Interestingly, the theoretical effect of this slow rate of entry of ^{3}H-uridine into the uridine triphosphate pool is to convert the expected curvilinear total labeling pattern (Fig. 5A) to a more nearly linear one (Fig. 5B). In fact, most experimental observations demonstrate apparent linearity of incorporation of ^{3}H-uridine during several hours (Cooper and Rubin, 1965; Kay, 1968a). Without an understanding of the complex interaction among the various RNA classes and precursor pools, measurements of such pseudo linear incorporation rates have little meaning. A PHA-induced change in the rate at which the uridine triphosphate pool is saturated with exogenous ^{3}H-uridine may alter the apparently linear rate of labeling of RNA without any change in RNA synthesis.

In fact, such changes in the rate of uridine entry into, and phosphorylation by the cell have been noted following PHA stimulation (Lucas, 1967; Hausen and Stein, 1968; Kay and Handmaker, 1970; Peters and Hausen,

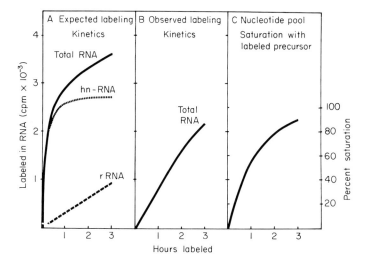

Fig. 5. Kinetics of incorporation of radioactive nucleosides into total RNA. (A) Expected accumulation of label in RNA, assuming saturation of precursor pool (e.g., UTP) with exogenous labeled nucleoside (e.g., ^{3}H-uridine) from time zero. Under this assumption, the precursor pool specific activity remains constant from the beginning of the labeling period. The curvilinear kinetics for total RNA labeling result from a combination of radioactivity entering a rapidly saturated pool of material which turns over quickly (hn-RNA) and a slowly synthesized pool of stable material (rRNA). (B) Typical linear labeling kinetics actually observed in experiments. This is due to the slow entry of exogenous radioactive nucleosides into the precursor pool. As a result, the specific activity of the precursor entering RNA rises continuously during the experiment (C). This will lower the amount of label entering RNA during the early period of the experiment relative to later periods, producing the pseudolinear kinetics seen in (B).

1971a). Uridine kinase activity rises roughly in parallel to the increase in rate of ^{3}H-uridine incorporation into RNA. This finding gave rise to the speculation that the rate of RNA labeling by ^{3}H-uridine was limited by the level of uridine kinase activity, and that rises in the rate of RNA labeling were due not to increased RNA synthesis but to increased uridine kinase activity (Lucas, 1967; Hausen and Stein, 1968). However, it has been shown that the rate of ^{3}H-uridine incorporation into RNA begins to rise before any increase in uridine kinase activity, indicating that RNA synthesis probably does increase in absolute rate (Kay and Handmaker, 1970). Moreover, dipyridamole, a drug that inhibits uridine transport, did not block PHA-induced lymphocyte growth (Peters and Hausen, 1971a). Increased uridine transport and uridine kinase activity, therefore, are probably secondary phenomena facilitating the acquisition of exogenous nucleosides for RNA synthesis. As a final caution, the rate of ^{3}H-uridine uptake may be affected by the concentration of the nucleoside in the culture medium (Hausen and Stein, 1968), making comparison of results among laboratories difficult.

Thus, quantitative inferences about changes in RNA synthesis drawn from observed alterations in labeling by exogenous nucleosides are perilous. Corrections for nucleoside triphosphate pool specific activities can be made theoretically. However, in practice, this quantitation in lymphocytes is difficult because of limitations in the number of cells available.

2. *Studies of Various Classes of RNA*

From the foregoing, it should be apparent that attempts to quantitate the rates of overall RNA synthesis in resting lymphocytes are difficult and may be of little value. A potentially more useful approach is the separation of the various classes of RNA so that they may be studied independently. This has been done, using the technique of phenol extraction and separation of molecules of various sizes by sedimentation in sucrose gradients or by electrophoresis in polyacrylamide gels.

a. *Heterogeneous Nuclear RNA.* Using this approach, it was shown that the most rapidly synthesized RNA in both resting and growing lymphocytes is the high molecular weight heterogeneous nuclear RNA (Cooper, 1968, 1969c; Neiman and Henry, 1971). This material, whose molecular weight ranges from 2×10^6 to possibly 10^7 daltons, is rapidly synthesized and turned over. It comprises no more than 1% of the total cellular RNA and its half-life may be less than 30 minutes (Darnell, 1968; Cooper, 1968; Scherrer et al., 1970). Nevertheless, it represents over 90% of the RNA being synthesized at any moment in the resting cell, and about 80% in the growing lymphocyte (Cooper, 1968, 1969c; Neiman and Henry, 1971). This class of RNA appears to contain copies of fairly large regions of the genome (Torelli *et al.,* 1968; Scherrer *et al.,* 1970). However, little, if any, of this RNA ever leaves the nucleus; most of it is apparently rapidly degraded within the nucleus shortly after its synthesis (Attardi et al., 1966; Cooper, 1968; Darnell, 1968; Neiman and Henry, 1971). Its biologic role in animal cells is currently unknown, and therefore the subject of intense speculation. It is probable that at least some of this heterogeneous RNA is messenger RNA which is transported to the cytoplasm. Possibly, much of it is potential messenger RNA which is degraded as part of a control process. Some studies suggest that it is quite different metabolically from messenger RNA, since its synthesis is little affected by the drug cordycepin, while messenger RNA production may be halted by the drug (Penman et al., 1970).

Following PHA stimulation, there is an increase in the rate of incorporation of ^{3}H-uridine into heterogeneous RNA within the first 6 hours (Cooper, 1969c). It is not certain to what degree changes in precursor pool specific activity may contribute to this change in labeling, so that the safest conclusion is that the early increase in heterogeneous nuclear RNA labeling can at best reflect only a modest early increase in the rate of synthesis of this class of RNA.

After about 20 hours of PHA stimulation, short labeling experiments suggest that heterogeneous nuclear RNA labeling by ^{3}H-uridine increases some 10- to 20-fold, on the average (Cooper, 1969c). Of this increase, perhaps two- to four-fold can be attributed to altered precursor pool specific activity (Cooper, 1972b), leaving a five-fold increase in synthesis rate to make up the difference. Thus, by the time cell growth has progressed to about the point where DNA synthesis begins in the most advanced cells, there is clearly an increased rate of heterogeneous nuclear RNA synthesis.

Since it is possible, though not certain, that heterogeneous nuclear RNA includes informational molecules (Darnell et al., 1971), it is of interest to know whether or not the material produced in PHA-stimulated cells includes molecules transcribed from previously repressed regions of the genome. Attempts to answer this question by RNA–DNA hybridization competition have given no evidence that any RNA molecules are produced in growing lymphocytes that are not also synthesized in resting cells (Torelli et al., 1968; Neiman and MacDonnell, 1970). It is probable that the RNA–DNA hybridization techniques employed would not detect the synthesis of a relatively small amount of new unique nonrepetitive RNA in stimulated cells. However, the data indicate that the bulk of heterogeneous nuclear RNA (and other RNA) molecules produced after PHA stimulation have base sequences similar to those molecules already produced by the resting cell.

This conclusion is hard to reconcile with certain widespread preconceptions, largely derived from the impressive results with bacteria, that antici-

7. EFFECTS OF MITOGENS ON THE MITOTIC CYCLE 161

pate "turning on" of many repressed genes when resting cells are stimulated to enter the mitotic cycle. Microscopic observations, which show a change from heterochromatic to euchromatic appearance of the nucleus following PHA treatment (Tokuyasu et al., 1968), suggest a more active state for much of the nuclear chromatin. Autoradiographic evidence indicates that chromatin-associated RNA synthesis occurs in relation to euchromatin rather than heterochromatin (Hay and Revel 1963; Littau et al., 1964; Tokuyasu et al., 1968). Rapid increases in acetylation of nuclear histones follow PHA treatment and are thought to reflect dissociation of inhibitory histones from DNA, allowing transcription of repressed regions (Pogo et al., 1966; Mukherjee and Cohen, 1969; Ono et al., 1969; Cross and Ord, 1970; Desai and Foley, 1970). Similarly, the PHA-induced increase in acridine orange staining of lymphocyte nuclei was thought to indicate a greater accessibility of regions of the genome previously involved in some type of inhibitory configuration (Killander and Rigler, 1969; Rigler and Killander, 1969). Nuclei isolated from PHA-treated lymphocytes are reported to act more effectively than controls as templates for bacterial RNA polymerase (R. Hirschhorn et al., 1969). All these observations support the expectation that transcription of new types of RNA molecules is induced by PHA treatment; yet these have not been found. To date, there is no direct evidence for the existence of gene activation of the sort found in bacteria during enzyme induction (Jacob and Monod, 1961a,b) in growth-stimulated lymphocytes.

b. *Transfer RNA and Precursor Molecules.* Transfer RNA comprises some 10% of total cellular RNA. Its synthesis was found to be increased within 90 minutes of addition of PHA (Kay and Cooper, 1969). A more rapid increase was noted in the rate of labeling of another low molecular weight RNA which appears in the cytoplasm (Kay and Cooper, 1969). The rapidity of this increase, especially relative to the rate of labeling of heterogeneous nuclear and transfer RNA's, suggests that changes in precursor pool specific activity are not the cause but that a true increase in synthesis occurs. There is some evidence that this material may be a precursor to transfer RNA, giving rise to the speculation that new species or increased quantities of transfer RNA are required to permit the accelerated protein synthesis found after growth stimulation.

c. *Ribosomal RNA.* As noted above, total lymphocyte RNA content rises about 1.5–2-fold during the first 24–48 hours of culture with PHA. Since the bulk of cellular RNA (at least 80%) is ribosomal RNA, this increase represents primarily an accumulation of new molecules of that type. This inference is supported by electron microscopic studies, which show a copious increase in the content of cytoplasmic ribosomes in PHA-stimulated cells (Douglas, 1972). It has been shown that progressive increase in total ribosomal RNA of the lymphocyte is a requirement if the PHA-treated cell is to progress normally through the mitotic cycle to DNA replication and division. If ribosomal RNA synthesis is inhibited specifically with very low doses of actinomycin D, most of the early responses of the cell to PHA stimulation still take place. Thus, the overall rate of protein synthesis rises and changes in phospholipid metabolism and membrane transport phenomena take place in relatively normal fashion. However, after about 24 hours, further increase in protein synthesis is halted, and DNA synthesis never begins (Kay et al., 1969). Thus, progressive accumulation of new ribosomes is an essential part of the normal mitogenic response in lymphocytes.

The metabolism of ribosomal RNA may be studied by labeling with radioactive nucleosides, as with other RNA forms. An additional feature of ribosomal RNA is that, unlike other forms of RNA in the same molecular weight range, it contains a fairly large number of methyl groups added to the molecule at the same time as or shortly after polymerization of the nucleotide chain (Attardi and Amaldi, 1970; Darnell, 1968). Ribosomal RNA may therefore be studied by radioactive labeling of these methyl

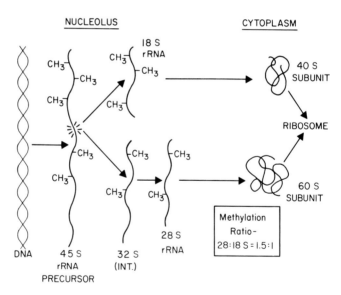

Fig. 6. Production of ribosomal RNA in animal cells. A large precursor (45 S) is transcribed from DNA (rDNA). During transcription the molecule is methylated at various points. Cleavage into 32 S and 18 S segments occurs. Further cleavage of the 32 S to 28 S rRNA follows. The 28 S and 18 S rRNA's are transported to cytoplasm, in combination with a large number of proteins, to form the ribosomal subunits which are active in protein synthesis.

7. EFFECTS OF MITOGENS ON THE MITOTIC CYCLE

groups. Using this technique, together with elaborate cell fractionation methods, the following scheme has been elaborated for the production of ribosomal RNA in animal cells (Fig. 6) (Darnell, 1968).

Ribosomal RNA is transcribed from a highly reduplicated region of the genome that is closely related to the nucleolus. It is synthesized first as a high molecular weight (45 S; $\sim 4.5 \times 10^6$ daltons) precursor which is found only in the nucleolus. During transcription of the precursor, methyl groups are attached to bases and to the 2′-OH position of ribosyl residues at specific positions along the polynucleotide chain. Methylation is virtually complete by the time the chain is fully transcribed. Following transcription of the 45 S precursor, the single large molecule is cleaved into a 32 S intermediate preribosomal RNA and an 18 S ribosomal RNA. Several transitory intermediate forms have also been found which we will not consider here. The 18 S moiety, in combination with some 20–30 proteins, is transported quickly to the cytoplasm as the smaller (40 S) ribosomal subunit. The 32 S RNA moiety is retained in the nucleolus for a period, where a further cleavage occurs, producing a 28 S ribosomal RNA molecule. This species is then transported to the cytoplasm in combination with perhaps 30–40 proteins, where it is recognized as the larger (60 S) ribosomal subunit. The two ribosomal subunits function together in protein synthesis, comprising the ribosome monomer (80 S). It should be noted that during maturation of the single 45 S RNA to the two mature ribosomal RNA forms about half the 45 S molecule is lost, or at least cannot be accounted for. Interestingly, only the methylated portions of the 45 S molecule survive as the mature 18 S and 28 S RNA's.

In the resting lymphocyte, during relatively short labeling periods, the rate of accumulation of new ribosomal RNA molecules in the cytoplasm, as measured by ^{3}H-uridine labeling, is only 2%–4% that of total RNA labeling (the latter consisting mainly of heterogeneous nuclear RNA labeling). After 20 hours of incubation with PHA, this figure rises to about 20% (Cooper, 1969c). Apparently, there is a differential stimulation of ribosomal RNA production relative to heterogeneous nuclear RNA production. This may be shown directly in sucrose gradient sedimentation profiles of ^{3}H-uridine-labeled RNA. Prominent labeling of the 45 S moiety is seen together with heterogeneous RNA after a pulse label in growing cells. The same length of pulse in resting cells, however, shows labeling primarily of heterogeneous RNA (Cooper, 1969c). Using methyl labeling, the increase in the rate of 45 S RNA synthesis can be detected by 1–2 hours after addition of PHA (Cooper, 1969b, 1970b). Thus, a characteristic of PHA-induced lymphocyte growth is a rapid differential stimulation of ribosomal RNA synthesis. After 20 hours of PHA treatment, the rate of ribosomal RNA accumulation in the cytoplasm is about 10–20-fold increased over resting cell levels, when corrections for precursor pool specific activity are considered.

It is of interest that the rate of 45 S RNA synthesis can be increased in the absence of concurrent protein synthesis, since the early effect of PHA on this process is seen in lymphocytes treated with cycloheximide (Cooper, 1970a,b). This indicates that at least initially no new polymerase or other enzymes need be synthesized to produce this increase. Apparently, RNA polymerase molecules are present which are not fully active, either because the polymerase itself is inhibited or because the DNA template is unavailable. One of the important unanswered questions in this regard is whether the PHA-induced increase in the rate of 45 S RNA synthesis is due to an increase in the number of templates available for transcription (i.e., derepression) or to an increased frequency of transcription initiation on the same templates previously available. The RNA polymerase (polymerase I; nucleolus associated) involved in ribosomal RNA synthesis appears to be different from that engaged in nonribosmal RNA synthesis (polymerase II; chromatin associated) (Roeder and Rutter, 1969). These can be distinquished by their different sensitivities to the drug, α-amanitin, which inhibits RNA polymerase II, but not polymerase I (Kedinger et al., 1970; Lindell et al., 1970). RNA polymerase I increases in activity rapidly after PHA addition to lymphocytes and reaches a plateau at about 24 hours. Polymerase II activity, however, rises more slowly at first but its increase accelerates after 12 hours of stimulation (Handmaker and Graef, 1970; Kay and Cooke, 1971; Cooke et al., 1971). These changes in polymerase activity roughly parallel the changes in rate of synthesis of 45 S ribosomal RNA precursor and heterogeneous nuclear RNA discussed above.

The maturation process by which 45 S RNA is converted to the final 28 S and 18 S ribosomal RNA forms is also accelerated following PHA stimulation. The onset of this acceleration is rapid and is observed within the initial 2 hours of growth stimulation (Cooper, 1969c; Rubin, 1970). While the factors involved in the maturation process are unknown, it is likely that various enzymes and other proteins are involved. Somewhat paradoxically, in resting cells, the apparent rate of conversion of 45 S RNA to 32 S and 18 S forms was increased when protein synthesis was abolished by cycloheximide. However, in growing lymphocytes, after PHA stimulation, the opposite was true: cycloheximide caused a decrease in the rate of conversion of 45 S RNA to 32 S and 18 S forms (Cooper, 1969d). The latter observation is true in other growing cells as well (Craig and Perry, 1970).

Another intriguing aspect of the control of ribosomal RNA production during lymphocyte growth stimulation concerns the survival of newly synthesized ribosomal RNA molecules. In resting lymphocytes, of all the ribosomal RNA synthesized in the nucleus, only half reaches the cytoplasm as mature 28 and 18 S RNA (Rubin, 1968, 1970; Cooper, 1969a,b; Cooper and Gibson, 1971). The remainder is apparently lost through degradation at some point between its synthesis in the nucleus and its appearance in the cytoplasm. This wastage of new RNA is quickly reversed by growth stimulation, allowing for an approximate doubling in the rate of accumulation of new ribosomal RNA in the cytoplasm even without an increase in actual ribosomal RNA synthesis (Cooper, 1969b, 1970b; Rubin, 1970).

The survival of new ribosomal RNA molecules in both resting and growing lymphocytes is dependent on continuous protein synthesis. If protein synthesis is abolished with cycloheximide virtually all RNA synthesized thereafter is rapidly degraded shortly after its transcription (Cooper, 1970b; Cooper and Gibson, 1971). This is true of other cell systems as well (Ennis, 1966; Willems *et al.*, 1969; Mandal, 1969; Cooper, 1970b). However, 45 S RNA precursor molecules synthesized under normal conditions can be processed to mature ribosomal RNA subsequently even if protein synthesis is abolished during the processing period (Cooper, 1970b). This indicates that some event requiring continuous protein synthesis takes place during the production of 45 S RNA which subsequently protects the ribosomal RNA molecules from degradation during the final maturation of the molecule.

The accumulated evidence suggests that a protein (or proteins) acts to protect newly synthesized ribosomal RNA from nuclease attack during the process of maturation from the 45 S precursor form (Cooper and Gibson, 1971; Hopper and Cooper, 1971). The amount of this ribosomal protective protein available appears to act as an upper limit on the number of new ribosomal RNA molecules which can survive during any period, all molecules in excess of this amount being degraded. The protective protein must be produced continuously, either because it is rapidly turned over, or because it forms a stable association with ribosomal RNA in order to protect it from nuclease attack. The reduced wastage of ribosomal RNA which follows PHA treatment may result from an increased production of the ribosomal protective protein.

These observations suggest that the rate of production of the protective protein acts as a control on the growth of the cell. Since normal lymphocyte growth and division do not occur if new ribosomal accumulation does not take place in the cytoplasm (Kay *et al.*, 1969), and since the survival of new ribosomal RNA molecules is sharply controlled in resting lymphocytes by the limiting rate of production of the protective protein, then normal growth stimulation must provide for an early increase in the rate of synthesis of ribosomal protective protein. An increase in 45 S RNA synthesis without a concomitant elevation of protective protein production would theoretically fail to provide an increased accumulation of mature ribosomal RNA and growth stimulation would fail. Interestingly, nature seems to have provided exactly this situation in the disease chronic lymphatic leukemia. In this condition, large numbers of small lymphocytes crowd the circulation. The cells appear no different from normal small lymphocytes. However, when treated with PHA, very little DNA synthesis occurs at 48–72 hours, as it does in normals (Oppenheim et al., 1965; Havemann and Rubin, 1968; Abell et al., 1970). For some time it was thought that the leukemic lymphocytes were unresponsive to PHA. Recently, Rubin observed that, as in normal cells, a rapid rise in 45 S RNA synthesis occurred after PHA treatment. However, this rise was not accompanied by a decrease in ribosomal RNA wastage, possibly because production of ribosomal protective protein did not increase. Accumulation of cytoplasmic ribosomal RNA was not increased, and cell growth did not ensue at the usual time (Rubin, 1971).

In this connection, it was reported that the surface of chronic lymphatic leukemia lymphocytes contains subnormal numbers of binding sites for erythroagglutinating PHA (S. Kornfeld, 1969). This finding may be reconciled with that described above by proposing that it requires fewer PHA binding events to accomplish enhancement of 45 S RNA synthesis than it does to promote the synthesis of sufficient protective protein to permit survival of the new ribosomal RNA molecules.

The foregoing discussion indicates that the regulation of ribosomal RNA production is both complex and of extreme importance in the mitogen-induced entry of resting lymphocytes into the mitotic cycle. It seems likely that the restrictions placed on RNA accumulation by these controls play an important role in maintaining the lymphocyte in its nongrowing condition, and their orderly modification may be essential to the controlled cell growth characteristic of normal tissues.

d. *Messenger RNA.* In my consideration of lymphocyte RNA metabolism I have not touched upon messenger RNA. This is for the good reason that no studies to date have shed any direct light on possible changes in the movement or production of messenger RNA in lymphocytes during the onset of cell growth. Some inferences concerning messenger RNA may be drawn from studies with protein synthesis, however, and will be considered below.

H. Protein Synthesis

1. *Overall Measurements*

Resting lymphocytes synthesize protein at a continuous baseline rate. This activity ceases within 2–3 hours on addition of large doses of actinomycin D, sufficient to inhibit nearly all RNA synthesis (5–10 μg/ml)

(Kay, 1967). This suggests a mean life of that magnitude for cytoplasmic messenger RNA in resting lymphocytes. New messenger RNA must, therefore, be continuously entering the cytoplasm in resting cells.

With low doses of actinomycin D (0.005–0.05 μg/ml), which inhibit specifically only ribosomal RNA synthesis, the baseline level of protein synthesis is affected only slightly, even with exposures as long as 24 hours (Kay et al., 1969). Apparently, preexisting cytoplasmic ribosomes are present in adequate quantity and are sufficiently stable to carry on protein synthesis without continuous replacement. These observations also suggest that in the resting lymphocyte the continuous entry of messenger RNA molecules into the cytoplasm is not dependent on movement of new ribosomal particles from nucleus to cytoplasm. Thus, the mode of transport of messenger RNA to the cytoplasm probably does not require it to be bound to newly emerging ribosomal subunits.

The overall rate of protein synthesis in lymphocytes increases shortly after exposure to PHA (K. Hirschhorn et al., 1963; Kay et al., 1969). Such an increase has been reported within minutes of mitogen treatment (Pogo et al., 1966). However, it has generally been possible to document reproducibly an increased amino acid incorporation into total cell protein after about 2 hours of treatment with PHA. Overall protein synthesis rises about tenfold during the first 24 hours of culture with PHA, and about 20-fold during 48 hours (Kay et al., 1971). Thereafter further increases in total protein synthesis are not generally seen. It should be observed that inability to detect overall increases in protein synthesis before about 2 hours does not eliminate the possibility of discrete elevations in production of some specific proteins which have not been detected to date.

Following PHA stimulation for 24–48 hours, high doses of actinomycin D still inhibit protein synthesis markedly. It has been suggested, however, that some of the messenger RNA in growing cells may have greater stability than that found in resting cells (Kay, 1967).

Inhibition of ribosomal RNA synthesis in resting lymphocytes with low concentrations of actinomycin D does not prevent the PHA-induced rise in total protein synthesis during the initial 24 hours of growth (Kay et al., 1969). Thereafter, the additional rise seen normally does not occur, and DNA synthesis does not begin.

2. Ribosome Activation

From this data, it is apparent that the preexisting ribosomes of the resting lymphocytes are not maximally active in protein synthesis, and that their activity is initially enhanced by growth stimulation.

The means by which preexisting ribosomes of the resting lymphocyte may function more actively in protein synthesis has been investigated (Kay et al., 1971). In the nongrowing lymphocyte, 70% of cytoplasmic ribosomes are found in the monomeric form (80 S) and their ease of dissociation into subunits by high salt concentrations suggests that they are free of any association with messenger RNA. The proportion of such inactive ribosomes falls to about 40% after 10–15 hours of PHA treatment, and to less than 30% by 24 hours. A parallel increase in the proportion of ribosomes in polysomes or as nondissociable (messenger-bound) monosomes occurs. Thus, enhanced protein synthetic activity of preexisting ribosomes is due to mobilization of the relatively large proportion of inactive ribosomes in resting lymphocytes. Studies with antigen-stimulated rat spleen cells suggest that activation of ribosomes is related to the appearance of a soluble cytoplasmic factor that enhances the ability of the ribosome to translate mRNA (Selawry and Starr, 1971a,b). Like the limitation in ribosome accumulation discussed in Section IV,G,2,c, the restriction of the protein synthetic activity of preexisting ribosomes may constitute a major growth regulating mechanism.

Preexisting ribosomes appear capable of handling the quantitative increase in protein synthesis during the initial 24 hours of stimulation. However, after 24 hours the cells must be competent to synthesize new ribosomes for growth to proceed normally. It is not clear whether this is due to a simple quantitative deficiency of ribosomes or whether new and perhaps functionally different ribosomes are required at that time to perform specific tasks. Evidence on this point is found in studies of DNA polymerase activity.

3. *DNA Polymerase*

DNA polymerase activity is very low in resting lymphocytes. There is no evidence that the enzyme is synthesized by nongrowing cells. After 20 hours in culture with PHA, the activity of this enzyme begins to rise parallel to the rate of DNA synthesis (Loeb et al., 1968; Loeb and Agarwal, 1971). Experiments with inhibitors of protein synthesis suggest that this increase is due to synthesis of new polymerase molecules (Loeb et al., 1970). When low doses of actinomycin D are added together with PHA, DNA polymerase activity does not begin its expected rise 20 hours later. The failure to produce DNA polymerase would, of course, explain the lack of DNA synthesis in lymphocytes treated with low doses of actinomycin D. If 6–8 hours of PHA treatment are given and actinomycin is then added (at a time when no production of the enzyme has yet occurred) there is only partial inhibition of enzyme synthesis some 12–18 hours later (Agarwal and Loeb, 1972). In this experiment, the effective concentrations of actinomycin D were low ($0.005 - 0.015$ μg/ml) and were unlikely to have prevented the synthesis of mRNA for DNA polymerase production, although the possibility cannot be ruled out entirely. It is possible, however, that the effects are related to the failure to enhance rRNA production caused by such low concentrations of the drug.

It appears that some early event, requiring production of new ribosomes, is essential for the translation of messenger RNA for DNA polymerase some 18–20 hours later. This event may be simply an increase in the total number of ribosomes available. It is difficult to see, however, how the cell detects this quantitative change so precisely as a signal to begin translation of a specific messenger RNA many hours later. One alternative possibility is that the new ribosomes are required to permit the messenger RNA for DNA polymerase to emerge from the nucleus or to be translated. Since it has already been seen that such a requirement does not exist for the messenger RNA which specify the proteins being synthesized by the resting cell (Section V,H,1), this possibility implies some special interaction between the new ribosomes and the messenger RNA for DNA polymerase. Thus, messengers for new activities which are "induced" as part of the onset of cell growth may require an association with ribosomes in order to emerge from the nucleus. In this case, the polymerase messenger would emerge soon after PHA stimulation but remain in a stable, unused state for a prolonged period. Its combination with the ribosome might play a role in this stabilization. The special interaction, mentioned above, between new ribosomes and the DNA polymerase messenger, may involve a special class of ribosomes with affinity for that messenger RNA. This hypothetical class of ribosomes might have similar affinity for a number of messenger RNAs related to growth-induced functions. Such messengers, as a class, might be identified by specific nucleotide sequences.

Variants of this hypothesis may easily be envisaged. Its advantage over the seemingly simpler proposition, that a quantitative change in the number of available ribosomes is the sufficient cause of the late onset of DNA polymerase synthesis, is that no mechanism can be postulated in terms of current knowledge to explain how the quantitative change would produce the final effect. The hypothesis of specific ribosome–messenger RNA interaction is both comprehensible in terms of current information, and leads to interesting possibilities for regulatory mechanisms.

Thus, this hypothesis predicts heterogeneity among ribosomes and/or among groups of messenger RNAs. The possibility of heterogeneity among ribosomes is supported by biochemical evidence from two directions. First, data being accumulated indicate that the protein components of ribosomes are not uniform for all ribosomes. This is demonstrated most clearly with bacterial ribosomes, where it has been shown that some ribosomal proteins are probably present on all ribosomes, while others are not found in sufficient amounts for them to be represented on all ribosomes (Kurland et al., 1969; Traut et al., 1969). Some of these may be ribosomal proteins, which attach to and detach from the ribosome during its normal function. Others, however, may be permanent components of a portion of the ribosome population, conferring specific characteristics on those particles. Findings of a similar nature, though less detailed, have been reported for animal cell ribosomes (Gould, 1970).

The second source of ribosome heterogeneity is found in the ribosomal RNA. Reports have appeared suggesting that some nucleotide sequences are not found in the rRNA of all ribosomes (Young, 1968). More recently, data has appeared suggesting that the high molecular weight preribosomal RNA in plant (Grierson and Loening, 1972) and animal (Tiollais et al., 1971) cells is not homogeneous within the same cell or among different tissues. Thus, there is a biochemical basis for potential functional heterogeneity among ribosomes, which might permit particular messenger RNAs to be stabilized and/or transported in the company of specific ribosomal particles.

The possibility of distinction among various messenger RNA's in terms of special nucleotide sequences is supported by the finding that many, but not all messenger RNA's contain extensive sequences of adenylate residues (Darnell et al., 1971; Edmonds et al., 1971; Lee et al., 1971).

A modification of this hypothesis would suggest that the requirement for the new ribosomes occurred only at the onset of translation, some 18 hours after the initial stimulatory event. This would require either continual emergence of polymerase messenger, with failure to translate it without some special signal, or presence of the messenger in a stable "masked" form, again awaiting some late signal for translation. That the late signal is ineffective without new ribosome synthesis again indicates the requirement for more, or new types of ribosomes. The same argument as given above may be advanced for favoring the qualitative rather than the quantitative hypothesis.

These observations raise the general possibility of stable or masked messenger RNAs which may exist in the resting lymphocyte but which are not used for protein synthesis until growth stimulation. The existence of masked messenger RNA seems well established as the explanation of events that occur during the initial stages of embryogenesis in sea urchins (Piatigorsky, 1968; Raff et al., 1972).

The diminished ability of actinomycin D to prevent subsequent DNA polymerase production when given 6–8 hours after stimulation is similar to the findings noted earlier (Section V, D, 2) relating to cAMP. Elevated cAMP levels were inhibitory to eventual DNA synthesis, but with decreasing effectiveness as cell growth proceeded. In both situations, an early event apparently occurs which determines the later onset of DNA synthesis. The inhibitory effect of cAMP may be mediated, therefore, through prevention of the same early step which provides for subsequent DNA polymerase production.

4. *Immunoglobulins*

Conflicting reports have appeared regarding immunoglobulin production by PHA-stimulated lymphocytes (reviewed by Ling, 1968). Several workers reported increased numbers of immunoglobulin-producing cells and elevated synthesis of immunoglobulin molecules after human lymphocytes were incubated with PHA (Bach and Hirschhorn, 1963; Ripps and Hirschhorn, 1967; Elves *et al.,* 1963b; Forbes and Turner, 1965; Forbes and Henderson, 1966). However, others were unable to confirm this finding (Parenti *et al.,* 1966). Instead, it was shown that the proportion of protein synthesis devoted to immunoglobulin production was no higher after PHA stimulation than in resting cells (Asofsky and Oppenheim, 1966). Thus, an elevated overall production of immunoglobulins in PHA-stimulated cultures might be seen without any specific stimulation of its production. The overall elevation of protein synthesis due to PHA stimulation would cause the observed increase, provided that the proportion of protein synthetic activity devoted to immunoglobulins remained constant.

Nevertheless, there may be other reasons for the conflicting findings. First, most of the above studies were performed before the division of the lymphocyte population into B and T cells was generally known. If it is true, as most immunologists currently believe it is, that immunoglobulins are produced primarily by B cells, then variability of numbers of B cells in the experimental population might influence the results. Most of the studies mentioned used human perpheral blood leukocytes. These were subjected to varying degrees and methods of purification to remove nonlymphoid elements. These purification techniques, however, sometimes remove B cells preferentially (Plotz and Talal, 1967; Shortman *et al.,* 1971; Rosenthal *et al.,* 1972). Since only about 10%–20% of the circulating human lymphocyte population are B cells (Heller *et al.,* 1971; Abdou, 1971), great variation in the numbers of potential antibody-producing cells may have been present in the cultures studied in different laboratories.

Finally, the ability of PHA to stimulate B cells may vary, depending on the way in which the mitogen is presented to the cells. This is indicated by recent studies using mouse cells, where relatively pure populations of B and T cells may be obtained. When PHA was added to the medium of purified mouse T cells, significant stimulation of DNA synthesis was observed 48 hours later, whereas B cells treated similarly showed no response. However, if the PHA was immobilized by covalent bonding to Sepharose beads, excellent B-cell stimulation was obtained (Greaves and Bauminger, 1972). While it is too soon to speculate on the nature of the difference in mitogen–lymphocyte interaction caused by this alteration in mode of presentation of the mitogen, it is clear that PHA will stimulate B cells, but only under special conditions. Since all other studies of immunoglobulin production have used soluble PHA, the variable results obtained may have resulted from differences in PHA preparations and their mode of use in cell culture. Stimulation of B cells may have occurred to the extent that the soluble PHA resembled Sepharose-immobilized PHA under the conditions employed.

Using such purified B- and T-cell populations it was shown that pokeweed mitogen stimulated specifically the production of immunoglobulin (IgM) by B cells or by mixtures of B and T cells, relative to total protein synthesis. PHA, however, failed to do so, as in earlier studies (Parkhouse et al., 1972). Both PHA and pokeweed mitogen used in this study were presented in soluble form. It will be of interest to note the results of similar experiments on immunoglobulin production where immobilized PHA is used.

It may be concluded from the available evidence that mitogens will specifically stimulate immunoglobulin production if they activate B cells present in the population. The appearance of plasma-cell-like cells in pokeweed-treated cultures (Chessin et al., 1966; Douglas et al., 1967), where specific stimulation of immunoglobulin production seems probable, suggests that the mechanism of such stimulation is through the development of such cells.

5. *Other Enzymes and Mediators*

A number of specific enzymes and functional materials of a presumptive protein nature are either produced in increased quantity or are first elaborated by lymphocytes following growth stimulation. These are summarized in Table I.

a. *Specific Enzymes.* In general, most of the rises in specific enzyme activity that may be attributable to increased or *de novo* protein synthesis are detected only after growth stimulation and increased overall protein synthesis are well established. The fairly late rises in thymidine kinase and thymidine monophosphate kinase activities, like that of DNA polymerase, are prevented if low doses of actinomycin D are added with the mitogen (Loeb et al., 1970). The synthesis of these enzymes, which function in the salvage pathway for thymidine and are thus concerned with DNA replication, may not occur until the above described period when ribosomal RNA synthesis is required. However, some essential RNA-dependent event which determines such subsequent production may occur shortly after growth induction. The low doses of actinomycin D, which are effective in this inhibition, again

TABLE I

Specific Factors or Enzymes Produced by Lymphocytes Whose Activities Are Increased or Induced by PHA or Other Mitogens

Mediators of cellular immunity		
Mediator	Time after mitogen when first detected (hour)	References
Macrophage migration inhibition factor	6	Thor and Dray (1968) Bennett and Bloom (1968)
Blastogenic factor	4	Kasakura and Lowenstein (1965); Wolstencroft (1971)
Lymphotoxin	2	Kolb and Granger (1968)
Chemotactic factor	24–72	Ward (1970)
Interferon	2–24	Wheelock (1965); Friedman and Cooper (1967)

Specific Enzymes			
Activity	Time of first increase (hour)	Baseline level in resting cells	References
Lactic dehydrogenase	24[a]	+	Rabinowitz and Dietz (1967)
Ornithine decarboxylase	6–14	±	Kay and Cooke (1971)
DNA polymerase	20	±	Loeb et al. (1968); Rabinowitz et al. (1969)
Thymidine kinase	48–72	+	Loeb et al. (1970)
TMP kinase	48–72	+	Loeb et al. (1970)
Deoxycytidine kinase	40–72	+	Pegoraro and Bernengo (1971)
Deoxycytidylate kinase	40–72	+	Pegoraro and Bernengo (1971)
tRNA methylase(s)	38–40	+[b]	Riddick and Gallo (1971)
Uridine kinase	2	+[c]	Kay and Handmaker (1970); Hausen and Stein (1968)
RNA polymerase I	Rapid increase	+[c]	Kay and Cooke (1971); Handmaker and Graef (1970)
RNA polymerase II	Slow increase	+	Kay and Cooke (1971); Handmaker and Graef (1970)
Adenyl cyclase	0.83–1.66	+[c]	Smith et al. (1971a)
Alkaline ribonuclease inhibitor	3	−	Kraft and Shortman (1970)

[a] Alteration in preexisting isozyme pattern.
[b] New class of enzyme may appear.
[c] Preexisting enzyme may be activated.

suggest that this determination is connected in some way with the early enhancement of ribosomal RNA production rather than with messenger RNA synthesis. The final effect, however, might result from failure of new messenger RNA to be stabilized or to reach the cytoplasm.

b. *Mediators.* The so-called mediators of cellular immunity produced by activated lymphocytes include those listed in Table I, and others which have not been listed (for more complete list, see Lawrence and Landy, 1969; Bloom and Glade, 1971). Those shown are all of moderately high molecular weight and in some cases (interferon) are almost certainly proteins. In others, the characterization of the mediator as a protein is less certain. With the exception of interferon and lymphotoxin, the mediators are not detected within the earliest period of stimulation. However, the sensitivity of the functional assays required for these factors may preclude recognition of the earliest time of production.

In each case, no evidence of the existence of mediators can be detected in lysates of unstimulated cells, suggesting *de novo* synthesis of new proteins following growth stimulation. These materials appear to be products related to the differentiated function of the lymphocyte and may be an example of true induction in an animal cell system. At present, however, it is not possible to attribute mediator production to derepression and transcription of specific genes even in cases where it can be shown that the onset of production is dependent upon RNA and protein synthesis. It is as likely in such cases that the apparent induction is due to enhanced use of messenger RNA, which was synthesized continuously in the resting cell. These messenger RNA's may not have been used prior to stimulation, and hence degraded rapidly, due to failure either of transport or of stabilization. The distinction between these possibilities is important in understanding the regulation of function in differentiated animal cells, since the two possibilities place the level of control at different points: Derepression places the control at the level of gene transcription, while enhanced use places the control at the level of protection and transport of messenger RNA to the cytoplasm. Naturally, the two modes of control are not mutually exclusive, and it is probable that both forms occur.

c. *Relevance of Mediator Production to the Mitotic Cycle.* The production of mediators of cellular immunity by lymphocytes following growth stimulation raises the question of the relevance of such production to the entrance and progress of the lymphocyte through the mitotic cycle. It seems most likely that the elaboration of these materials reflects the behavior of the lymphocyte as a differentiated cell and may not represent an essential biochemical step in the sequence leading to cell division. However, the phenomenon of growth response to various agents that bind to the lymphocyte surface may have evolved as a means of facilitating exactly that production of mediators. Mitotic activity itself may be regarded as one component of the differentiated function of the lymphocyte, since it appears that part of the normal immune response is induction of multiplication of a population of T cells that are specifically reactive to the inducing antigen (Burnet, 1959; Sprent and Miller, 1972). This multiplication enhances the degree and specificity of the immune response, which in most cases benefits the host.

Nevertheless, some phenomena can be apparently divorced from the purely mitotic aspects of lymphocyte growth stimulation. Thus, the production of interferon was found to be related less to the state of cell growth than to the presence or absence of PHA in the culture medium (Friedman and Cooper, 1967). Also, its elaboration seems to depend strongly upon the presence of macrophages in PHA cultures, since removal of those cells abolished interferon production, while DNA synthesis was not affected (Epstein *et al.,* 1970).

I. DNA Synthesis

1. *General Considerations*

The hallmark of the lymphocyte mitogenic response, and a primary attraction for investigators, is the onset of DNA synthesis. Elevation of ^{3}H-thymidine incorporation above resting cell levels (expressed either as numerical increment or as ratio, depending on the purposes and persuasion of the investigator) is used widely to measure the degree of stimulation of resting lymphocyte populations by various agents and conditions. A rough calculation indicates that an incorporation of 1000 cpm of ^{3}H-thymidine during 2 hours, using specific activities, labeling, and counting methods currently in vogue, will correspond to about 10^2–10^3 cells engaged in DNA synthesis during the labeling period. Many workers have based conclusions on increases over baseline levels in that range, using cultures containing 10^5–10^6 cells. Thus, evidence of mitogenic activity may be claimed for agents which cause one in 10^4 cells to be in S phase at the time studied, a mere doubling of the baseline level. While this may have considerable immunologic significance and indicates the great sensitivity of the technique, it is doubtful if much biochemical information relating to DNA synthesis can be obtained from such studies.

The use of DNA synthesis as a measurement of lymphocyte activation, despite the advantage of extreme sensitivity, suffers from the disadvantage of requiring a 24–48 hour interval between application of the stimulus and measurement of the effect. As we have seen, numerous phenomena occur rapidly after adding the mitogen. Little information about these early and often essential steps is gained from studies that measure only the incorporation of ^{3}H-thymidine 1–2 days (and sometimes 5–7) after stimulation has begun.

2. *DNA Repair*

The replication of DNA during the mitotic cycle must be distinguished from synthesis of DNA during repair processes. The resting lymphocyte possesses enzymes capable of repairing damage to DNA caused by chemical or physical agents. Their activity is shown by a small stimulation of ^{3}H-thymidine incorporation into DNA following treatment of resting lymphocytes with agents such as nitrogen mustard. The repair enzymes are different from the DNA polymerase involved in DNA replication, since only the latter is inhibited by hydroxyurea and cycloheximide. In fact, in order to lower the background sufficiently to detect the activity of the repair enzymes, the very small numbers of cells normally engaged in DNA replication in the resting cell population must be suppressed by treatment with hydroxyurea (Lieberman *et al.*, 1971a,b).

3. *Drugs Which Affect DNA Synthesis*

Most of the studies that have yielded information about S phase in this system have involved the use of various drugs or chemicals that inhibit DNA synthesis. Some of these agents may be added to lymphocyte cultures together with the mitogen without apparent effect on the initial activation of the lymphocyte or its passage through the prereplicative phase. However, the cells do not synthesize DNA at the appropriate time. Instead, they accumulate at the dividing point between G_1 and S where they remain viable for up to 12 hours. The block may be circumvented by the removal of the inhibitor or by the addition of a required metabolite. The technique is well known as a means of partially synchronizing the mitotic cycles of cells in continuous growth (Mueller, 1971). Agents which have been used in this way are methotrexate (plus adenosine), 5-fluorodeoxyuridine, and thymidine excess (Rueckert and Mueller, 1960; Xeros, 1962; Bootsma *et al.*, 1964; Studzinski and Lambert, 1969). Addition of thymidine reverses the block in the first two cases, while removal of excess thymidine reverses the third. Methotrexate and fluorodeoxyuridine act by inhibiting the endogenous pathway for thymidylate production, while excess thymidine inhibits the production of deoxycytidine. There is no apparent requirement for these pyrimidines by the cell during the resting or prereplicative phases, so all growth processes preceding the onset of S phase seem to occur as usual. The rapidity with which the block may be reversed suggests that the essential early steps leading to production of DNA polymerase are also accomplished. However, it has not been determined whether or not the polymerase is actually synthesized on schedule in the presence of these inhibitors. It would be desirable to have this information because, if polymerase is synthesized on schedule, one might anticipate that the initiating steps in DNA replication may occur as well, only to be aborted when the blocked pyrimidines were required to permit the reaction to continue. In such a system it might be possible to learn more about the process of initiation of DNA replication. Conversely, if DNA polymerase is not synthesized on schedule, it would be rewarding to study the means by which the lack of deoxypyrimidines might inhibit the synthesis of this protein.

4. *Cell Cycle Measurements*

The length of S phase or S + G_2 has been estimated in stimulated lymphocytes using inhibitors to produce partial synchrony. A brief pulse with ^{3}H-thymidine was given at the time of release of the block, to label cells entering S phase. At various times, a 3-hour accumulation of mitotic cells was obtained with colcemide and the number of first-round-labeled mitoses scored with radioautography. The results suggested that, after release of the metabolic block, there was a biphasic movement of lymphocytes through S phase. The minimal length of S + G_2 was 6–9 hours, with a median S + G_2 duration of 16–18 hours (Steffen and Stolzmann, 1969).

Earlier studies with nonsynchronized cells had suggested that S phase lasted 15–20 hours and G_2 lasted 3–5 hours (Kikuchi and Sandberg, 1964; Bianchi and deBianchi, 1965). These values are rather longer than those generally found for human cells. However, the biphasic nature of the curve for appearance of labeled mitoses suggests nonhomogeneity of the population being studied. The earliest wave of mitoses may, indeed, come from cells moving through S phase in about 6–8 hours, followed by 2–3 hour G_2 phase. These values are typical of human cell cycle measurements (Lipkin, 1971). However, the secondary peak suggests that some cells, entering S phase at the time of removal of the metabolic block, proceed through S more slowly, or remain in G_2 for a long period before entering mitosis. The cause of this delay is not known, but artifacts due to cell culture conditions cannot be entirely ruled out. Conflicting reports exist regarding the length of S phase in the second round of mitotic cells. In one study the same biphasic behavior was noted (Steffen and Stolzmann, 1969), while another revealed a uniform wave of mitoses with S-phase duration of about 8–9 hours (Cave, 1966).

VI. Discussion and Conclusions

It is evident that a great many events are initiated nearly simultaneously by the action of mitogens on resting lymphocytes. It also seems apparent that many of these phenomena are not encountered or perceived in studies of continuously growing cells, very probably for the simple reason that they relate particularly to the maintenance of the nongrowing condition and are involved mainly in the orderly conversion to the growing state. These events either do not occur in continuously growing cells, or they occur too rapidly to be detected and studied.

As the stimulated lymphocyte progresses through G_1 toward S, it becomes biochemically less distinct from other growing cell systems in regard to those processes which may relate specifically to the control of cell growth. Thus, after 20 hours of PHA stimulation, the overall aspect of lymphocyte RNA metabolism differed little from that of the HeLa cell (Cooper, 1968; Darnell, 1968). From the viewpoint of biological activity, however, this is evidently not the case, since a variety of factors and mediators, presumably related to immunological activity, are produced by the lymphocyte after partial passage through the mitotic cycle.

A. Sequence of Events in Lymphocyte Activation

The following attempt to organize the many phenomena that have been reviewed is designed to emphasize possible regulatory interrelationships. At worst, this will be merely a speculative exercise; at best, it will suggest specific questions to be resolved. It is unlikely to afford any new insights which have not already occurred to workers in this field and is open to immediate revision upon the appearance of new data.

First, an initial binding event occurs between the mitogen and a specific receptor site on the cell surface. Recognition may depend on specific carbohydrate configurations in either the receptor site or the mitogen, or both.

Second, an alteration of the cell surface occurs, which is likely to be of a steric nature. This alteration activates a great variety of preexisting, inactive transport sites; activates preexisting, inactive enzymes; and causes a redistribution of cellular hydrolyses, probably by altering the lysosomal membrane.

These phenomena seem to occur directly in response to the binding reaction and do not require the synthesis of new enzymes, proteins, or RNA molecules. Only the particular inactive sites in the neighborhood of each mitogen binding point may be activated by binding at that point. Thus, a number of mitogen-binding events may be required fully to activate all of the necessary functions.

Third, as a result of the preceding steps, new active enzymes, substrates, and ionic conditions are provided that accelerate the synthesis and turnover of cell membrane constituents such as phospholipids and glycoproteins.

Phosphorylation of nucleotides is accelerated through activation of previously inactive nucleoside kinases.

A number of nuclear changes occur that parallel these cell membrane changes. These also involve alteration of preexisting proteins by acetylation and phosphorylation, reducing their binding to DNA. The nuclear DNA thereby becomes more easily available to act as template for RNA synthesis. Again, RNA and protein synthesis are not required, the effect stemming from activation of preexisting enzymes. Preexisting RNA polymerase molecules are activated either by a direct action on the enzyme, or through making new template available. Currently unknown cytoplasmic factors are activated that will permit an increase in protein synthesis without the production of new mRNA or rRNA. These may be ribosome dissociation or initiation factors.

All the above phenomena have in common the activation of a preexisting inactive or minimally active system. In virtually every case, the activation may be explained by the conversion of inactive enzyme molecules of various types to active ones. (The term "enzyme" is used in the broad sense of a protein with a specific combining site which alternately binds and releases substrate molecules, leading to some alteration of the state or location of the substrate following its release). The presence of this array of inactive enzymes, and their activation as a result of a steric change in the lymphocyte membrane, I suppose, is the central fact of lymphocyte growth induction by mitogens. This has been previously suggested by others (Quastel *et al.,* 1970, 1972).

Such widespread activation of diverse systems suggests generally nonspecific activation mechanisms.

The most direct activating effect of a change in membrane configuration, requiring no other chemical step, would be simple exposure of sites that were physically blocked by the organization of the cell membrane. Akin to this would be juxtaposition or separation of sites that require propinquity or its lack to permit function. This may be the mode of activation of preexisting transport sites. Subsequent generalized activation of inactive enzymes might be accomplished through equally nonspecific biochemical means, themselves initiated by the above process.

Evidence for two such mechanisms has been given previously. Changes in cAMP content may dissociate enzymes from repressor proteins, and thereby activate protein kinase; this, in turn may activate or inactivate other proteins through phosphorylation. Proteolytic activation of enzyme precursors is a second means.

Fourth, the stage is now set for an increase in all the major cellular metabolic activities. (1) The synthesis of proteins is increased through increased activity of preexisting ribosomes. This may provide increased amounts of a few critical proteins, like ribosomal protective protein, which function in a regulatory capacity. (2) The synthesis of ribosomal RNA, transfer RNA, and probably heterogeneous nuclear RNA is increased through the changes mentioned in Section V,A,3,c. The directly stimulated increase in ribosomal protective protein production permits enhanced survival of new ribosomal RNA, which in turn, is conducive to continued elevation of protein synthesis. Also, new classes of ribosomes may begin to enter the cytoplasm at this time. (3) Turnover and net synthesis of cell membrane constituents is increased (already noted) through direct activation of enzymes governing the relevant pathways. (4) Similar activation of enzymes regulating pathways of glucose utilization produce increased glycolysis and pentose–phosphate pathway activity. Increasing macromolecular synthesis furnishes more enzyme for these pathways, providing for further acceleration of these processes.

Fifth, new messenger RNA's may be used in protein synthesis. These new messages would arise through (1) transcription of previously repressed genetic regions (classical derepression); (2) stabilization, transport to the cytoplasm, and translation of messenger RNA's previously synthesized but rapidly degraded or otherwise not used. This may be accomplished through the activity of new classes of ribosomal particles. Alternatively, specific protective proteins may be produced which achieve the same end. Another mechanism, currently widely investigated, may involve the terminal addition of a polyadenylate sequence to certain messenger RNA molecules (Darnell et al., 1971; Edmonds et al., 1971; Lee et al., 1971). (3) "Masked" messenger RNA that was previously present in the cytoplasm in an inactive form may be activated.

Sixth, certain messenger RNA's may be made early in the course of growth stimulation that are stabilized in the cytoplasm for translation at a later time. In particular, the messenger RNA's for synthesis of DNA polymerase and enzymes involved in thymidine metabolism may be handled in this way.

At this point, probably all the essential changes have occurred which determine the shift of the lymphocyte from the resting state to progress through the mitotic cycle. The next problems are those involved in comprehending the orderly passage of cells through G_1 to S phase. These are not specific to the lymphocyte and will not be considered further in this chapter.

B. Restoration of the Resting State

To conclude, we may consider the restortion of the nongrowing state after a period of mitotic activity. We know only that this does occur in stimulated lymphocytes (Polgar et al., 1968b; Cooper, 1969b) and that the secondary resting cells often remain responsive to mitogens (Ling and Holt, 1967; Cooper, 1969b).

The simplest view of this question is that following mitosis, the cell will return to the resting state if mitogen is not present to cause new surface receptor sites to assume the active configuration, thereby restimulating the daughter cells. Since the daughter cell membranes are synthesized in the presence of whatever mitogen remained in the cell culture, then it follows that any newly formed mitogen receptor sites will be activated as soon as they are competent to participate in the binding reaction. When the mitogen level begins to fall through use or degradation, daughter cells may be produced that have parental sites still activated and new sites which are unbound. Eventually, an insufficient number of sites will be bound to permit reactivation. However, since some subthreshold number of sites may remain bound, restimulation with secondary addition of mitogen may proceed more rapidly. This expectation is consistent with experimental data (Ling and Holt, 1967; Cooper, 1969b). The secondarily resting cell must restore itself to the state where previously active enzyme molecules remain present in an inactive form so that second-round growth stimulation may occur. The mechanisms that have been considered to explain initial activation of these same enzymes may be invoked. Thus, if many of these proteins are produced normally in an inactive form requiring proteolytic activation, then reduction of over-all proteolytic activity might permit nonactivated proteins to accumulate. If such nonactivated proteins were inhibitory to further synthesis of the same protein, then the system would "wind down," but remain primed for further synthesis as soon as proteolysis was resumed.

Similarly, newly synthesized enzyme molecules may require phosphorylation to become active. Here, the level of protein kinase activity, governed by cAMP concentrations, may fall, permitting accumulation of inactive enzyme. Finally, if cAMP is generally inhibitory to lymphocyte growth, the action of protein kinase may be to inactivate various enzymes through phosphorylation. Reactivation would then require either dephosphorylation or partial proteolysis.

References

Abdou, N. I. (1971). Immunoglobulin (Ig) receptors on human peripheral leukocytes. II. Class restriction of Ig receptors. *J. Immunol.* **107**, 1637–1642.
Abell, C. W., Kemp, C. W., and Johnson, L. D. (1970). Effects of phytohemagglutinin and isoproterenol on DNA synthesis in lymphocytes from normal donors and patients with chronic lymphatic leukemia. *Cancer Res.* **30**, 717–723.

Adinolfi, M., Gardner, B., Gianelli, F., and McGuire, M. (1967). Studies on human lymphocytes stimulated *in vitro* with anti-γ and anti-μ antibodies. *Experientia* **23**, 271–272.

Agarwal, S. S., and Loeb, L. A. (1972). Studies on the induction of DNA polymerase during transformation of human lymphocytes. *Cancer Res.* **32**, 107–113.

Alford, R. H. (1970). Metal cation requirements for phytohemagglutinin-induced transformation of human peripheral blood lymphocytes. *J. Immunol.* **404** 698–703.

Asofsky, R., and Oppenheim, J. J. (1966). Protein synthesis by human leukocytes transformed *in vitro*. *J. Reticuloendothel. Soc.* **3**, 373–374.

Attardi, G., and Amaldi, F. (1970). Structure and synthesis of ribosomal RNA. *Annu. Rev. Biochem.* **39**, 183–226.

Attardi, G., Parnas, H., Hwang, M., and Attardi, B. (1966). Giant-size rapidly labeled nuclear ribonucleic acid and cytoplasmic messenger ribonucleic acid in immature duck erythrocytes. *J. Mol. Biol.* **80**, 145–182.

Bach, F., and Hirschhorn, K. (1963). γ-globulin production by human lymphocytes *in vitro*. *Exp. Cell Res.* **32**, 592–595.

Bach, F., and Hirschhorn, K. (1964). Lymphocyte interaction: A potential histocompatibility test *in vitro*. *Science* **143**, 813–814.

Bain, B., and Lowenstein, L. (1964). Genetic studies on the mixed leukocyte reaction. *Science* **145**, 1315–1316.

Bain, B., Vas, M. R., and Lowenstein, L. (1964). The development of large immature mononuclear cells in mixed leukocyte cultures. *Blood* **23**, 108–116.

Barker, B. E., Lutzner, M., and Farnes, P. (1969). Ultrastructural properties of pokeweed-stimulated leucocytes *in vivo* and *in vitro*. In "Proceedings of the Third Leucocyte Culture Conference (W. Rieke, ed.), pp. 587–606. Appleton, New York.

Bender, M. A., and Prescott, D. M. (1962). DNA synthesis and mitosis in cultures of human peripheral leucocytes. *Exp. Cell Res.* **27**, 221–229.

Bennett, B., and Bloom, B. R. (1968). Reactions *in vivo* and *in vitro* produced by a soluble substance associated with delayed type hypersensitivity. *Proc. Nat. Acad. Sci. U. S.* **59**, 756–762.

Bianchi, N. O., and deBianchi, M. (1965). DNA replication sequence of human chromosomes in blood cultures. *Chromosoma* **17**, 273–290.

Biberfeld, P., Biberfeld, G., and Perlmann, P. (1971). Surface immunoglobulin light chain determinants on normal and PHA-stimulated human blood lymphocytes studied by immunofluorescence and electronmicroscopy. *Exp. Cell Res.* **66**, 177–189.

Bloom, B. R., and Glade, P. R., eds. (1971). "*In Vitro* Methods in Cell-Mediated Immunity." Academic Press, New York.

Bootsma, D., Budke, L., and Vos, O. (1964). Studies on synchronous division of HeLa cells initiated by excess thymidine. *Exp. Cell Res.* **33**, 301–309.

Borberg, H., Yesner, I., Gesner, B., and Silber, R. (1968). The effect of N acetyl D-galactosamine and other sugars on the mitogenic activity and attachment of PHA to tonsil cells. *Blood* **31**, 747–757.

Börjeson, J., Reisfeld, R., Chessin, L. N., Welsh, P. O., and Douglas, S. D. (1966). Studies on human peripheral blood lymphocytes *in vitro*. I. Biological and physicochemical properties of the pokeweed mitogen. *J. Exp. Med.* **124**, 859–872.

Burnet, F. M. (1959). "The Clonal Selection Theory of Acquired Immunity." Cambridge Univ., London and New York.

Butcher, R. W. (1968). Role of cyclic AMP in hormone actions. *N. Engl. J. Med.* **279**, 1378–1384.
Cave, M. (1966). Incorporation of tritium-labeled thymidine and lysine into chromosomes of cultured human leukocytes. *J. Cell Biol.* **29**, 209–222.
Chessin, L. N., Börjeson, J., Welsh, P. D., Douglas, S. D., and Cooper, H. L. (1966). Studies on human peripheral blood lymphocytes *in vitro*. II. Morphological and biochemical studies on the transformation of lymphocytes by the pokeweed mitogen. *J. Exp. Med.* **124**, 873–884.
Claman, H. N., and Brunstetter, F. H. (1968). Effects of antilymphocyte serum and phytohemagglutinin upon cultures of human thymus and peripheral blood lymphoid cells. I. Morphologic and biochemical studies of thymus and blood lymphoid cells. *Lab. Invest.* **18**, 757–762.
Cooke, A., Kay, J. E., and Cooper, H. L. (1971). RNA polymerase activity as a measure of RNA synthesis. *Biochem. J.* **125**, 74p.
Cooper, E. H., Barkham, P., and Hale, A. J. (1963). Observations on the proliferation of human leucocytes cultured with phytohemagglutinin. *Brit. J. Haematol.* **9**, 101–110.
Cooper, H. L. (1968). Ribonucleic acid metabolism in lymphocytes stimulated by phytohemagglutinin. II. Rapidly synthesized ribonucleic acid and the production of ribosomal ribonucleic acid. *J. Biol. Chem.* **243**, 34–43
Cooper, H. L. (1969a). Ribosomal ribonucleic acid production and growth regulation in human lymphocytes. *J. Biol. Chem.* **244**, 1946–1952.
Cooper, H. L. (1969b). Ribosomal ribonucleic acid wastage in resting and growing lymphocytes. *J. Biol. Chem.* **244**, 5590–5596.
Cooper, H. L. (1969c). Alterations in RNA metabolism in lymphocytes during the shift from resting state to active growth. *In* "Biochemistry of Cell Division" (R. Baserga, ed.), pp. 91–112. Thomas, Springfield, Illinois.
Cooper, H. L. (1969d). Unpublished data.
Cooper, H. L. (1970a). Early biochemical events in lymphocyte transformation and their possible relationship to growth regulation. *In* "Proceedings of the Fifth Leucocyte Culture Conference" (J. Harris, ed.), pp. 15–30. Academic Press, New York.
Cooper, H. L. (1970b). Control of synthesis and wastage of ribosomal RNA in lymphocytes *Nature (London)* **227**, 1105–1107.
Cooper, H. L. (1972a). Purification of lymphocytes from peripheral blood. *In* "Biomembranes: Cells, Organelles and Membranous Components" (S. Fleischer, L. Packer, and R. Estabrook, eds.). Academic Press, New York (in press).
Cooper, H. L. (1972b). Lymphocyte RNA-labeling with ³H-uridine: correction of data by analysis of UTP pool saturation kinetics. *In* "Proceedings of the Seventh Leukocyte Culture Conference" (F. Daguillard, ed.). Academic Press, New York (in press).
Cooper, H. L., and Gibson, E. M. (1971). Control of synthesis and wastage of ribosomal ribonucleic acid in lymphocytes. II. The role of protein synthesis. *J. Biol. Chem.* **246**, 5059–5506.
Cooper, H. L., and Rubin, A. D. (1965). RNA metabolism in lymphocytes stimulated by phytohemagglutinin: Initial responses to phytohemagglutinin. *Blood* **25**, 1014–1027.
Coulson, A. S., and Chalmers, D. G. (1967). Response of human blood lymphocytes to tuberculin PPD in tissue culture. *Immunology* **12**, 417–429.
Craig, N. C., and Perry, R. P. (1970). Aberrant intranuclear maturation of ribosomal precursors in the absence of protein synthesis. *J. Cell Biol.* **45**, 554–564.

Cross, M. E., and Ord, M. G. (1970). Changes in the phosphorylation and thiol content of histones in phytohemagglutinin-stimulated lymphocytes. *Biochem. J.* **118**, 191–192.
Darnell, J. E. (1968). Ribonucleic acids from animal cells. *Bacteriol. Rev.* **32**, 262–290.
Darnell, J. E., Wall, R., and Tushinski, R. (1971). An adenylic-rich sequence in messenger RNA of HeLa cells and its possible relationship to reiterated sites in DNA. *Proc. Nat. Acad. Sci. U.S.* **68**, 1321–1325.
Desai, L. S., and Foley, G. E. (1970). Studies on the nucleic acids of human lymphocytic cells: Acetylation of histones. *Arch. Biochem. Biophys.* **141**, 552–556.
Douglas, S. D. (1972). Human lymphocyte growth *in vitro*: Morphologic, biochemical and immunologic significance. *Int. Rev. Exp. Pathol.* **10**, 41–114.
Douglas, S. D., and Fudenberg, H. H. (1969). *In vitro* development of plasma cells from lymphocytes following pokeweed mitogen stimulation: A fine structural study. *Exp. Cell Res.* **54**, 277–279.
Douglas, S. D., Hoffman, P. F., Borjeson, H., and Chessin, L. N. (1967). Studies on human peripheral blood lymphocytes *in vitro*. III. Fine structural features of lymphocyte transformation by pokeweed mitogen. *J. Immunol.* **98**, 17–30.
Edmonds, M., Vaughan, M., and Nakazato, H. (1971). Polyadenylic acid sequences in heterogeneous nuclear RNA and rapidly-labeled polyribosomal RNA of HeLa cells: Possible evidence of a precursor relationship. *Proc. Nat. Acad. Sci. U.S.* **68**, 1336–1340.
Elves, M. W., Roath, S., and Israels, M. C. G. (1963a). The response of lymphocytes to antigen challenge *in vitro*. *Lancet* **1**, 806–807.
Elves, M. W., Roath, S., Taylor, G., and Israels, M. C. G. (1963b). The *in vitro* production of antibody lymphocytes. *Lancet* **1**, 1292–1293.
Ennis, H. L. (1966). Synthesis of ribonucleic acid in L cells during inhibition of protein synthesis by cycloheximide. *Mol. Pharmacol.* **2**, 543–557.
Epstein, L. B., Cline, M. J., and Merigan, T. C. (1970). Macrophage-lymphocyte interaction in the PHA-stimulated production of interferon *in vitro*. In "Proceedings of The Fifth Leucocyte Culture Conference." (J. Harris, ed.), pp. 501–513. Academic Press, New York.
Erlichman, J., Hirsch, A. H., and Rosen, O. M. (1971). Interconversion of cyclic nucleotide-activated and cyclic nucleotide independent forms of a protein kinase from beef heart. *Proc. Nat. Acad. Sci. U.S.* **68**, 731–735.
Everett, N. B., and Tyler, R. W. (1970). Quantitative aspects of lymphocyte formation and destruction. In "Formation and Destruction of Blood Cells" (T. Greenwalt and G. Jamieson, eds.), pp. 264–283. Lippincott, Philadelphia, Pennsylvania.
Farnes, P., Barker, B. E., Brownhill, L. E., and Fanger, H. (1964). Mitogenic activity in *Phytolacca americana* (Pokeweed). *Lancet* **2**, 1100–1101.
Fawcett, D. W. (1966). "An Atlas of Fine Structure. The Cell: Its Organelles and Inclusions." pp. 152–155. Saunders, Philadelphia, Pennsylvania.
Fisher, D. B., and Mueller, G. C. (1968). An early alteration in phospholipid metabolism of lymphocytes by phytohemagglutinin. *Proc. Nat. Acad. Sci. U.S.* **60**, 1396–1402.
Fisher, D. B., and Mueller, G. C. (1969). The stepwise acceleration of phosphatidyl choline synthesis in phytohemagglutinin treated lymphocytes. *Biochim. Biophys. Acta* **176**, 316–323.
Fitzgerald, P. H. (1964). The immunological role and long life span of small lymphocytes. *J. Theor. Biol.* **6**, 12–25.

Forbes, I. J., and Henderson, D. W. (1966). Globulin synthesis by human peripheral lymphocytes. *In vitro* measurements using lymphocytes from normals and patients with disease. *Ann. Intern. Med.* **65**, 69–79.

Forbes, I. J., and Turner, K. J. (1965). Synthesis of protein by human lymphocytes *in vitro*. I. Clinical studies. *Australas. Ann. Med.* **14**, 304–310.

Forsdyke, D. R. (1967). Quantitative nucleic acid changes during phytohaemagglutinin-induced lymphocyte transformation *in vitro*. Dependence of the response on phytohemagglutinin/serum ratio. *Biochem. J.* **105**, 679–684.

Forster, J., Lamelin, J. P., Green, I., and Benacerraf, B. (1969). A quantitative study of the stimulation of DNA synthesis in lymph node cell culture by antilymphocyte serum, specific antigen and phytohemagglutinin. *J. Exp. Med.* **129**, 295–313.

Friedman, R. M., and Cooper, H. L. (1967). Stimulation of interferon production in human lymphocytes by mitogens. *Proc. Soc. Exp. Biol. Med.* **125**, 901–905.

Ginsburg, H., Hollander, N., and Feldman, M. (1971). The development of hypersensitive lymphocytes in cell culture. *J. Exp. Med.* **134**, 1062–1082.

Goldberg, M. L., Rosenau, W., and Burke, G. C. (1969). Fractionation of phytohemagglutinin: Purification of the RNA and DNA synthesis-stimulating substances and evidence that they are not proteins. *Proc. Nat. Acad. Sci. U.S.* **64**, 283–289.

Goldstein, I. J., Hollerman, C. E., and Smith, E. E. (1965). Protein-carbohydrate interaction. II. Inhibition studies on the interaction of concanavalin-A with polysaccharides. *Biochemistry* **4**, 876–883.

Gordon, J., and MacLean, L. D. (1965). A lymphocyte-stimulating factor produced *in vitro*. *Nature (London)* **208**, 795–796.

Gould, H. J. (1970). Proteins of rabbit reticulocyte ribosomal subunits. *Nature (London)* **227**, 1145–1147.

Gowans J. L., and MacGregor, D. D. (1965). The immunological activities of lymphocytes. *Progr. Allergy* **9**, 1–78.

Gräsbeck, R., Nordman, C., and de la Chapelle, A. (1963). Mitogenic action of antileukocyte immune serum on peripheral leukocytes *in vitro*. *Lancet* **2**, 385–386.

Greaves, M., and Bauminger, S. (1972). Activation of T and B lymphocytes by insoluble phytomitogens. *Nature (London), New Biol.* **235**, 67–70.

Grierson, D., and Loening, U. E. (1972). Distinct transcription products of ribosomal genes in two different tissues. *Nature (London), New Biol.* **235**, 80–82.

Handmaker, S. D., and Graef, J. W. (1970). The effect of phytohemagglutinin on the DNA-dependant RNA polymerase activity of nuclei isolated from human lymphocytes. *Biochim. Biophys. Acta* **199**, 95–102.

Handmaker, S. D., Cooper, H. L., and Leventhal, B. G. (1969). The kinetics of PHA-stimulation of human lymphocytes. *In* "Proceedings of the Third Leukocyte Culture Conference" (W. Rieke, ed.), pp. 53–67. Appleton, New York.

Hausen, P., and Stein, H. (1968). On the synthesis of RNA in lymphocytes stimulated by phytohaemagglutinin. I. Induction of uridine kinase and the conversion of uridine to UTP. *Eur. J. Biochem.* **4**, 401–406.

Havemann, K., and Rubin, A. D. (1968). The delayed response of chronic lymphatic leukemia leukocytes to phytohemagglutinin *in vitro*. *Proc. Soc. Exp. Biol. Med.* **127**, 668–671.

Hay, E. D., and Revel, J. P. (1963). The fine structure of the DNA component of the nucleus. An electron microscopic study utilizing autoradiography to localize DNA synthesis. *J. Cell Biol.* **16**, 29–51.

Hayden, G. A., Crowley, G. M., and Jamieson, G. A. (1970). Studies on glycoproteins.

V. Incorporation of glucosamine into membrane glycoproteins of phytohemagglutinin-stimulated lymphocytes. *J. Biol. Chem.* **245**, 5827–5832.

Hedeskov, C. J. (1968). Early effects of phytohemagglutinin on glucose metabolism of normal human lymphocytes. *Biochem. J.* **110**, 373–380.

Heller, P., Bhoopalam, N., Yakulis, V. J., and Costea, N. (1971). Kappa and lambda receptor sites on single lymphocytes. *Clin. Exp. Immunol.* **9**, 637–643.

Henney, C. S., and Lichtenstein, L. M. (1971). The role of cyclic AMP in the cytolytic activity of lymphocytes. *J. Immunol.* **107**, 610–612.

Hirschhorn, K., Bach, F., Kolodny, R. L., Firschein, I. L., and Hashem, N. (1963). Immune response and mitosis of human lymphocytes *in vitro*. *Science* **142**, 1185–1187.

Hirschhorn, K., Schreibman, R., Verbo, S., and Gruskin, R. (1964). The action of streptolysin-S on peripheral lymphocytes of normal subjects and patients with acute rheumatic fever. *Proc. Nat. Acad. Sci. U.S.* **52**, 1151–1157.

Hirschhorn, R., Brittinger, G., Hirschhorn, K., and Weissmann, G. (1968). Studies on lysosomes. XII. Redistribution of acid hydrolases in human lymphocytes stimulated by phytohemagglutinin. *J. Cell Biol.* **37**, 412–423.

Hirschhorn, R., Troll, W., Brittinger, G., and Weissmann, G. (1969). Template activity of nuclei from stimulated lymphocytes. *Nature (London)* **222**, 1247–1250.

Hirschhorn, R., Grossman, J., and Weissmann, G. (1970). Effect of cyclic 3',5'-adenosine monophosphate and theophyllin on lymphocyte transformation. *Proc. Soc. Exp. Biol. Med.* **133**, 1361–1365.

Hirschhorn, R., Grossman, J., Troll, W., and Weissmann, G. (1971). The effect of epsilon amino caproic acid and other inhibitors of proteolysis upon the response of human peripheral blood lymphocytes to phytohemagglutinin. *J. Clin. Invest.* **50**, 1206–1217.

Holm, G., and Perlmann, P. (1965). Phytohemagglutinin-induced cytotoxic action of unsensitized immunologically competent cells on allogeneic and xenogeneic tissue culture cells. *Nature (London)* **207**, 818–821.

Holt, L. J., Ling, N. R., and Stanworth, D. R. (1966). The effect of heterologous antisera and rheumatoid factor on the synthesis of DNA and protein by human peripheral lymphocytes. *Immunochemistry* **3**, 359–372.

Hopper, G. D. K., and Cooper, H. L. (1971). Nucleolar proteins in cells exhibiting rRNA wastage. *Fed. Proc., Fed. Amer. Soc. Exp. Biol.* **30**, 1257.

Jacob, F., and Monod, J. (1961a). Genetic regulatory mechanisms in the synthesis of proteins. *J. Mol. Biol.* **3**, 318–356.

Jacob, F., and Monod, J. (1961b). On the regulation of gene activity. *Cold Spring Harbor Symp. Quant. Biol.* **26**, 193–211.

Johnson, L. D., and Abell, C. (1970). The effects of isoproterenol and cyclic adenosine 3',5'-phosphate on phytohemagglutinin stimulated DNA synthesis in lymphocytes obtained from patients with chronic lymphocytic leukemia. *Cancer Res.* **30**, 2718–2723.

Kasakura, S., and Lowenstein, L. (1965). A factor stimulating DNA synthesis derived from the medium of leukocyte cultures. *Nature (London)* **288**, 749–795.

Kay, J. E. (1966). RNA and protein synthesis in lymphocytes stimulated with phytohemagglutinin. *In* "The Biological Effects of Phytohemagglutinin," Report of the Orthopedic Hospital Management Committee, pp. 37–52. Oswestry, England.

Kay, J. E. (1967). Effect of actinomycin on protein synthesis by lymphocytes. *Nature (London)* **215**, 77–78.

Kay, J. E. (1968a). Early effects of phytohemagglutinin on lymphocyte RNA synthesis. *Eur. J. Biochem.* **4**, 225–232.
Kay, J. E. (1968b). Phytohemagglutinin: An early effect on lymphocyte lipid metabolism. *Nature (London)* **219**, 172–173.
Kay, J. E. (1970). The role of the stimulant in the activation of lymphocytes by PHA. *Exp. Cell Res.* **58**, 185–188.
Kay, J. E. (1971a). Interaction of lymphocytes and phytohemagglutinin. Inhibition by chelating agents. *Exp. Cell Res.* **68**, 11–16.
Kay, J. E. (1971b). The binding of PHA to lymphocytes. *In* "Proceedings of The Fourth Leukocyte Culture Conference" (O. McIntyre, ed.), pp. 21–35. Appleton, New York.
Kay, J. E., and Cooke, A. (1971). Ornithine decarboxylase and ribosomal RNA synthesis during the stimulation of lymphocytes by phytohemagglutinin. *FEBS Lett.* **16**, 9–12.
Kay, J. E., and Cooper, H. L. (1969). Rapidly labeled cytoplasmic RNA in normal and phytohemagglutinin-stimulated human lymphocytes. *Biochim. Biophys. Acta* **186**, 62–84.
Kay, J. E., and Handmaker, S. D. (1970). Uridine incorporation and RNA synthesis during stimulation of lymphocytes by PHA. *Exp. Cell Res.* **63**, 411–421.
Kay, J. E., Leventhal, B. G., and Cooper, H. L. (1969). Effects of inhibition of ribosomal RNA synthesis on the stimulation of lymphocytes by phytohemagglutinin. *Exp. Cell Res.* **54**, 94–100.
Kay, J. E., Ahern, T., and Atkins, M. (1971). Control of protein synthesis during the activation of lymphocytes by phytohemagglutinin. *Biochim. Biophys. Acta* **247**, 322–334.
Kedinger, C., Gniazdowski, M., Mandel, J. L., and Chambon, P. (1970). α-amanitin: A specific inhibitor of one of two DNA-dependent RNA polymerase activities from calf thymus. *Biochem. Biophys. Res. Commun.* **38**, 165–171.
Kikuchi, Y., and Sandberg, A. A. (1964). Chronology and pattern of human chromosome replication. I. Blood leukocytes of normal subjects. *J. Nat. Cancer Inst.* **32**, 1109–1144.
Killander, D., and Rigler, R. (1969). Activation of deoxyribonucleoprotein in human leucocytes stimulated by phytohemagglutinin. I. Kinetics of the binding of acridine orange to deoxyribonucleoprotein. *Exp. Cell Res.* **54**, 163–170.
Kincade, P. W., Lawton, A. R., and Cooper, M. D. (1971). Restriction of surface immunoglobulin determinants to lymphocytes of the plasma cell line. *J. Immunol.* **106**, 1421–1423.
Kirchner, H., and Rühl, H. (1970). Stimulation of human peripheral lymphocytes by Zn^{2+} *in vitro*. *Exp. Cell Res.* **61**, 229–230.
Kleinsmith, L., Allfrey, V., and Mirsky, A. E. (1966). Phosphorylation of nuclear proteins early in the course of gene activation in lymphocytes. *Science* **154**, 780–781.
Kolb, W. P., and Granger, G. A. (1968). Lymphocyte *in vitro* cytotoxicity: Characterization of human lymphotoxin. *Proc. Nat. Acad. Sci. U.S.* **61**, 1250–1255.
Kornfeld, R., and Kornfeld, S. (1970). The structure of a phytohemagglutinin receptor site from human erythrocytes. *J. Biol. Chem.* **245**, 2536–2545.
Kornfeld, S. (1969). Decreased phytohemagglutinin receptor sites in chronic lymphatic leukemia. *Biochim. Biophys. Acta* **192**, 542–545.
Kornfeld, S., and Kornfeld, R. (1969). Solubilization and partial characterization of a phytohemagglutinin receptor site from human erythrocytes. *Proc. Nat. Acad. Sci. U.S.* **63**, 1439–1446.

Kraft, N., and Shortman, K. (1970). A suggested control function for the animal tissue ribonuclease-ribonuclease inhibitor system, based on studies of isolated cells and phytohaemagglutinin-transformed lymphocytes. *Biochim. Biophys. Acta* **217**, 164–175.

Kurland, C. G., Voynow, R., Hardy, S. J. S., Randall, L., and Lutter, L. (1969). Physical and functional heterogeneity of *E. coli* ribosomes. *Cold Spring Harbor Symp. Quant. Biol.* **34**, 17–24.

Lamelin, J.-P. (1971). Inhibition of macrophage migration. *In* "Cell-mediated Immunity. In Vitro Correlates" (J. P. Revillard, ed.), pp. 75–102. Karger, Basel.

Landy, M., and Chessin, L. N. (1969). The effect of plant mitogens on humoral and cellular immune responses. *Antibiot. Chemother. (Basel)* **15**, 199–212.

Lawrence, H. S., and Landy, M., eds. (1969). "Mediators of Cellular Immunity." Academic Press, New York.

Lee, S. Y., Mendecki, J., and Brawerman, G. (1971). A polynucleotide segment rich in adenylic acid in the rapidly-labeled polyribosomal RNA component of mouse sarcoma 180 ascites cells. *Proc. Nat. Acad. Sci. U.S.* **68**, 1331–1335.

Leiken, S., Whang-Peng, J., and Oppenheim, J. (1970). *In vitro* transformation of human cord blood lymphocytes. *In* "Proceedings of the Fifth Leukocyte Culture Conference" (J. Harris, ed.), pp. 389–402. Academic Press, New York.

Levis, W. R., and Robbins, J. H. (1970). Effect of glass-adherent cells on the blastogenic response of 'purified' lymphocytes to phytohemagglutinin. *Exp. Cell Res.* **61**, 153–158.

Lieberman, M. W., Baney, R. N., Lee, R. E., Sell, S., and Farber, E. (1971a). Studies on DNA repair in human lymphocytes treated with proximate carcinogens and alkylating agents. *Cancer Res.* **31**, 1297–1306.

Lieberman, M. W., Sell, S., and Farber, E. (1971b). Deoxyribonusleoside incorporation and the role of hydroxyurea in a model lymphocyte system for studying DNA repair in carcinogenesis. *Cancer Res.* **31**, 1307–1312.

Lindahl-Kiessling, K. M., and Böök, J. A. (1964). Effects of phytohemagglutinin on leukocytes. *Lancet* **2**, 591.

Lindahl-Kiessling, K., and Mattson, A., (1971). Mechanism of phytohemagglutinin (PHA) action. IV. Effect of some metabolic inhibitors on binding of PHA to lymphocytes and the stimulatory potential of PHA-pretreated cells. *Exp. Cell Res.* **65**, 307–312.

Lindahl-Kiessling, K. M., and Peterson, R. D. A. (1969). The mechanism of phytohemagglutinin action. II. The effect of certain enzymes and sugars. *Exp. Cell Res.* **55**, 81–84.

Lindell, T. J., Weinberg, F., Morris, P. W., Roeder, R. G., and Rutter, W. J. (1970). Specific inhibition of nuclear RNA polymerase ii by γ-amanitin. *Science* **170**, 447–449.

Ling, N. R. (1968). "Lymphocyte Stimulation." North-Holland Publ., Amsterdam.

Ling, N. R., and Holt, P. J. L. (1967). The activation and reactivation of peripheral lymphocytes in culture. *J. Cell Sci.* **2**, 57–70.

Ling, N. R., and Husband, E. M. (1964). Specific and non-specific stimulation of peripheral lymphocytes. *Lancet* **1**, 363–365.

Ling, N. R., Spicer, E., James, K., and Williamson, N. (1965). The activation of human peripheral lymphocytes by products of staphylococci. *Brit. J. Haematol.* **11**, 421–431.

Lipkin, M. (1971). The proliferative cycle of mammalian cells. *In* "The Cell Cycle and Cancer" (R. Baserga, ed.), pp. 6–26. Dekker, New York.

Littau, V. C., Allfrey, V. G., Frenster, J. H., and Mirsky, A. E. (1964). Active and inactive regions of nuclear chromatin as revealed by electron microscopic autoradiography. *Proc. Nat. Acad. Sci. U.S.* **52**, 93–100.

Lloyd, K. O., and Kabat, E. A. (1968). Immunochemical studies on blood groups. XLI. Proposed structures for the carbohydrate portions of blood group A, B, H, Lewis[a] and Lewis[b] substances. *Proc. Nat. Acad. Sci. U.S.* **61**, 1470–1477.

Loeb, L. A., and Agarwal, S. S. (1971). DNA polymerase: Correlations with DNA replication during transformation of human lymphocytes. *Exp. Cell Res.* **66**, 299–304.

Loeb, L. A., Agarwal, S. S., and Woodside, A. M. (1968). Induction of DNA polymerase in human lymphocytes by phytohemagglutinin. *Proc. Nat. Acad. Sci. U.S.* **61**, 827–834.

Loeb, L. A., Ewald, J. L., and Agarwal, S. S. (1970). DNA polymerase and DNA replication during lymphocyte transformation. *Cancer Res.* **30**, 2514–2520.

Lucas, D. O., Shohet, S. B., and Merler, E. (1971). Changes in phospholipid metabolism which occur as a consequence of mitogenic stimulation of lymphocytes. *J. Immunol.* **106**, 768–772.

Lucas, Z. J. (1967). Pyrimidine nucleotide synthesis: Regulatory control during transformation of lymphocytes *in vitro*. *Science* **156**, 1237–1240.

McDonald, J. W. D. (1971). Lymphocyte and platelet adenylcyclase. *Can. J. Biochem.* **49**, 316–319.

MacHaffie, R. A., and Wang, C. H. (1967). The effect of phytohemagglutinin upon glucose catabolism in lymphocytes. *Blood* **29**, 640–646.

MacKinney, A. A., Stohlman, F., and Brecher, G. (1962). The kinetics of cell proliferation in cultures of human peripheral blood. *Blood* **19**, 349–358.

McIntyre, O. R., and Ebaugh, F. G. (1962). The effect of phytohemagglutinin on leukocyte cultures as measured by P^{32} incorporation in the DNA, RNA and acid soluble fractions. *Blood* **19**, 443–453.

Maini, R. N., Bryceson, A. D. M., Wolstencroft, R. A., and Dumonde, D. C. (1969). Lymphocyte mitogenic factor in man. *Nature (London)* **224**, 43–44.

Mandal, R. K. (1969). RNA synthesis in ascites tumor cells during inhibition of protein synthesis. *Biochim. Biophys. Acta* **182**, 375–381.

Mendelsohn, J., Skinner, A., and Kornfeld, S. (1971). The rapid induction by phytohemagglutinin of increased α-aminoisobutyric acid uptake by lymphocytes. *J. Clin. Invest.* **50**, 818–826.

Möller, G. (1970). Induction on DNA synthesis in human lymphocytes: Interaction between non-specific mitogens and antigens. *Immunology* **19**, 583–598.

Monjardino, J. P. P. V., and MacGillivray, A. J. (1970). RNA and histone metabolism in small lymphocytes stimulated by phytohemagglutinin. *Exp. Cell Res.* **60**, 1–15.

Mueller, G. C. (1971). Biochemical perspectives of the G1 and S intervals in the replication cycle of the animal cell: A study in the control of cell growth. *In* "The Cell Cycle and Cancer" (R. Baserga, ed.), pp. 269–307. Dekker, New York.

Mukherjee, A. B., and Cohen, M. M. (1969). Histone acetylation: Cytological evidence in human lymphocytes stimulated with phytohemagglutinin. *Exp. Cell Res.* **54**, 257–260.

Neiman, P. E., and Henry, P. H. (1971). An analysis of the rapidly synthesized ribonucleic acid of the normal human lymphocyte by agarose-polyacrylamide gel electrophoresis. *Biochemistry* **10**, 1733–1740.

Neiman, P. E., and MacDonnell, D. M. (1970). Studies on the mechanism of increased protein synthesis in human phytohemagglutinin stimulated lymphocytes. In "Proceedings of the Fifth Leukocyte Culture Conference" (J. Harris, ed.), pp. 61-74. Academic Press, New York.

Novogrodsky, A., and Katchalski, E. (1970). Effect of phytohemagglutinin and prostaglandins on cyclic AMP synthesis in rat lymph node lymphocytes. *Biochim. Biophys. Acta* **215**, 291-296.

Novogrodsky, A., and Katchalski, E. (1971a). Lymphocyte transformation induced by concanavalin-A and its reversion by methyl-α-D mannopyranoside. *Biochim. Biophys. Acta* **228**, 759-583.

Novogrodsky, A., and Katchalski, E. (1971b). Induction of lymphocyte transformation by periodate. *FEBS Lett.* **12**, 297-300.

Nowell, P. C. (1960). Phytohemagglutinin: An initiator of mitosis in cultures of normal human leukocytes. *Cancer Res.* **20**, 462-466.

Ono, T., Terayama, H., Takaku, F., and Nakao, K. (1969). Hydrocortisone effect upon the phytohemagglutinin-stimulated acetylation of histones in human lymphocytes. *Biochim. Biophys. Acta* **179**, 214-220.

Oppenheim, J. J. (1968). Relationship of *in vitro* lymphocyte transformation to delayed hypersensitivity in guinea pigs and man. *Fed. Proc., Fed. Amer. Soc. Exp. Biol.* **27**, 21-28.

Oppenheim, J. J., Whang, J., and Frei, E. (1965). Immunologic and cytogenetic studies of chronic lymphatic leukemia cells. *Blood* **26**, 121-132.

Oppenheim, J. J., Leventhal, B. G., and Hersh, E. M. (1968). The transformation of column-purified lymphocytes with nonspecific and specific antigenic stimuli. *J. Immunol.* **101**, 262-270.

Oppenheim, J. J., Rogentine, G. N., and Terry, W. D. (1969). The transformation of human lymphocytes by monkey antisera to human immunoglobulins. *Immunology* **16**, 123-138.

Pachman, L. M. (1967). The carbohydrate metabolism and respiration of isolated small lymohocytes. *In vitro* studies of normal and phytohemagglutinin stimulated cells. *Blood* **30**, 691-706.

Papermaster, B. W., Condie, R. M., Findstad, J., and Good, R. A. (1964). Evolution of the immune response. I. The phylogenetic development of adaptive immunological responsiveness in vertebrates. *J. Exp. Med.* **119**, 105-130.

Pardee, A. B. (1968). Membrane transport proteins. *Science* **162**, 632-637.

Parenti, F., Franceschini, P., Forti, G., and Ceppelini, R. (1966). The effect of phytohemagglutinin on the metabolism and γ-globulin synthesis of human lymphocytes. *Biochim. Biophys. Acta* **123**, 181-187.

Parkhouse, R. M. E., Janossy, G., and Greaves, M. F. (1972). Selective stimulation of IgM synthesis in mouse B lymphocytes by pokeweed mitogen. *Nature (London), New Biol.* **235**, 21-23.

Pastan, I., and Perlman, R. L. (1971). Cyclic AMP in metabolism. *Nature (London), New Biol.* **229**, 5-7

Pauly, J. L., Caron, G. A., and Suskind, R. R. (1969). Blast transformation of lymphocytes from guinea pigs, rats and rabbits induced by mercuric chloride *in vitro*. *J. Cell Biol.* **40**, 847-850.

Pearmain, G., Lycette, R. R., and Fitzgerald, P. H. (1963). Tuberculin-induced mitosis in peripheral blood leukocytes. *Lancet* **1**, 637-638.

Pegoraro, L., and Bernengo, M. G. (1971). Thymidine kinase, deoxycytidine kinase and deoxycytidylate deaminase activities in phytohemagglutinin stimulated human lymphocytes. *Exp. Cell Res.* **68**, 283–290.

Penman, S., Rosbush, M., and Penman, M. (1970). Messenger and heterogeneous nuclear RNA in HeLa cells: Differential inhibition by cordycepin. *Proc. Nat. Acad. Sci. U.S.* **67**, 1878–1885.

Peters, J. H., and Hausen, P. (1971a). Effect of phytohemagglutinin on lymphocyte membrane transport. I. Stimulation of uridine uptake. *Eur. J. Biochem.* **19**, 502–508.

Peters, J. H., and Hausen, P. (1971b). Effect of phytohemagglutinin on lymphocyte membrane transport. 2. Stimulation of "facilitated diffusion" of 3-o-methyl-glucose. *Eur. J. Biochem.* **19**, 509–513.

Piatigorsky, J. (1968). Ribonuclease and trypsin treatment of ribosomes and polyribosomes from sea urchins. *Biochim. Biophys. Acta* **166**, 142–155.

Plotz, P. H., and Talal, N. (1967). Fractionation of splenic antibody forming cells on glass bead columns. *J. Immunol.* **99**, 1236–1242.

Pogo, B. G. T., Allfrey, V. G., and Mirsky, A. E. (1966). RNA synthesis and histone acetylation during the course of gene activation in lymphocytes. *Proc. Nat. Acad. Sci. U.S.* **55**, 805–812.

Polgar, P. R., Foster, J. M., and Cooperband, S. R. (1968a). Glycolysis as an energy source for stimulation of lymphocytes by phytohemagglutinin. *Exp. Cell Res.* **49**, 231–237.

Polgar, P. R., Kibrick, S., and Foster, J. M. (1968b). Reversal of PHA-induced blastogenesis in human lymphocyte cultures. *Nature (London)* **218**, 596–597.

Powell, A. E., and Leon, M. A. (1970). Reversible interaction of human lymphocytes with concanavalin-A. *Exp. Cell Res.* **62**, 315–325.

Quastel, M. R., and Kaplan, J. G. (1970). Lymphocyte stimulation: The effect of ouabain on nucleic acid and protein synthesis. *Exp. Cell Res.* **62**, 407–420.

Quastel, M. R., and Kaplan, J. G. (1971). Early stimulation of potassium uptake in lymphocytes treated with PHA. *Exp. Cell Res.* **63**, 230–233.

Quastel, M. R., Dow, D. S., and Kaplan, J. G. (1970). Stimulation of K^{42} uptake into lymphocytes by phytohemagglutinin and role of intracellular K^+ in lymphocyte transformation. In "Proceedings of the Fifth Leukocyte Culture Conference" (J. Harris, ed.), pp. 97–123. Academic Press, New York.

Quastel, M. R., Wright, P., and Kaplan, J. G. (1972). Potassium uptake and lymphocyte activation: Generality of the effect of ouabain and a model of events at the lymphocyte surface induced by phytohemagglutinin. In "Proceedings of the Sixth Leukocyte Culture Conference" (O. R. Schwarz, ed.), pp. 185–212. Academic Press, New York.

Rabinowitz, Y. (1964). Separation of lymphocytes, polymorphonuclear leukocytes and monocytes on glass columns, including tissue culture observations. *Blood* **23**, 811–828.

Rabinowitz, Y., and Dietz, A. (1967). Genetic control of lactate dehydrogenase adn malate dehydrogenase isozymes in cultures of lymphocytes and granulocytes: Effect of addition of phytohemagglutinin, actinomycin-D or puromycin. *Biochim. Biophys. Acta* **139**, 254–264.

Rabinowitz, Y., McCluskey, I. S., Wong, P., and Wilhite, B. A. (1969). DNA polymerase activity of cultured normal and leukemic lymphocytes. Response to phytohemagglutinin. *Exp. Cell Res.* **57**, 257–262.

Raff, M. C. (1970). Two distinct populations of peripheral lymphocytes in mice distinguishable by immunofluorescence. *Immunology* 19, 637–650.

Raff, R. A., Colot, H. V., Selvig, S. E., and Gross, P. R. (1972). Oogenetic origin of messenger RNA for embryonic synthesis of microtuble proteins. *Nature (London)* 235, 211–214.

Reimann, E. M., Brostrom, C. O., and Corbin, J. D. (1971). Separation of regulatory and catalytic subunits of the cyclic 3',5' adenosine monophosphate-dependent protein kinase(s) of rabbit skeletal muscle. *Biochem. Biophys. Res. Commun.* 42, 187–194.

Riddick, D. H., and Gallo, R. C. (1971). The transfer RNA methylases of human lymphocytes. I. Induction by PHA in normal lymphocytes. *Blood* 37, 282–292.

Rigas, D. A., and Osgood, E. E. (1955). Purification and properties of the phytohemagglutinin of Phaseolus vulgaris. *J. Biol. Chem.* 212, 607–609.

Rigler, R., and Killander, D. (1969). Activation of deoxyribonucleoprotein in human leucocytes stimulated by phytohemagglutinin. II. Structural changes of deoxyribonucleoprotein and synthesis of RNA. *Exp. Cell Res.* 54, 171–180.

Ripps, C., and Hirschhorn, K. (1967). The production of immunoglobulin by human peripheral blood lymphocytes *in vitro*. *Clin. Exp. Immunol.* 2, 377–390.

Robison, G. A., Sutherland, E. W. and Butcher, E. (1971). "Cyclic AMP." Academic Press, New York.

Roeder, R. G., and Rutter, W. J. (1969). Multiple forms of DNA-dependent RNA polymerase in eukaryotic organism. *Nature (London)* 224, 234–237.

Rosenthal, A. S., Davie, J., Rosenstreich, D. L., and Blake, J. T. (1972). Depletion of antibody-forming cells and their precursors from complex lymphoid populations. *J. Immunol.* 108, 279–281.

Rubin, A. D. (1968). Possible control of lymphocyte growth at the level of ribosome assembly. *Nature (London)* 229, 196–197.

Rubin, A. D. (1970). Ribosome biosynthesis in cultured lymphocytes. II. The role of ribosomal RNA production in the initiation and maintenance of lymphocyte growth. *Blood* 35, 708–720.

Rubin, A. D. (1971). Defective control of ribosomal RNA processing in stimulated leukemic lymphocytes. *J. Clin. Inves.* 50, 2485–2497.

Rubin, A. L., Stenzel, K. H., Hirschhorn, K., and Bach, F. (1964). Histocompatibility and immunologic competence in renal homotransplantation. *Science,* 143, 815–816.

Rueckert, R. R., and Mueller, G. C., (1960). Studies on imbalanced growth in tissue culture. I. Induction and consequences of thymidine deficiency. *Cancer Res.* 20, 1584–1591.

Rühl, H., Kirchner, H., and Bochert, G. (1971). Kinetics of the Zn^{2+} -stimulation of human peripheral lymphocytes *in vitro*. *Proc. Soc. Exp. Biol. Med.* 137, 1089–1092.

Sasaki, M. S., and Norman, A. (1966). Proliferation of human lymphocytes in culture. *Nature (London)* 210, 913–914.

Scherrer, K., Spohr, G., Granboulan, N., Morel, C., Grosclaude, J., and Chezzi, C. (1970). Nuclear and cytoplasmic messenger-like RNA and their relation to active messenger RNA in polyribosomes of HeLa cells. *Cold Spring Harbor Symp. Quant. Biol.* 35, 539–554.

Selawry, H. S., and Starr, J. L. (1971a). Protein biosynthesis in the spleen. I. Effect of primary immunization on microsomal and ribosomal function *in vitro*. *J. Immunol.* 106, 349–357.

Selawry, H. S., and Starr, J. L. (1971b). Protein biosynthesis in the spleen. II. Effect of primary immunization on the mechanism of protein synthesis. *J. Immunol.* **106**, 358–363.

Sell, S. (1967). Studies on rabbit lymphocytes *in vitro*. V. Induction of blast transformation with sheep antisera to rabbit IgG subunits. *J. Exp. Med.* **125**, 289–301.

Sell, S., and Gell, P. G. H. (1965). Studies on rabbit lymphocytes *in vitro*. I. Stimulation of blast transformation with an antiallotype serum. *J. Exp. Med.* **122**, 423–440.

Sell, S., Rowe, D. S., and Gell, P. G. H. (1965). Studies on rabbit lymphocytes *in vitro*. III. Protein, RNA and DNA synthesis by lymphocyte cultures after stimulation with staphylococal filtrate, with antiallotype serum and with heterologous antiserum to rabbit whole serum. *J. Exp. Med.* **122**, 823–839.

Shortman, K., Williams, N., Jackson, H., Russell, P., and Byrt, P. (1971). The separation of different cell classes from lymphoid organs. IV. The separation of lymphocytes from phagocytes and antibody-forming cells. *J. Cell Biol.* **48**, 566–579.

Smith, J. W., Steiner, A. L., Newberry, W. M., and Parker, C. W. (1971a). Cyclic adenosine 3′,5′-monophosphate in human lymphocytes. Alterations after phytohemagglutinin stimulation. *J. Clin. Invest.* **50**, 432–448.

Smith, J. W., Steiner, A. L., and Parker, C. W. (1971b). Human lymphocyte metabolism. Effects of cyclic and non-cyclic nucleotides on stimulation by phytohemagglutinin. *J. Clin. Invest.* **50**, 442–448.

Sprent, J., and Miller, J. F. A. P. (1972). Activation of thymus cells by histocompatibility antigens. *Nature (London), New Biol.* **234**, 195–198.

Steffen, J. A., and Stolzmann, W.M. (1969). Studies on *in vitro* lymphocyte proliferation in cultures synchronized by the inhibition of DNA synthesis. I. Variability of S plus G2 periods of first generation cells. *Exp. Cell. Res.* **56**, 453–460.

Stewart, C. C., and Breitner, J. (1969). Quantitative measurement of hemagglutinating, leucoagglutinating, and mitogenic activities of PHA. *In* "Proceedings of the Third Leukocyte Culture Conference" (W. Rieke, ed.), pp. 31–39. Appleton, New York.

Strobo, J. D., Rosenthal, A. S., and Paul, W. E. (1972). Functional heterogeneity of murine lymphoid cells. I. Responsiveness to and surface binding of concanavalin-A and phytohemagglutinin. *J. Immunol.* **108**, 1–17.

Studzinski, G. P., and Lambert, W. C. (1969). Thymidine as a synchronizing agent. I. Nucleic acid and protein formation in synchronous HeLa cultures treated with excess thymidine. *J. Cell. Physiol.* **73**, 109–118.

Takahashi, T., Ramachandramurthy, P., and Liener, I. E. (1967). Some physical and chemical properties of a phytohemagglutinin isolated from Phaseolus vulgaris. *Biochim. Biophys. Acta* **133**, 123–133.

Thor, D. E., and Dray, S. (1968). A correlate of human delayed hypersensitivity: Specific inhibition of capillary tube migration of sensitized human lymph node cells by tuberculin and histoplasmin. *J. Immunol.* **101**, 51–61.

Tiollais, P., Gailibert, F., and Boiron, M. (1971). Evidence for the existence of several molecular species in the "45s fraction" of mammalian precursor RNA. *Proc. Nat. Acad. Sci. U.S.* **68**, 1117–1120.

Tokuyasu, K., Madden, S. C. and Zeldis, L. J. (1968). Fine structural alterations of interphase nuclei of lymphocytes stimulated to growth activity *in vitro*. *J. Cell Biol.* **39**, 630–660.

Torelli, U. L., Henry, P. H., and W sman, S. M. (1968). Characteristics of the RNA synthesized *in vitro* by the normal human small lymphocyte and the changes induced by phytohemagglutinin stimulation. *J. Clin. Invest.* **47**, 1083–1095.

Tormey, D. C., and Mueller, G. C. (1965). An assay for the mitogenic activity of phytohemagglutinin preparations. *Blood* **26**, 569–578.

Traut, R. R., Delius, H., Ahmad-Zadeh, C., Bickle, T. A., Pearson, P., and Tissières, A. (1969). Ribosomal proteins of *E. coli*. Stoichiometry and implications for ribosome structure. *Cold Spring Harbor Symp. Quant. Biol.* **34**, 25–38.

Ward, P. A. (1970). Neutrophil chemotactic factors and related clinical disorders. *Arthritis Rheum.* **13**, 181–186.

Watkins, W. M. (1966). Blood-group substances. *Science* **152**, 172–181.

Weber, T., Nordman, C. T., and Gräsbeck, R. (1967). Separation of lymphocyte-stimulating and agglutinating activities in phytohemagglutinin (PHA) from *Phaseolus vulgaris*. *Scand. J. Haematol.* **4**, 77–80.

Wheelock, E. F. (1965). Interferon-like virus-inhibitor induced in human leukocytes by phytohemagglutinin. *Science* **149**, 310–311.

Wigzell, H. (1970). Specific fractionation of immunocompetent cells. *Transplant. Rev.* **5**, 76–104.

Willems, M., Penman, M., and Penman, S. (1969). The regulation of RNA synthesis and processing in the nucleolus during inhibition of protein synthesis. *J. Cell Biol.* **41**, 177–187.

Wolstencroft, R. A. (1971). Lymphocyte mitogenic factor in relation to mediators of cellular immunity. *In* "Cell-mediated Immunity. *In Vitro* Correlates" (J. P. Revillard, ed.), pp. 130–153. Karger, Basel.

Xeros, N. (1962). Deoxyriboside control and synchronization of mitosis. *Nature (London)* **194**, 682–683.

Young, R. J. (1968). The fractionation of quaternary ammonium complexes of nucleic acids. Evidence for heterogeneity of ribosomal ribonucleic acids. *Biochemistry* **7**, 2263–2272.

8

Intestinal Cytodynamics: Adductions from Drug Radiation Studies

RONALD F. HAGEMANN and S. LESHER

I. Introduction 195
II. Nature of the Crypt Progenitor Cell 196
III. Factors Influencing Intestinal Tolerance to Cytotoxins 200
IV. Control Aspects of Intestinal Cell Renewal 205
V. Summary 213
    References 215

I. Introduction

The epithelial cells lining the adult mammalian small intestine exemplify a steady-state cell renewal system. The progenitor compartment is contained in the crypts of Lieberkühn, wherein cell replication occurs (Leblond and Stevens, 1948). Approximately two-thirds of the distance from the base of the crypt a rather abrupt transition occurs; proliferation ceases and differentiative activity commences. As the cells migrate from the crypt to the base of the villi, differentiation to the mature, striated border, columnar absorbing cell is completed. These serve as fully functional cells until the villous apex is attained, whereupon they are exfoliated into the intestinal lumen and degraded. In the mouse, the size of the various compartments, as well as cell fluxes between them, is known (Hagemann et al., 1970a,b,c).

The kinetic interaction between elements catenated to form a steady-state system may be analyzed by instituting a transient perturbation at some point and examining the subsequent compartmentalized events that lead either to reestablishment of preexisting conditions or do not, in which case failure of the entire system may ensue. The intestinal epithelium at steady state provides such an opportunity. The unidirectional progression of cells from the circumscribed crypt proliferative region to the villous extrusion zone represents an orderly series of compartments conducive to perturbational analysis. Fortunately, certain agents (of which various drugs and ionizing radiation may be considered class representatives) are known to exhibit selective efficacy for proliferating cells. These may be used to affect a single vulnerable intestinal compartment with reasonable specificity, after which a systems analysis provides insight as to the nature of events normally occurring at steady state.

In this chapter, we wish to consider the intestinal response to cytotoxic agents from the standpoint of the basic biology of this *in vivo* renewal system. The following will be examined: (1) What constitutes the intestinal stem cell? (2) Which cytokinetic factors are the determinants of system survival (or failure)? (3) What is the nature of the feedback loop controlling cell proliferation at steady state?

II. Nature of the Crypt Progenitor Cell

We may consider a crypt progenitor cell (CPC) as a cell having the capacity to repopulate, through successive division, an intestinal crypt. It is entirely analogous to a stem cell, which is usually considered to possess the capacity for unlimited divisions. The CPC, however, is bound by the constraints of *in vivo* proliferative control mechanisms and cannot divide indefinitely for that reason. Radiation survival curves for mammalian intestinal crypts have been obtained (Hagemann *et al.,* 1971). The curves are characterized by a board shoulder followed by an exponential decrease in survival with increasing radiation dose. The existence of an exponential portion on the radiation survival curve implies that, in this dose region, surviving crypts originate from a single surviving cell on a probability basis. It has been reported that, in the mouse, the number of cells in the proliferative compartment is about 160 per crypt (Lesher and Bauman, 1969; Hagemann *et al.,* 1970c). If a single surviving CPC can repopulate a crypt, and if it is *assumed that* there are 160 CPC per crypt, the number of surviving CPC can be calculated for any given exposure (Withers and Elkind, 1969) and the number of surviving crypts measured for that exposure. One would anticipate no crypt attrition until an average of between one and three surviving CPC per crypt is reached (from Poisson statistics). In the actual analysis, this prediction is amply borne out (Hagemann *et al.,* 1971). Thus, the assumption that there are about 160 CPC per crypt appears substantiated. This indicates that the number of CPC approximates that of proliferative

8. DRUG AND RADIATION STUDIES

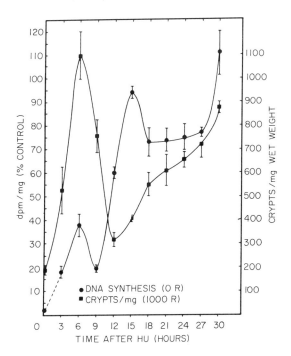

Fig. 1. Intestinal crypt survival following irradiation given at various times after partial synchronization with hydroxyurate (HU) (squares). DNA synthesis indicates relative number of cells in the S phase (circles).

crypt cells. Furthermore, since the majority of cells in the crypt are proliferative, the evidence suggests that this cell population constitutes the CPC compartment.

Hydroxyurea is a cycle-specific drug, lethal to cells in the S phase of the cell cycle (Sinclair, 1967), which has been used to partially synchronize cells *in vitro* (e.g., Sinclair, 1968) and *in vivo* (Mauro and Madoc-Jones, 1969; Hagemann and Lesher, 1971a). In the intestine, the drug (1.5 gm/kg body weight) blocks entry of G_1 cells into S for about 2–3 hours; a second dose is equally effective. It is well known that proliferating mammalian cells exhibit marked alterations in radiosensitivity as a function of cell age (i.e., position in the cell cycle). The effect of irradiation at various times after partial synchronization of the proliferative compartment (obtained by two injections of hydroxyurea separated by 2 hours), then, should indicate whether or not the CPC is in cycle. The results are shown in Fig. 1. The age response following partial synchronization with hydroxyurea suggests that the CPC is, in fact, an actively proliferating cell. This may be examined in an alternative manner using split-dose irradiation. In this instance, the first

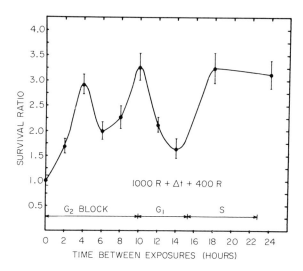

Fig. 2. Intestinal crypt survival following split-dose irradiation (relative to a single total dose). Definite progression effects are apparent.

exposure is chosen on the exponential portion of the single dose survival curve, and is followed at various times thereafter by a second exposure. Typically, in the case of actively proliferating cells, there is an initial rapid rise in survival as the interval between exposures is lengthened (due to re-

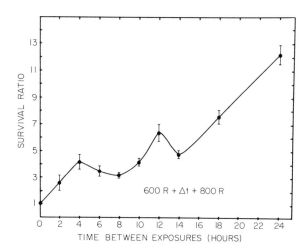

Fig. 3. Similar to Fig. 2 except for a smaller initial exposure that results in a significant residual multiplicity following the first dose. Progression effects are markedly damped.

pair of sublethal damage; Elkind et al., 1967), followed by fluctuations due to progression of cells, surviving the first exposure, through phases of the cell cycle each with characteristic radiosensitivity (Elkind and Whitemore, 1967). Split dose survival for the CPC is illustrated in Fig. 2. Both repair and progression components of the curve are evident. Thus the CPC progresses through its cell cycle in the interval between exposures as detected radiobiologically. A lower first exposure may be chosen so that a residual CPC multiplicity greater than one will result. In this instance, progression effects would be expected to be damped out because of variations in cell cycle transit of individual cells. In Fig. 3, an initial exposure (600 R) was employed, which results in about 20 surviving CPC per crypt. Progression effects are evident, but appreciably less marked than in Fig. 2. These split dose radiation studies demonstrating CPC progression through the cell cycle in the interval between exposures, provide additional evidence that CPC are normally in the division cycle.

The foregoing suggests that a cell in the crypt proliferative compartment is a CPC. If this is indeed the case, then an alteration in the size of the form-

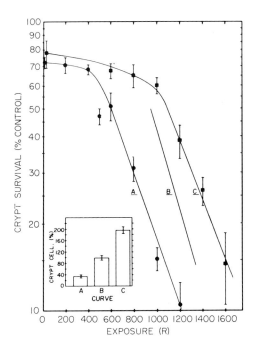

Fig. 4. Radiation survival of crypts of varying multiplicity of proliferative cells at the time of exposure. The insert indicates relative proliferative cellularity existing at the time of exposure for each survival curve.

er should be reflected in the number of the matter; crypt survival should be consonant with CPC per crypt existing at the time of irradiation. The number of proliferative cells per crypt may be altered by a conditioning radiation exposure (Hagemann et al., 1972). This is shown in Fig. 4. In these studies, an initial radiation exposure of 900 R was followed at 1 day (low proliferative cellularity per crypt) or 3 days (high cellularity) by a graded series of exposures. A control survival curve (no prior irradiation) is given for comparison. Estimates of proliferative cellularity per crypt existing at the time of the graded series of exposures is shown in the insert. Alterations in proliferative cellularity per crypt are attended by similar alterations in crypt survival and, by inference, by CPC per crypt.

The foregoing suggests that the stem cell (CPC) compartment of the mammalian small intestine is indistinct from the proliferative compartment. Thus, each crypt cell in the division cycle has the potential for entirely repopulating that crypt. In addition, recent evidence suggests that a surviving crypt (originating from a single CPC) can, after a time, undergo replication (Cairnie, 1971; Hagemann, 1972). This probably occurs by budding and eventual separation to form a new crypt (Cairnie, 1971). Hence, even after damage sufficiently severe to result in crypt attrition, the normal crypt-to-villus ratio may eventually be reestablished. The consequence of these findings in terms of the intestinal response to antiproliferative agents will be considered in the next section.

III. Factors Influencing Intestinal Tolerance to Cytotoxins

The long-term function of the lining of the small bowel is principally to absorb essential substances from the chyme, secrete various digestive aids, act as a barrier against bacterial invasion, and to prevent entrance of noxious and loss of vital materials. One notes that these activities are all within the physiological realm of the mature, striated, columnar villous cell. The majority of cytotoxic agents act at the level of the crypt proliferative cell, which is subsequently translated into a functional deficit via a depletion of villous epithelial cells. As already pointed out, surviving CPC's can completely repopulate a crypt and, given sufficient time, crypts can "replicate." Thus the temporal course of events attending damage to the proliferative compartment becomes extremely important. It follows that with respect to intestinal injury, one must be concerned mainly with the acutely critical function of villous cells, i.e., the maintenance of an intact barrier epithelium. If the system as a whole survives, the villous epithelium will have been restored prior to manifestations of nutritional imbalance. In the case of irradiation, this acute interval occurs within 5 days of exposure for many mammals. Hence the $LD_{50/5}$ days is commonly used as an indicator for failure primarily of the intestinal renewal system (Quastler, 1956).

In the normal animal there exists a given number of crypts for each villus. The product of cell production rate per crypt and crypts per villus determines the cell input rate into the villous compartment (see Hagemann et al., 1970c). At steady state, this is a constant value and equals the rate of cell exfoliation at the villous apex. Cytotoxic agents such as hydroxyurea and relatively low doses of radiation result in partial depletion of the CPC (proliferative) compartment. This is followed by a compensatory hyperplasia of surviving CPC, which rapidly repopulates and even overpopulates the crypt (Lesher and Lesher, 1970; Hagemann and Lesher, 1971b; *vide post*). As the level of initial damage is increased, e.g., by repeated doses of hydroxyurea, some crypts will be left with no surviving CPC and crypt attrition results. Failure of the system (i.e., denudation of villous epithelium) occurs when the crypt-to-villus ratio is sufficiently reduced so that the rate of cell production per surviving crypt, although proceeding maximally, is not adequate to counterbalance villous cell loss. From this, it may be adduced that a rather discrete level of crypt survival should be compatible with survival of the system (meaning animal as well) (Hagemann et al., 1969). Figure 5 shows crypt survival curves for irradiation with fission neutrons, X-rays, and ^{60}Co γ-rays. The dashed lines indicate the level of crypt

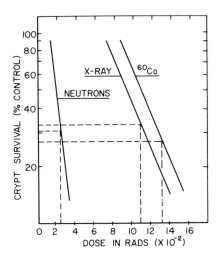

Fig. 5. Crypt survival curves for X-rays, ^{60}Co γ-rays, and fission spectrum neutrons. The dashed lines intersect the abcissa at the $LD_{50/5}$ dose for each radiation. The intersect of these lines with the ordinate indicates the level crypt survival at the $LD_{50/5}$ dose.

survival at the animal $LD_{50/5}$ dose. Clearly, the $LD_{50/5}$ radiation dose corresponds to a reasonably narrow range of crypt survival.

Crypt survival is naturally a consequence of proliferative cell (CPC) survival, and events which affect the latter will affect the former. However, CPC are of practical use only after having reformed a proliferative unit (crypt). For example, two surviving CPC in one crypt would not be as effective in total cellular repopulation as would one CPC in each of two crypts. One can destroy about 1397 of the 1400 CPC surrounding each villus (in the mouse; Hagemann et al., 1970c, 1971) before reducing crypt survival to the critical 30% level (approximately three per villus). None-

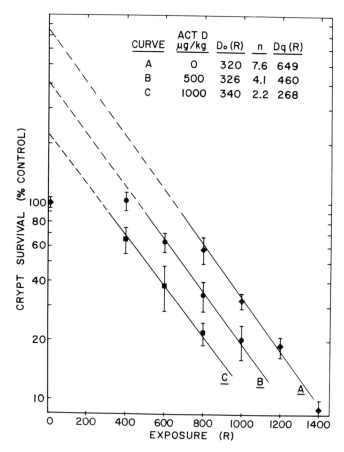

Fig. 6. Crypt survival curves following X-irradiation and actinomycin D (Act D). The drug was given immediately after exposure and crypt survival measured 4 days thereafter.

theless, crypt survival is a direct consequence of CPC survival, and thus proliferative cytotoxins may have profound intestinal influence once the CPC level is sufficiently reduced. It is noteworthy that (predictable?) interactions between different agents may occur. For example, a single injection of hydroxyuria cannot cause crypt attrition, simply because of the multiplicity of CPC per crypt and their random distribution through the cell cycle. It is known that cells in the late S phase are most resistant to irradiation (Gilete et al., 1970; Hagemann and Lesher, 1971a), hence the survivors of a radiation dose on the exponential portion of the CPC survival curve will be those cells that were in S at the time of exposure. If, however, one injects hydroxyurea immediately after irradiation, a marked decrease in the $LD_{50/5}$ is observed (Hagemann and Lesher, 1971a). In those cases wherein the independent mechanisms of action of agents is known, such interactions may be anticipated. The synergism observed between hydroxyurea and radiation could be negated by giving the drug several hours after irradiation when the surviving middle-late S phase cells had moved into G_2.

It has been well documented that crypts surviving cytotoxic agents such as hydroxyurea or irradiation undergo a marked compensatory hyperplasia

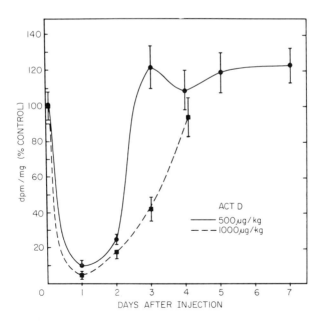

Fig. 7. Pulse tritiated thymidine incorporation by jejunum as a function of time after injection of actinomycin D (Act D). The ^{3}H-thymidine was given at daily intervals and the animals sacrificed 30 minutes later.

(Lesher and Lesher, 1970; Hagemann and Lesher, 1971b). In the case of irradiation at least, this compensatory response must be well underway within about 3 days of exposure to be effective. Subsequent increases occur too late to ameliorate epithelial denudation (Hagemann et al., 1971). This suggests that accelerated cell production per surviving crypt, as well as the crypt-to-villus ratio, is an important aspect in determining the efficacy of intestinal cytotoxins. We have investigated the interaction between actinomycin D and X-radiation on the intestine. The effect of the drug on crypt survival when given immediately after irradiation is shown in Fig. 6. As in cultured cells (Elkind et al., 1968), the drug primarily reduces the width of the shoulder. The finding relevant to the present discussion, however, is that the $LD_{50/5}$ occurs at a level of crypt survival of about 70% instead of the usual approximately 30%. This suggests that the drug has deleterious action in addition to decreasing CPC survival. The affect of actinomycin D on intestinal DNA synthesis was measured. Two amounts of actinomycin D were

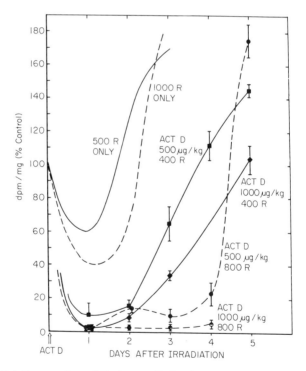

Fig. 8. Relative number of S phase cells in the jejunum as a function of time after X-irradiation only, or irradiation plus actinomycin D (Act D). The drug is seen to delay the onset of compensatory proliferative hyperplasia.

injected into different groups of mice at the beginning of the experiment and ^{3}H-thymidine was injected at daily intervals. Thirty minutes after ^{3}H-thymidine, the animals were killed and incorporated radioactivity determined by liquid scintillation spectrometry. The results are shown in Fig. 7. The drug causes a marked reduction in net DNA synthesis; recovery commences about 2 days later. A parallel experiment was performed wherein labeled nuclei (LN; S phase cells) per crypt were measured under identical conditions. The curves for LN/crypt were found to closely follow those for dmp/mg* following actinomycin D. This indicates that the drug is affecting cell proliferation (i.e., the number of S phase cells per crypt) and is not, in this instance, a late manifestation of inhibited DNA synthesis per se (Baserga et al., 1966). The data in Fig. 8 show that the interruption of intestinal cell proliferation is nearly additive to that resulting from irradiation. Hence, the drug is seen to delay the commencement of the compensatory proliferative response to irradiation, which, if it is to be effective, must occur within about 3 days (vide infra). From this we may surmise that agents which retard the onset of rapid cell proliferation, which normally follows partial depletion of the CPC compartment, will result subsequently in a critical shortage of functional villous cells. This may occur at levels of crypt survival greatly exceeding about 30%, in those instances wherein the agent interferes with the compensatory hyperplasia of surviving crypts. It again stresses the importance of cell input rate per villus (the product of production rate per crypt and crypts per villus) within the critical 5-day posttreatment interval.

From the foregoing, it appears that the intestine combats the effects of cytotoxic agents by accelerated proliferation of surviving CPC's which results, some days later, in a rapid repopulation of the villous epithelium. The efficacy of this process is a direct function of the crypt-to-villus ratio (which reflects the crypts with at least one surviving CPC), and the compensatory proliferative response of each surviving crypt. Treatments, which adversely affect either or both of these parameters, may be expected to inflict severe damage to the intestinal cell renewal system.

IV. Control Aspects of Intestinal Cell Renewal

The rapid compensatory cell proliferation observed in crypts surviving treatment with cytotoxic agents may be viewed as a temporary failure of the normal control mechanisms governing this process. Since proliferative compartment kinetics subsequently return to steady-state levels, one may surmise

---

* dpm/mg = disintegrations per minute per milligram.

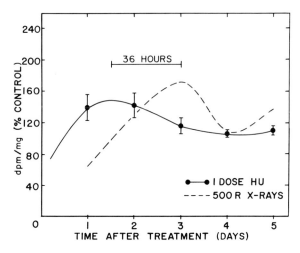

Fig. 9. A comparison between the jejunal proliferative response elicited by a single dose of hydroxyurea (HU) and 500 R of X-radiation.

that the defect is not intracellular (i.e., is not caused by the cell's indifference to a control influence that is still present). This provides, in part, the rationale for a compartmentalized analysis of the intestinal cell renewal system following perturbation by antiproliferative agents, with an eye toward the elucidation of controlling factors operational at steady state.

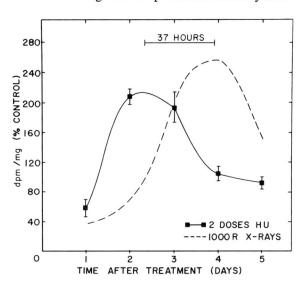

Fig. 10. A comparison between the jejunal proliferative responses elicited by 2 doses of hydroxyurea (HU) given 8 hours apart and 1000 R of X-radiation.

8. DRUG AND RADIATION STUDIES

We have found that cell killing by drugs as well as irradiation results in intestinal compensatory hyperplasia. In each case, the extent is dependent on the degree of cell killing (Hagemann, 1972). It is evident that a temporary breakdown in proliferative control occurs, which is manifest as supra-control levels of proliferative activity, and is dependent on the degree of damage. The proliferative reaction to hydroxyurea is compared with that occurring following doses of X-radiation, which results in a similar maxima of response, in Figs. 9 and 10. The two doses of hydroxyurea were separated by 8 hours for maximum effectiveness in total cell killing. In each instance, the two curves are similar in shape but clearly out of phase. The peak in proliferative activity following X-irradiation occurs about 36 hours later than the peak for hydroxyurea. This suggests that the breakdown in the

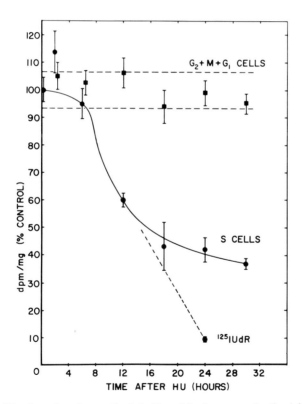

Fig. 11. Kinetics of cycle-specific lethality of hydroxyurea in the jejunum. Cells were prelabeled with ^{3}H-thymidine and either 30 minutes (S cells) or 6 hours (G_2 + M + G_1 cells) subjected to hydroxyurea (HU). The curve labeled S cells contains a component of reused ^{3}H as evidenced by the point labeled ^{125}I-iododeoxyuridine (^{125}IUdR) (which is not appreciably reused).

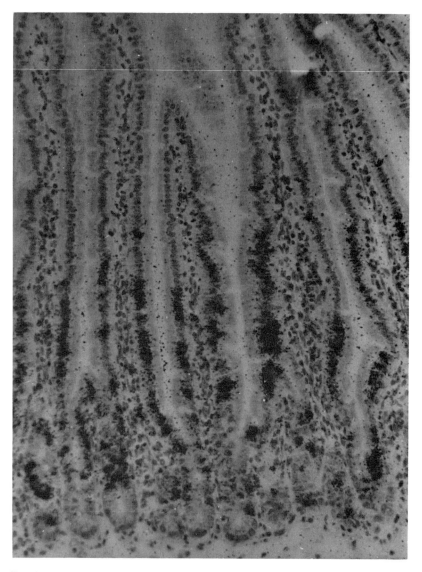

Fig. 12. Photomicrograph of a radioautograph of cells labeled with ^{125}I-iododeoxyuridine. The cells were labeled with the precursor, 30 minutes later irradiated with 1200 R, and 24 hours later the sample was taken. Note a fairly discrete leading edge of migrating cells, the vast majority of which are lethally damaged with respect to reproductive integrity.

normal control mechanism occurs earlier in the case of hydroxyurea than for irradiation.

The kinetics of hydroxyurea cytotoxicity *in vivo* may be assessed by following the tissue content of proliferative cells prelabeled in S phase with ^{3}H-thymidine or ^{125}I-iododeoxyuridine (to minimize reuse of label; Heiniger *et al.*, 1971) and exposed to the drug in either S or the remaining portions of the cell cycle. This is illustrated in Fig. 11. The well-established cycle-specific lethality (Sinclair, 1967) is apparent in this *in vivo* system as is the partial reuse of ^{3}H-thymidine from disintegration cells (Heiniger *et al.*, 1971). For the present purposes, the salient feature of the curves is this: In excess of 90% of the S phase cells, lethally damaged by the drug, die and disintegrate *in situ*. Radiation-induced lethality may be analyzed in a similar experiment. In the case of this agent, more than 50% of the proliferative cells, lethally damaged with respect to clonogenic potential, nevertheless enter the villous compartment (Hagemann, 1972), as illustrated by the autoradiograph in Fig. 12. In this instance, S cells were labeled with ^{125}I-iododeoxyuridine and exposed to 1200 R (which results in a clonogenic fraction of about 3×10^{-3}; Withers and Elkind, 1970); the autoradiograph was prepared from jejunum 24 hours thereafter. It is apparent that the majority of cells have migrated to the villus and exhibit a distinct leading edge of traversing cells. Both cells in S at the time of irradiation (labeled) and those in the more radiosensitive G_2, M, and G_1 phases (Gillette *et al.*, 1970; Hagemann and Lesher, 1971a; unlabeled cells) have ascended the villus. It follows directly that the villous compartment will be depleted earlier in the case of hydroxyurea (after which the lethally damaged cells disintegrate *in situ*) than in the case of irradiation. Quantitatively, this difference would be expected to approximate one villus transit time, which, in the mouse, is about 30–32 hours (Fry *et al.*, 1963). Returning to Figs. 9 and 10, it is seen that the measured difference in peak proliferative compensatory response is about 36 hours. These findings strongly suggest that the villous compartment exerts a regulative influence on cell division in the crypt.

The relationship between villous cellularity and rate of crypt cell proliferation may be examined in an alternative fashion. The size of the villous compartment can be decreased by reducing the crypt-to-villus ratio instead of cell production rate per crypt *(vide infra)*. This is readily accomplished by X-irradiation (Hagemann *et al.*, 1972), as shown in Fig. 13. An exposure of 1100 R initially reduces the crypt-to-villus ratio from 8.7 to 2.7 (Hagemann *et al.*, 1971). This is followed by gradual crypt repopulation. Cell proliferation per surviving crypt is indicated as well. Clearly, crypt cell replication is maximal when the crypt-to-villus ratio is least; as the ratio returns toward steady-state levels, proliferative activity diminishes. A similar pat-

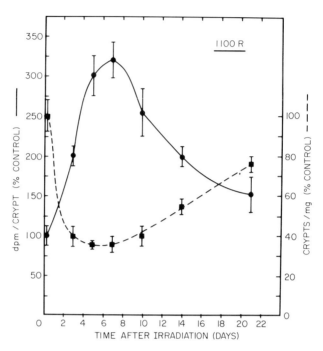

Fig. 13. Relative number of S cells per crypt (circles) and crypts/mg jejunum (squares) as a function of time after irradiation with 1100 R. Note reciprocal relationship.

tern emerges following irradiation of a crypt population partially synchronized by hydroxyurea (Hagemann et al., 1971). In this case, intervals of maximal CPC sensitivity (i.e., lowest crypt-to-villus ratios) result in the maximal proliferative response per surviving crypt. This is again indicative of a negative feedback loop, from villus to crypt, controlling intestinal cell proliferation.

We have investigated the causative factors involved in intestinal cell migration from crypt to villous apex to see if this would provide a clue as to the nature of villus-to-crypt interaction. If cell extrusion at the end of the villous effects a pulling influence on successively deeper cells, then cell migration should continue in spite of cessated cell division in the crypt. This was examined in two ways. First, cell division in the crypt was markedly reduced, or completely blocked, by successive relatively small exposures of X-radiation. The movement of ^{3}H-thymidine-labeled cells from the crypt was followed by periodic radioassay of crypts. The loss of radioactivity per crypt (exodus of labeled cells) was found to continue when cell division was slowed or blocked by small successive X-ray exposures. It appears that cells continue to egress the crypts, for a time, when cell division is severely abated. Second, cells were labeled in the crypts and allowed to migrate approximately three-fourths of the way up the sides of the villi. At this time, cell division was abruptly interrupted in the crypts by either a large exposure of X-rays or injection of colchicine. Intestinal radioassay then indicated the temporal course of cell attrition at the villous apexes, as shown in Fig. 14. Again, cell migration is apparent despite the temporary absence of crypt cell division. From these results, there seems to exist an influential pathway in the direction of villus to crypt. Further work will be required to decide if

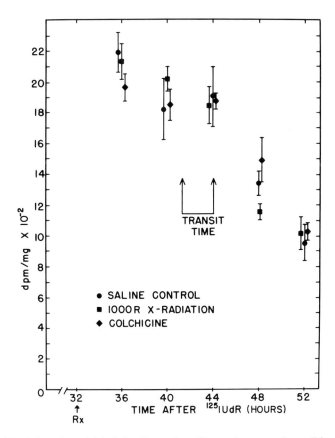

Fig. 14. Migration of labeled cells on the villus under normal conditions or when crypt cell division has been abruptly cessated (32 hours after labeling) by either 1000 R of X-radiation or colchicine. Exfoliation occurs at about the same time irrespective of treatment.

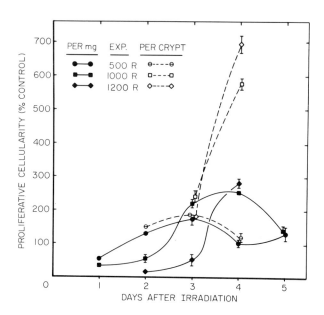

Fig. 15. Temporal course of changes in proliferative cellularity per milligram jejunum and per crypt following varied radiation exposures. The maximum response per crypt occurs at exposures sufficient to reduce the crypt-to-villus ratio.

this is related to control of intestinal cell proliferation or represents a spurious finding in this regard.

It is conceivable that the size of the crypt proliferative compartment itself affects cell proliferation rate owing to recognition of a "critical mass." The data in Fig. 15 show crypt proliferative activity as a function of time after various X-ray exposures. It is apparent that proliferative activity per single crypt mirrors that per milligram wet weight intestine for exposures which do not result in crypt attrition. The maximal rate per crypt occurs when the crypt-to-villus ratio is reduced (1000 and 1200 R), a condition that results in maximal depletion of the villous compartment. In addition, surviving crypts are capable of attaining extremely large proportions without cessation of proliferative activity (see also Fig. 13). It is unlikely, therefore, that a critical crypt cell mass exists that, when surpassed, depresses cell division. It is possible that the region of the crypt immediately above the proliferative zone (crypt differentiating compartment) constitutes a region innately inhospitable to cell division (e.g., via a chemical gradient). However, autoradiographic analysis of this region several days after irradiation clearly indicates upward extension of the proliferative compartment, which may even encompass the base of the villus (Lesher and Bauman, 1969). The CPC could contain a limiting amount of some substance (e.g., messenger) that, when exhausted, negates further replication. This seems unlikely in view of the fact that following drug or radiation treatment surviving CPC remain in the proliferative compartment and undergo numerous divisions.

At the crypt level, it is likely that cell production is governed by (1) the size of the proliferative compartment and (2) the duration of the generation cycle (Tg). During the compensatory response to both drugs and irradiation, a marked increase in the size of the CPC compartment, as well as a shortening of Tg, occurs (Lesher, 1967). Under conditions of reduced cell division, such as observed in the geriatric animal, these directions are reversed. Regulation, at the CPC level, is instituted primarily between mitosis and DNA synthesis. Since a mature villous cell has the DNA content of a postmitotic G_1 cell (Quastler and Sherman, 1959; Baserga, 1965; Hagemann et al., 1970c), the decision as to whether to remain in the proliferative compartment, or to differentiate after mitosis, probably occurs soon after division. In the former case, the cell enters G_1 proper and biochemically prepares, *inter alia,* for a succeeding round of DNA replication. In the latter, the cell hypertrophies in the functional direction and irreversibly exits the proliferative compartment. These decisions may be symetrical or asymetrical (Cairnie et al., 1965; Thrasher, 1970). Both shortening of the cell cycle, such as seen following irradiation (Lesher, 1967), as well as lengthening, as observed in aged animals (Lesher et al., 1961), is primarily a consequence of the duration of the G_1 phase. This implies that in the normal intestine, the duration of S, G_2, and M are relatively static (and proceeding at a near maximal rate), whereas reaction rates in G_1 are regulated and, given the appropriate stimulus, may either accelerate or deccelerate according to the demand for cell production. We may, therefore, adduce that the villous compartment regulates CPC proliferation both by influencing the differentiative decision of cells shortly after mitosis, and by governing the duration of the G_1 phase of the cell cycle.

V. Summary

In the intestine, antiproliferative agents act at the level of the crypt progenitor cell (CPC). Lethally damaged CPC (with respect to reproductive integrity) may either disintegrate *in situ* or ascend the villus, depending on the nature of the cytotoxic agent. In the former case, depletion of the functional villous epithelial compartment will occur relatively soon after treatment; in the latter, depletion occurs later. A reduction in the villous compartment size, a consequence of continued apical cell loss coupled with diminished cell input rate, influences the surviving CPC. This is manifest by a reduction in the duration of the cell cycle (primarily G_1) and by an enhanced predelection to remain in the proliferative cycle. These two processes cause a marked, rapid expansion of the CPC compartment and, as a subsequent consequence, restoration of villous epithelial cellularity. As the villous compartment is repopulated, negative feedback influences on the CPC are reinstated and the duration of the cell cycle and CPC compartment size return to steady-state levels.

In general, the intestinal cell renewal system is not overly sensitive to proliferative cytotoxins (compare, for instance, the hematopoietic system) for three reasons, two of which are morphological.

(1) The sizable multiplicity of CPC per crypt (e.g., 160 in the mouse) any one of which surviving treatment, can entirely repopulate its crypt, indicates that the efficacy of an agent is not directly proportional to its lethal action on the CPC. Rather, it is a reflection of the affect upon the last surviving CPC per crypt. On a probability basis, this CPC will have been in a state most resistant to the particular agent. By way of example, a single injection of hydroxyurea cannot entirely deplete crypts of CPC due to their multiplicity and distribution throughout the cell cycle. The crypts will be repopulated by resistant cells in G_2, M, and G_1 at the time of treatment. This may be contrasted with the situation following irradiation, in which the surviving CPC are likely to be late S phase cells. One could predict then, that hydroxyurea would be extremely effective, in terms of intestinal damage, when given immediately after radiation exposure. This logical interaction is realized experimentally (Hagemann and Lesher, 1971a).

(2) There is a multiplicity of crypts per villus (8.7 in the mouse). In those cases wherein surviving CPC are unaffected in replicative capacity, substantially less than this number is required to ensure maintenance of the villous epithelial barrier. Following treatment with cytotoxic agents, the villous compartment becomes depleted due to continued cell loss in the face of diminished cell input rate. This becomes critical when cellular discontinuities arise, resulting in intracellular gaps that permit unregulated bidirectional movement of elements between intestinal lumen and blood. Only about 2–3 surviving CPC per villus are sufficient to obviate epithelial denudation.

(3) Surviving CPC have the capacity to greatly accelerate cell production by shortening Tg and by remaining in cycle, which occur in response to partial depletion of the differentiated villous compartment. This results in a marked expansion of the size of the proliferative compartment, and, subsequently, in an increased cell input into the differentiated compartment when measured on a per crypt basis. The efficacy of this proliferative response in circumventing epithelial denudation depends on the total cell production per villus, i.e., the product of cell production rate per crypt and crypts per vil-

lus, and on the temporal course of these events. Thus, there is a requisite number of cells that must be fed into the differentiated compartment, and this must occur prior to epithelial denudation to be effective. We have suggested that animal survival following the acute phase of intestinal damage is dependent on continued maintenance of an essentially intact epithelial barrier, rather than absorptive efficacy. Hence, the adaptive response of the differentiated compartment to partial cell depletion (shortening of villi and progression from columnar to squamous cell morphology) delays epithelial denudation, and so provides a temporal extension for the compensatory response by the crypt proliferative compartment.

From the foregoing, it should be possible to predict, in many cases, the outcome of administering antiproliferative agents insofar as the intestine is concerned, and perhaps more importantly, possible synergistic interactions between agents whose cellular action is known. The aspects to be considered are: (1) CPC survival (including dose response, cycle specificity, and intracellular repair of damage); (2) the crypt-to-villus ratio (i.e., crypt survival); and (3) the extent and temporal course of compensatory hyperplasia resulting from partial depletion of the functional compartment. These are the constituants of the *sine qua non* determinant of the outcome of intestinal damage caused by proliferative cytotoxins, *viz.,* total epithelial cell production per villus within approximately 4 days of exposure to the agent.

Acknowledgments

This work was supported in part by AEC Contracts AT(30-1)-4247 and AT(11-1)-3098 and NCI Grants 1P02 CA10438-02 and 5T01 CA05224-03. The technical assistance of J. Reiland, G. Thomas, and D. Benko is gratefully recognized.

References

Baserga, R. (1965). The relationship of the cell cycle to tumor growth and control of cell division: A review. *Cancer Res.* **25**, 581–595.

Baserga, R., Estensen, R. D., and Petersen, R. O. (1966). Delayed inhibition of DNA synthesis in mouse jejunum by low doses of Actinomycin D. *J. Cell Physiol.* **68**, 177–184.

Cairnie, A. B. (1971). Intestinal cell renewal. *19th Annu. Meet., Radiat. Res. Soc.* p. 45.

Cairnie, A. B., Lamerton, L. F., and Steel, G. G. (1965). Cell proliferation studies in the intestinal epithelium of the rat. Theoretical aspects. *Exp. Cell Res.* **39**, 539–553.

Elkind, M. M., and Whitmore, G. F. (1967). "The Radiobiology of Cultured Mammalian Cells," p. 243. Gordon & Breach, New York.

Elkind, M. M., Kamper, C., Moses, W. B., and Sutton-Gilbert H. (1967). Sublethal-lethal radiation damage and repair in mammalian cells. *Brookhaven Symp. Biol.* **20**, 135–160.

Elkind, M. M., Sakamoto, K., and Kamper, C. (1968). Age-dependent toxic properties of Actinomycin D and x-rays in cultured Chinese hamster cells. *Cell Tissue Kinet.* **1**, 209–221.

Fry, R. J. M., Lesher, S., Kisieleski, W. E., and Sacher, G. (1963). *In* "Cell Proliferation in the Small Intestine" (L. E. Lamerton and R. J. M. Fry, eds.), p. 213. Blackwell, Oxford.

Gillette, E. L., Withers, H. R., and Tannock, I. F. (1970). The age sensitivity of epithelial cells of the mouse small intestine. *Radiology* **96**, 639–643.

Hagemann, R. F. (1972). On control of intestinal cell proliferation. In preparation.

Hagemann, R. F., and Lesher, S. (1971a). Intestinal crypt survival and total and per crypt levels of proliferative cellularity following irradiation: Age response and animal lethality. *Radiat. Res.* **47**, 159–167.

Hagemann, R. F., and Lesher, S. (1971b). Irradiation of the G. I. tract: Compensatory response of stomach, jejunum and colon. *Brit. J. Radiol.* **44**, 599–602.

Hagemann, R. F., Sigdestad, C. P., and Lesher, S. (1969). Radiation $LD_{50/5}$ and its relation to surviving intestinal stem cells. *Int. J. Radiat. Biol.* **16**, 291–292.

Hagemann, R. F., Sigdestad, C. P., and Lesher, S. (1970a). A method for quantitation of proliferative intestinal mucoral cells on a weight basis: Some values for C57BL/6. *Cell Tissue Kinet.* **3**, 21–26.

Hagemann, R. F., Sigdestad, C. P., and Lesher, S. (1970b). Quantitation of intestinal villus cells on a weight basis: Values for C57BL/6 mice. *Amer. J. Physiol,* **218**, 637–640.

Hagemann, R. F., Sigdestad, C. P., and Lesher, S. (1970c). A quantitative description of the intestinal epithelium of the mouse. *Amer. J. Anat.* **129**, 41–52.

Hagemann, R. F., Sigdestad, C. P. and Lesher, S. (1971). Intestinal crypt survival and total and per crypt levels of proliferative cellularity following irradiation: Single x-ray exposures. *Radiat. Res.* **46**, 533–546.

Hagemann, R. F., Sigdestad, C. P. and Lesher, S. (1972). Intestinal crypt survival and total and per crypt levels of proliferative cellularity following irradiation: Role of crypt cellularity. *Radiat. Res.* **50**, 583–591.

Heiniger, H. J., Friedrich, G., Feinenclegen, L. E., and Cantelmo, F. (1971). Reutilization of 5-^{125}I-iodo-2'-deoxyuridine and ^{3}H-thymidine in regenerating livers of mice. *Proc. Soc. Exp. Biol. Med.* **137**, 1381–1384.

Leblond, C. P., and Stevens, C. E. (1948). The constant renewal of the intestinal epithelium in the albino rat. *Anat. Rec.* **100**, 357–78.

Lesher, S. (1967). Compensatory reactions in intestinal crypt cells after 300 Roentgens of cobalt-60 gamma irradiation. *Radiat. Res.* **32**, 510–519.

Lesher, S., and Bauman, J. (1969). Cell kinetic studies of the intestinal epithelium: Maintenance of the intestinal epithelium in normal and irradiated animals. *Nat. Cancer Inst., Monogr.* **30**, 185–198.

Lesher, S., and Lesher, J. (1970). Effects of single-dose, whole-body, ^{60}Co gamma irradiation on number of cells in DNA synthesis and mitosis in mouse duodenal epithelium. *Radiat. Res.* **43**, 429–438.

Lesher, S., Fry, R. J. M., and Kohn, H. (1961). Aging and the generation cycle of intestinal epithelial cells in the mouse. *Gerontologia* **5**, 176–181.

Mauro, F., and Madoc-Jones H. (1969). Age response to x-radiation of murine lymphoma cells synchronized *in vivo*. *Proc. Nat. Acad. Sci. U.S.* **63**, 686–691.

Quastler, H. (1956). The nature of intestinal radiation death. *Radiat. Res.* **4**, 303–320.
Quastler, H., and Sherman, F. G. (1959). Cell population kinetics in the intestinal epithelium of the mouse. *Exp. Cell Res.* **17**, 420–438.
Sinclair, W. K. (1967). Hydroxyurea: Effects on Chinese hamster cells grown in culture. *Cancer Res.* **27**, 297–308.
Sinclair, W. K. (1968). Cyclic x-ray responses in mammalian cells *in vitro*. *Radiat. Res.* **33**, 620–643.
Thrasher, J. (1970). The relationship between cell division and cell specialization in the mouse intestinal epithelium. *Experientia* **26**, 74–76.
Withers, H. R., and Elkind, M. M. (1969). Radiosensitivity and fractionation response of crypt cells of mouse jejunum. *Radiat. Res.* **38**, 598–613.
Withers, H. R., and Elkind, M. M. (1970). Microcolony survival assay for cells of mouse intestinal mucosa exposed to radiation. *Int. J. Radiat. Biol.* **17**, 261–267.

9

The Effects of Antitumor Drugs on the Cell Cycle

JOSEPH HOFFMAN and JOSEPH POST

I. Introduction	219
II. Antitumor Drugs	221
A. Alkaloids	221
B. Alkylating Agents	225
C. Antifols	228
D. Antibiotics	230
E. Steroids	231
F. Pyrimidine Analogs	233
G. Miscellaneous Compounds	238
III. General Discussion	239
IV. Summary	242
References	242

I. Introduction

The spectacular successes of chemotherapy in the management of antimicrobial infections have provided a therapeutic model that has stimulated an intensive search for effective agents in the control of neoplastic disease. Many compounds were designed to arrest cells during the replication cycle. It is of interest that while the synthesis of these antimetabolites and their clinical application have been in progress for years, studies of their actions on cell multiplication are relatively recent. This sequence of events has come about because, in many instances, the biochemistry essential to the development of antimetabolites has been known, but the cell cycle kinetics and the interrelated metabolic sequences involved were still to be worked out. Indeed, the study of the replication of human tumor cells is less than 10 years old.

The background tools for the investigation of cell cycle kinetics began when Doniach and Pelc (1950) reported the invention of autoradiography. This was followed by the publication of Howard and Pelc (1953) in which the replication of the plant form, *Vicia faba,* was studied using ^{32}P as the macromolecular label. The cell cycle was divided into four separate intervals: postmitotic gap (G_1) ⟶ DNA synthesis (S) ⟶ postsynthetic, or premitotic gap (G_2) ⟶ mitosis (M). Subsequently, Quastler and Sherman (1959) reported on the proliferation of the mouse intestinal crypt cells *in vivo* using tritiated thymidine (^{3}H-TdR). This DNA precursor provided the high resolution of labeling that permitted the study of cell replication *in vitro* and *in vivo*. Since then the investigations of the cell cycle have been refined by the introduction of radioactive macromolecular precursors of RNA as well as of protein.

The syntheses of DNA, RNA, and protein comprise the major metabolic events of the cell cycle, and they involve a complex series of interdependent and interrelated biochemical reactions leading to the duplication of cellular constituents preparatory to cell division. At each step in the cell cycle those enzymes that are involved in macromolecular synthesis are formed in increased amounts. The levels of activity of thymidine kinase, dihydrofolate reductase, deoxycytidylate deaminase, and thymidylate synthetase have all been related to the particular cell cycle intervals (Baserga, 1968). The timing of the occurrence of these synthetic steps in the phases of the cycle have been made more precise by the use of synchronization techniques *in vitro*. Most of the studies of the effects of cytotoxins on the cell cycle have also been done *in vitro,* and thus the locus of action of an agent may be defined with considerable accuracy.

A useful method for cell study has been that of synchronization by colchicine metaphase arrest followed by release and the introduction of the cytotoxin. One can then study cell killing as well as the rate of entry into S and the next mitotic interval. Recently, it has been found that the removal of isoleucine from the medium will stop cell replication in G_1 and induce a stationary phase from which cells may be induced to exit by the addition of isoleucine (Tobey and Ley, 1971). Another synchronization method used may be that of contact-inhibition arrest and release by medium change (Todaro *et al.,* 1965).

In contact-inhibited stationary or lag phase (G_1) 3T3 cells, e.g., RNA synthesis goes on at a low level (one-tenth that of log phase cells). The addition of fresh medium to stimulate lag phase growth is followed within 30 minutes by a tenfold increase in RNA synthesis. Protein synthesis occurs at a low level in stationary phase cells. Within 2 hours of the medium change this too increases fivefold by 4 to 6 hours. The level falls to a plateau but remains elevated during the next 36 hours. DNA synthesis begins about

9. ANTITUMOR DRUGS ON THE CELL CYCLE

16–18 hours after the medium change and continues for 6–7 hours. It is interesting that the "commitment" to DNA synthesis occurs within a few minutes after exposure to medium change, long before the actual event begins (Todaro et al., 1965). In the system of the Chinese hamster cells synchronized with Colcemid, RNA synthesis increased following mitosis and reached a plateau after 1 hour. A second increase was found 3 hours later, about 1 hour after the beginning of DNA synthesis. The second increase in RNA synthesis could be prevented by blocking DNA synthesis (Klevecz and Stubblefield, 1967). In contact-inhibited W1-38 human fibroblasts, which were induced to replicate by change of medium, a biphasic increase in RNA synthesis could be observed. This was associated with an early synthesis of acidic nuclear proteins (Farber et al., 1971). Later studies have shown that these proteins are synthesized in greatest abundance during late G_1 and reach a peak just before DNA synthesis begins, following which they decline. The acidic nuclear proteins differ from the histone polypeptides, which are synthesized during S and are dependent on the synthesis of DNA. Gel electrophoresis of the acidic nuclear proteins has been reported to show different profiles in the G_1, S, and G_2 portions of the cycle (Stein and Brown, 1972). The synthesis and turnover of these proteins undergo cell cycle stage-specific changes (Rovera and Baserga, 1971). These findings have suggested that they may have regulatory roles in the control of cell replication.

Antimetabolites have been important in the delineation of the metabolic events of the cell cycle and, in turn, these biochemical studies have suggested ways of interrupting the progress of cells through the cycle. Dosage plays an important role, since in all instances these agents may be lethal for cells at appropriate concentrations. The description of the various agents and their influence on replication kinetics have been considered according to the generic class of the chemical compound. However, in Table I the information has been summarized within the framework of the cell cycle phases. Since new compounds are constantly being developed, data will be included only from studies of the more commonly investigated cytotoxins. It is to be emphasized that the action of many of these compounds may be only partly defined at present and that future studies may reveal additional effects. Furthermore, the interruption of the cycle at one point may have effects on other parts of the cycle.

II. Antitumor Drugs

A. ALKALOIDS

The two most important compounds in this group, vincristine and vinblastine, are derived from extracts of the periwinkle plant, *vinca rosea* Linn

TABLE I
Summary of Effects of Some Antitumor Drugs on Cell Cycle

Phase of cell cycle	Chemical class	Drug	Metabolic reaction	In vivo	In vitro	References
Mitosis	Antibiotic	Mitomycin C	Alkylation		√	Thomson and Biggers (1966)
	Amino acid analog	p-Fluorophenyl-alanine	Inhigition of protein synthesis		√	Sisken and Iwasaki (1969)
	Enzyme	L-Asparaginase	Inhibition of RNA synthesis		√	Becker and Broome (1967)
G_1	Nitrosourea	1, 3-bis (2-Chloro-ethyl)-1-nitro-sourea	Alkylation	√		Wheeler et al. (1970b); DeVita (1971); DeVita et al. (1972)
	Antibiotic	Actinomycin D (small dose)	Inhibition of DNA synthesis	√		Baserga et al. (1965); Frankfurt (1968); Kim et al. (1968)
	Antibiotic	Puromycin	Inhibition of protein synthesis		√	Baserga (1968)
	Antibiotic	Cycloheximide	Inhibition of DNA, RNA and protein synthesis	√		Estensen and Baserga (1966); Kim et al. (1968); Verbin and Farber (1967)
	Steroid	Hydrocortisone	Inhibition of DNA synthesis		√	Kollmorgen and Griffin (1969)
	Analog of hydroxyl-amine	Hydroxyurea	Inhibition of DNA synthesis		√	Sinclair (1965)
G_2	Mustard	HN_2	Presumably by alkylation (ileum)	√		Hoffman and Post (1972)
	Mustard	Cyclophosphamide	Alkylation	√		Hampton (1967); Layde and Baserga (1964); Wheeler (1967); Wheeler and Alexander (1969); Wheeler et al. (1970a)

9. ANTITUMOR DRUGS ON THE CELL CYCLE

	Class	Drug	Mechanism	Phase	References
	Nitrosourea	1,3-bis (2-Chloroethyl)-1-nitrosourea	Alkylation	√	Wheeler et al. (1970b)
	Antibiotic	Actinomycin D (large dose)	Inhibition of DNA synthesis	√	Tobey et al. (1966)
	Antibiotic	Puromycin D (large dose)	Inhibition of protein synthesis	√	Estensen and Baserga (1966); Tobey et al. (1966)
	Antibiotic	Cycloheximide	Inhibition of protein synthesis	√	Verbin and Farber (1967); Verbin et al. (1969)
	Antibiotic	Daunomycin	Complex with DNA. Inhibition of DNA synthesis and DNA-dependent RNA synthesis	√	Brehaut (1969); Wilkoff et al. (1971)
	Antifol	Methotrexate	Inhibition of DNA synthesis	√	Dalen et al. (1965); Hoffman and Post (1972)
	Steroid	Hydrocortisone	Undetermined	√	Hoffman and Post (1972)
	Pyrimidine analog	Cytosine arabinoside	Inhibition of DNA synthesis	√	Brewen (1965); Karon and Benedict (1970); Karon and Shirakawa (1969, 1970)
	Pyrimidine analog (halogenated)	Iodo-deoxyuridine	Inhibition of DNA synthesis	√	Post and Hoffman (1969a)
	Mustard	HN$_2$	Presumably by alkylation (spleen)	√	Hoffman and Post (1972); Levis et al. (1964)
	Nitrosourea	1,3-bis (2-Chloroethyl)-1-nitrosourea	Alkylation	√	DeVita (1971); DeVita et al. (1972)
	Antibiotic	Cycloheximide	Inhibition of protein synthesis	√	Verbin et al. (1969)
S	Antifol (antimetabolite)	Methotrexate	Inhibition of DNA synthesis by blocking of deoxyuridylate methylation	√	Dalen et al. (1965); Hoffman and Post (1972); Margolis et al. (1971); Stock (1966a)

TABLE I Cont'd.

Phase of cell cycle	Chemical class	Drug	Metabolic reaction	In vivo	In vitro	References
	Steroid	Hydrocortisone	Inhibition of DNA synthesis	√		Frankfurt (1968)
	Analog of hydroxylamine	Hydroxyurea	Inhibition of DNA synthesis		√	Sinclair (1965)
	Pyrimidine analog	Cytosine arabinoside	Inhibition of DNA synthesis by (1) Blocking of deoxycytidine formation (2) Inhibition of DNA polymerase		√ √	Chu and Fisher (1962); Karon and Shirakawa (1969, 1970); Kimball and Wilson (1968)
	Pyrimidine analog (halogenated)	Iododeoxyuridine	Inhibition of DNA synthesis	√		Post and Hoffman (1969a)

(Baserga, 1968; Henderson, 1969; Stock, 1966b). Their mechanisms of action and their effects on the cell cycle are discussed in detail elsewhere in this volume (Taylor, Chapter 2).

B. ALKYLATING AGENTS

Several members of this group are used clinically. They include nitrogen mustard (HN_2), chlorambucil, cyclophosphamide, thio-TEPA, melphalan, busulfan, and the nitrosoureas (Ochoa and Hirschberg, 1967; DeVita, 1971; DeVita *et al.*, 1972; Ochoa, 1969).

The mustards are compounds that contain at least one β-chloroethyl group attached to a sulfur or nitrogen atom. They are classified as monofunctional or polyfunctional depending on the number of alkyl groups present. Alkylation is effected through a cyclic sulfonium ion or a cyclic immonium ion, which reacts at a negative or nucleophilic center (Wheeler, 1962). Nitrogen mustard has two alkyl radicals and acts by alkylating DNA at the nitrogen atom in the 7 position on the imidazole ring of the purine, guanine. Cross-linking of the DNA strands occurs and this is presumably the chemical reaction responsible for the biological effects of this family of drugs.

In vitro, HN_2 did not interfere with the progression of human heteroplastic epitheliallike cells (H. Ep. No. 2) through G_1 and into S but did retard their advancement to mitosis. Cells initially in G_2 when HN_2 was introduced into the culture continued to mitosis (Wheeler *et al.*, 1970a). Wheeler and Alexander (1969) reported similar results with HN_2 and cyclophos-

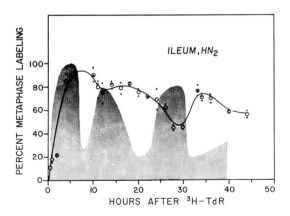

Fig. 1. Ileum, after HN_2. ^{3}H-TdR, 1μCi/gm body weight. Closed circles, individual rat; open circles mean for the group. The shaded background is the control. (From Hoffman and Post, 1972.)

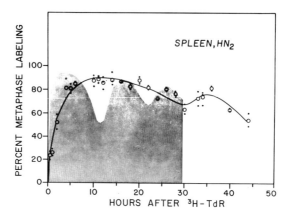

Fig. 2. Spleen, after HN_2. ^{3}H-TdR, 1 μCi/gm body weight. Symbols as in Fig. 1. The shaded background is the control. (From Hoffman and Post, 1972.)

phamide *in vivo* with hamster plasmacytomas. Experiments *in vivo* with Ehrlich ascites tumor cells also demonstrated that HN_2 blocked cell replication in the premitotic phase (Layde and Baserga, 1964). Hampton (1967) compared the effects of HN_2 and of radiation in the intact animal and suggested that the drug inhibited the cell cycle in G_2. Sisken *et al.* (1960) showed that 0.32 mg/gm body weight of cyclophosphamide, given

TABLE II
LABELING AND MITOTIC INDEXES AFTER SOME ANTITUMOR AGENTS

Group	Dose	Spleen		Ileum	
		Interphase labeling (%)	Mitotic index (%)	Interphase labeling (%)	Mitotic index (%)
^{3}H-TdR[a] (control)	1 μCi/gm	51.5	2.5	55.7	8.4
HN_2[b]	0.4 mg/kg[c]	17.4	0.8	58.3	2.0
MTX[b]	0.5 mg/kg[c]	19.5	1.2	67.9	3.8
Hydrocortiscone[b]	100 mg/kg[c]	46.2	1.6	53.7	5.3
^{3}H-IUdR[a]	1 μCi/gm	43.0	3.1	56.1	6.8
IUdR[a]	0.5 mg/kg[c]	53.2	2.4	53.3	9.0
IUdR[a]	150 mg/kg[c]	32.4	1.9	63.0	3.6

[a] From the work of Post and Hoffman (1969a).

[b] From the work of Hoffman and Post (1972).

[c] Daily injection for 5 days.

^{3}H-TdR (^{3}H-thymidine); HN_2 (nitrogen mustard); MTX (methotrexate); ^{3}H-IUdR (^{3}H-5-iodo-2'-deoxyuridine); IUdR (5-iodo-2'-deoxyuridine).

to mice 6 hours before ^{3}H-TdR, did not affect DNA synthesis but depressed the mitotic index.

The effects of sublethal doses of HN_2 on the cell cycle kinetics of coexistent ileal cells and spleen lymphocytes of the normal growing rat were investigated in this laboratory (Hoffman and Post, 1972). The dose of HN_2, 0.4 mg/kg, was given for 5 days. Four hours after the last injection of HN_2, the animals were pulse labeled with 1.0 μCi/gm ^{3}H-TdR. There was disruption of the mitotic labeling curves of the ileal crypt cells and the splenic lymphocytes (Figs. 1 and 2). The interphase labeling index of ileal crypt cells remained unchanged, but there was a decrease in the mitotic index suggesting a delay in S or a block in G_2 (Table II). However, among spleen lymphocytes there was a marked fall in the labeling index to one-third that of the control animals. This was accompanied by a drop in the mitotic index (Table II). These findings indicated that fewer cells were synthesizing DNA and advancing to mitosis. The experiments of Levis et al. (1964) with guinea pig kidney cells, also pointed to inhibition of DNA synthesis by HN_2.

The nitrosourea compounds act as alkylating agents (Young and Burchenal, 1971). The member of this group now being used is 1,3-bis(2-Chloroethyl)-1-nitrosourea (BCNU). Wheeler and co-workers (1970b), using cultures of H. Ep. No. 2 cells synchronized by Colcemid block, showed that BCNU delayed their progression from G_1, S, or early G_2 into metaphase. Late G_2 cells were not inhibited.

DeVita (1971; DeVita et al., 1972) reported that BCNU produced

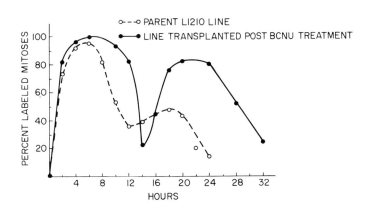

Fig. 3. Percent labeled mitoses curves for leukemia L1210 (ascites form) on day 6 of tumor growth. Dotted line is the untreated parent line; solid line represents the line following treatment, retransplantation, and growth. (From the work of DeVita, 1971; DeVita et al., 1972.)

changes in the mitotic labeling curves of transplanted ascites leukemia L1210 cells that were consistent with prolongation of the cell cycle and S times (Fig. 3).

C. ANTIFOLS

The folic acid antagonists have been used as anticancer agents since 1948 when Farber et al. (1948) demonstrated the effectiveness of aminopterin in the treatment of childhood leukemia. Subsequently the N^{10}-methyl derivative, amethopterin (methotrexate, MTX), was synthesized, and was found to be a more effective drug than aminopterin (Lin and Bruce, 1971; Stock, 1966a). Since MTX is the major antifol in clinical use, the actions of this drug will be discussed as representative of the group.

The metabolic effect of MTX is specific. It inhibits the enzyme, dihydrofolate reductase, preventing the reduction of folic acid to tetra-hydrofolic acid. This in turn limits the conversion of deoxyuridylic acid to thymidylic acid. Thus, DNA synthesis is blocked. The effects of MTX on the cell cycle have been studied in cell cultures, in tumor transplants, in intact animals, and in bacteria (Stock, 1966a). Chang strain human liver cells incubated in culture with MTX exhibited a marked decrease in mitotic index about 6 hours after treatment, indicating probable delay in S or G_2 (Dalen et al., 1965). Borsa and Whitmore (1969), investigating the viability of L cells exposed to MTX, found that the lethal effects could be reversed by the addition of cold thymidine to the medium and that cells in log phase culture were more susceptible to the drug than were those in the stationary phase.

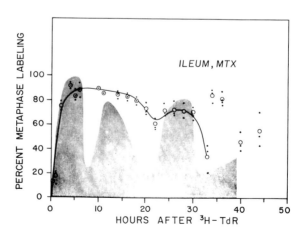

Fig. 4. Ileum after methotrexate (MTX). ^{3}H-TdR, 1 μCi/gm body weight. Symbols as in Fig. 1. The shaded portion is the control. (From Hoffman and Post, 1972.)

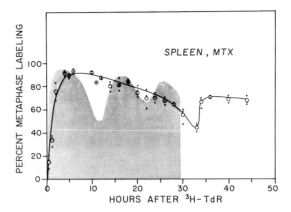

Fig. 5. Spleen, after methotrexate (MTX). ³H-TdR, 1 μCi/gm body weight. Symbols as in Fig. 1. Shaded background is the control. (From Hoffman and Post, 1972.)

Similar results were found with L5178Y cultures (Hryniuk and Bertino, 1971). The cells undergoing active proliferation were more susceptible to the action of the antimetabolite. Cell death resulted from the inhibition of DNA synthesis secondary to the interference with thymidylate production.

Margolis et al. (1971) showed that MTX (0.5 mg/kg body weight) reduced the incorporation of ³H-deoxyuridine into the DNA of mouse intestinal cells but had no effect on ³H-TdR incorporation. However, they demonstrated a concomitant fall in the uptake of ³²P-labeled orthophosphate, indicating that DNA synthesis was depressed and that there was a selective block of deoxyuridylate methylation by MTX. The mitotic index was also decreased after treatment with MTX. When MTX was given to growing rats, 0.5 mg/kg body weight, for five days, the wavelike character of the mitotic labeling curves of the coexistent ileal crypt cells and spleen lymphocytes was lost (Hoffman and Post, 1972). Estimations of generation and S times could not be determined from these curves (Figs. 4 and 5). There was a slight increase in the interphase labeling index of the ileal cells and the mitotic index was depressed, suggesting a delay in G_2. However, in the spleen lymphocytes of these animals there was a sharp reduction in both parameters (Table II). The results obtained in the rat ileal crypt cells were consistent with those of Margolis et al. (1971).

The incorporation of ³H-TdR by the gut cells represents the use of an exogenous source of the end product of thymidylate synthesis and thus need not be a measure of the DNA synthesis suppression induced by MTX. In the spleen, however, where ³H-TdR incorporation is markedly lowered, the effect of MTX is wider and more than one pathway of synthesis may be blocked.

D. ANTIBIOTICS

The success with antimicrobial compounds in the treatment of infectious disease has stimulated intensive investigations of soil organisms and other antibiotic sources for active antitumor drugs. To date, the results have been less spectacular than those experienced with penicillin, tetracycline, or streptomycin.

Most of the antitumor antibiotics are derived from the *Streptomyces* genus of soil bacteria. Actinomycin D (Act D) has been the most widely investigated of these drugs. It has also been used as a metabolic inhibitor in the study of the cell cycle. The basic biochemical event involves the formation of a DNA–Act D complex with the resultant inhibition of DNA-dependent RNA synthesis and, ultimately, DNA synthesis (Young, 1969). Small doses of Act D, 0.016 mg/gm body weight, inhibited Ehrlich ascites tumor cells in G_1 and produced a prompt fall in RNA synthesis and a delayed decrease in DNA synthesis (Baserga et al., 1965). Kim et al. (1968), using synchronized cultures of HeLa cells demonstrated that Act D acting in G_1 blocked DNA synthesis. However, Kishimoto and Lieberman (1964), using larger doses of Act D and rabbit kidney cortex cells observed a block in G_2. This work was confirmed by Tobey and co-workers (1966) in synchronized Chinese hamster cells. In an *in vivo* experiment with mouse forestomach, Frankfurt (1968) reported that Act D, 1 μg/gm body weight, produced a diminished labeling index and a reduction in the RNA synthesis requisite for DNA synthesis.

Puromycin, another derivative of the *Streptomyces* groups, is an inhibitor of protein synthesis (Baserga, 1968; Jackson and Studzinski, 1968; Stock, 1966a), and its effects on the cell cycle have varied with the dosage employed. Synchronized L cells exposed to 10 μg/ml were arrested in G_1 (Baserga, 1968). Larger doses (50 μg/ml) blocked synchronized Chinese hamster ovary cells in G_2 (Tobey et al., 1966). Estensen and Baserga (1966) demonstrated that 0.083 mg/gm body weight of puromycin, given hourly for four doses, induced necrosis in the crypt cells of mouse small intestine. This was associated with reduced synthesis of protein and DNA. Later, RNA synthesis fell and there was delay of cells in G_2.

Cycloheximide (Actidione) is derived from *Streptomyces griseus* (Stock, 1966a). Although it is not an effective antitumor agent it has been used extensively in the study of cell kinetics. An inhibitor of protein synthesis (Estensen and Baserga, 1966), cycloheximide also influences RNA and DNA synthesis. Estensen and Baserga (1966) observed that at a dosage of 0.3 mg/gm of body weight in the intact mouse, the incorporation of ^{3}H-thymidine and of ^{3}H-cytidine was depressed as effectively as that of ^{3}H-leucine.

Verbin and Farber (1967) reported that 1.3 mg/kg cycloheximide inhibited protein and DNA synthesis in the crypt cells of rat small intestine *in vivo*, and produced a block in G_2; RNA synthesis was unaffected. In partially hepatectomized rats Verbin and co-workers (1969) found that this drug arrested cells in DNA synthesis and in G_2. However, Kim *et al.* (1968), employing synchronous and asynchronous cultures of HeLa cells, reported that cycloheximide interfered with DNA, RNA, and protein synthesis.

Daunomycin (rubidomycin) is an antibiotic that has been isolated from *Streptomyces peucetius* (Stock, 1966a), and it has been reported to be useful in the treatment of acute leukemia (Henderson, 1969). The basic biochemical reaction of this compound is its formation of a stable complex with preformed DNA. This interferes with DNA synthesis and DNA-dependent RNA synthesis (Brehaut, 1969; Wilkoff *et al.*, 1971). In experiments with cultured human leukocytes Brehaut (1969) indicated that daunomycin blocked the cell cycle in G_2. Wilkoff *et al.* (1971) showed that the lethal effects of this drug were dose dependent.

Mitomycin C, derived from *Streptomyces caespitosus*, acts as an alkylating agent (Scott, 1970; Stock, 1966a). Thomson and Biggers (1966), using preimplanted mouse embryos in culture, showed that the drug blocked mitosis at the two cell and eight cell stages and that the effect was dose dependent.

E. STEROIDS

The glucosteroids have been used in the treatment of cancer since their introduction into clinical medicine. Cardinale and co-workers (1964) showed that high doses of cortisone depressed the mitotic index of Syrian hamster bone marrow cells. Mitosis was also reduced in the regenerating liver after CCl_4 injury in young rats pretreated with cortisone (Hoffman *et al.*, 1955). The steroids lowered DNA synthesis in these regenerating livers (Hoffman *et al.*, 1955) and in the livers of rats following partial hepatectomy (Rizzo *et al.*, 1971). DNA synthesis in the livers of growing rats was reduced by the administration of cortisone (Henderson *et al.*, 1971). Ioachim (1971) produced the wasting "runt disease" in newborn rats with one injection of cortisone and found a marked decrease in the labeling indexes of spleen, liver, kidney, lung, muscle, and skin. Frankfurt (1968), in experiments with mouse forestomach, showed that hydrocortisone effected a delay in the onset of DNA synthesis. Thymidine incorporation was depressed in fibroblast cultures by high concentrations of prednisolone (10^{-4} and 10^{-5} M) but it was enhanced by lesser concentrations of 10^{-7} and 10^{-8} M (Achenbach *et al.*, 1970). Kollmorgen and Griffin (1969) reported a G_1

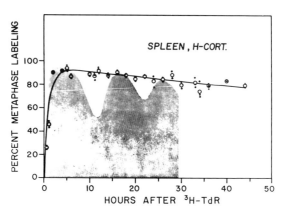

Fig. 6. Spleen, after hydrocortisone. ³H-TdR, 1 μCi/gm body weight. Symbols as in Fig. 1. Control is the shaded portion. (From Hoffman and Post, 1972.)

block in HeLa Chessen cells following hydrocortisone treatment. The doubling time changed from 18 to 35 hours. The drug was ineffective against HeLa S_3 cells.

In this laboratory (Hoffman and Post, 1972) hydrocortisone disrupted the mitotic-labeling curve of rat spleen lymphocytes (Fig. 6) so that the generation and S times could not be estimated therefrom. The interphase-labeling index of these cells remained unaffected but the mitotic index was reduced, suggesting a G_2 block. The interphase- and mitotic-labeling responses were similar in the coexistent ileum. The mitotic-labeling curve (Fig. 7) showed a slight lengthening of the generation time, and only one cycle was defined before the curve was dampened.

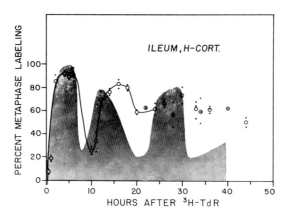

Fig. 7. Ileum, after hydrocortisone. ³H-TdR, 1 μCi/gm body weight. Symbols as in Fig. 1. Background is the control. (From Hoffman and Post, 1972.)

F. PYRIMIDINE ANALOGS

Although the purine and pyrimidine analogs are discussed elsewhere in this volume, mention will be made here of the pyrimidine analogs, 1-β-D-arabinofuranosylcytosine (Ara-C) and 5-iodo-2′-deoxyuridine (IUdR).

Ara-C is related to cytidine, and its biochemical action is the inhibition of DNA synthesis (Bertalanffy and Gibson, 1971; Chu and Fischer, 1962; Karon and Shirakawa, 1970; Kimball and Wilson, 1968; Livingston and Carter, 1968; Sartorelli and Creasy, 1969). The mechanisms of action may be via more than one pathway. Chu and Fischer (1962) demonstrated in L5178Y leukemia cells, that Ara-C blocked the conversion of cytidine di-

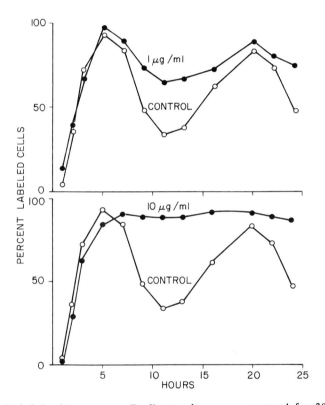

Fig. 8. Labeled mitoses curves. Replicate cultures were exposed for 30 minutes to tritiated thymidine and cold thymidine added, 10 μg/ml. The cells were treated with 1 and 10 μg/ml of 1-β-D-arabinofuranosyl cytosine (Ara-C) for 1 hour, before the drug was washed away and replaced with fresh media. Open circles, show the distribution of labeled mitoses for untreated cells (control); closed circles, cells treated with 1-β-D-arabinofuranosyl cytosine. Ordinate denotes percent labeled mitotic cells. (From Karon and Shirakawa, 1970.)

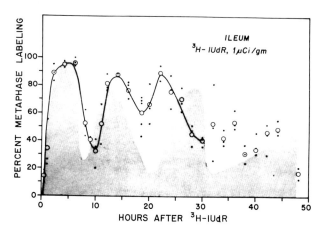

Fig. 9. Ileum after ³H-iododeoxyuridine (³H-IUdR). Symbols as in Fig. 1. Shaded portion is the control. (From Post and Hoffman, 1969a.)

phosphate to deoxycytidine diphosphate. More recent studies have indicated that Ara-C blocked DNA synthesis by inhibiting DNA polymerase (Graham and Whitmore, 1970; Kimball and Wilson, 1968). Kimball and Wilson (1968) reported that deoxycytidine 5′-triphosphate reversed this inhibition. It has also been noted that cell death after Ara-C may be correlated with its incorporation into RNA (Sartorelli and Creasey, 1969). The major effect of Ara-C on the cell cycle seems to be in late S and G_2, as shown by Karon

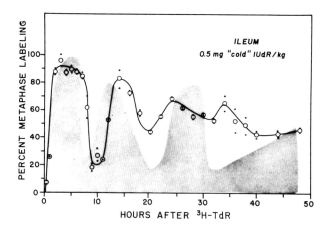

Fig. 10. Ileum after "cold" iododeoxyuridine (IUdR), 0.5 mg/kg body weight. ³H-TdR, 1 μCi/gm body weight. Symbols as in Fig. 1. Shaded background is the control. (From Post and Hoffman, 1969a.)

and Shirakawa (1969, 1970) in Don-C cells. The rate of progression from S to G_2 was slowed. The effect was related to the dose of the drug, as may be seen from the labeled mitosis curves (Fig. 8). An amount of 1 µg/ml dampens the curve after one cycle, but 10 µg/ml produces disruption of the wavelike character of the curve. However, the G_2 time in both cases remains the same. Others have shown that Ara-C-induced chromosome aberrations in cultured human leukocytes and in hamster fibroblasts. This effect was on cells which had been arrested in the G_2 phase of the cell cycle by Ara-C (Brewen, 1965; Karon and Benedict, 1970).

The replacement of the methyl group on carbon-5 of thymine with iodine transforms thymidine into IUdR (Prusoff, 1963). The halogenated analog is incorporated into nuclear DNA producing a false polymer (Clifton et al., 1963; Prusoff, 1960).

In vitro studies indicated that there was an alteration in mammalian cell replication after the first cycle (Erikson and Szybalski, 1963; Morris and Cramer, 1966). DNA synthesis was reduced because only about one-half the hybrid DNA replicated (Cramer and Morris, 1966). It has also been found that IUdR inhibits thymidine kinase and that the monophosphate inhibits thymidylate kinase (Delamore and Prusoff, 1962). In addition, IUdR increased the sensitivity of cells to radiation injury (Erikson and Szybalski, 1963).

Sublethal amounts of IUdR altered the cell cycles of the ileal crypt cells and of the spleen lymphocytes of the growing rat (Post and Hoffman, 1969a). Experiments were performed with (1) tritiated IUdR (^{3}H-IUdR),

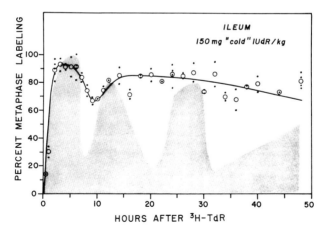

Fig. 11. Ileum after "cold" iododeoxyuridine (IUdR), 150 mg/kg body weight. ^{3}H-TdR, 1 µCi/gm body weight. Symbols as in Fig. 1. Shaded portion is the control. (From Post and Hoffman, 1969a.)

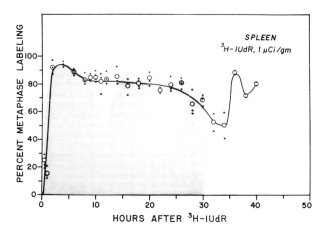

Fig. 12. Spleen after ³H-IUdR. Symbols as in Fig. 1. Shaded area is the control. (From Hoffman and Post, 1972.)

1 μCi/gm body weight; (2) "cold" IUdR, 0.5 mg/kg body weight daily for 5 days (the amount of carrier present in the dose of ³H-IUdR); and (3) cold IUdR, 150 mg/kg body weight daily for 5 days (Table II). Labeled mitosis curves were derived, and labeling and mitotic indexes were determined along with grain counts. The labeling and mitotic indices of the intestinal cells after ³H-IUdR were similar to those of the control animals (Table II).

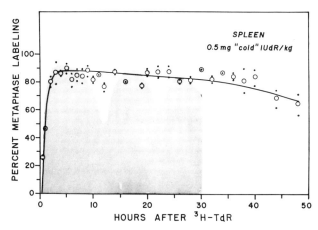

Fig. 13. Spleen after cold IUdR, 0.5 mg/kg body weight. ³H-TdR, 1 μCi/gm body weight. Symbols as in Fig. 1. Shaded portion is the control. (From Hoffman and Post, 1972.)

However, the mitotic labeling curve was altered (Fig. 9). There was slight prolongation of the generation and DNA synthesis times. The wavelike pattern of the labeled mitosis curve was lost after one cycle. Following IUdR, 0.5 mg/kg body weight, the results were similar to those with ^{3}H-IUdR (Table II) (Fig. 10). But when IUdR was administered at 150 mg/kg body weight the mitotic labeling curve was disrupted and the S and cell cycle times could not be measured (Fig. 11). The interphase labeling rate remained unchanged but there was a fall in the mitotic index (Table II). Thus, the administration of IUdR changed the wavelike sequence of the mitotic labeling curve of ileal crypt cells. The fall in the mitotic index and the unchanging labeling index suggested that there was a delay in the passage of cells through DNA synthesis and G_2 and that the effects were dose-dependent. A diminished grain count one hour after ^{3}H-TdR, as compared with controls, was consistent with a reduction in the rate of DNA synthesis.

The mitotic labeling curves of the spleen lymphocytes in each of the above experiments showed loss of the wavelike patterns of the controls (Figs. 12–14). Cell cycle times could not be measured. The labeling and mitotic indices after ^{3}H-IUdR and after 0.5 mg/kg IUdR were similar to those of the controls. However, 150 mg/kg body weight produced a 50% fall in the interphase labeling index and about a 25% decline in the mitotic index (Table II).

These results together with the initial and serial grain count changes, supported a reduction in the size of the proliferating population, as well as a delay in S and in G_2.

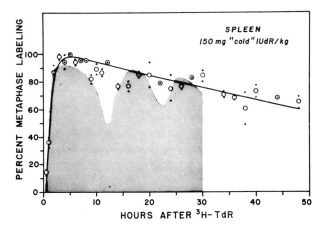

Fig. 14. Spleen after IUdR, 150 mg/kg. ^{3}H-TdR, 1 μCi/gm body weight. Symbols as in Fig. 1. Shaded portion is the control. (From Post and Hoffman, 1969a.)

G. Miscellaneous Compounds

It has been shown that protein synthesis is essential for cells to go from S to M (Sisken and Wilkes, 1967). In addition the inhibition of protein synthesis with cycloheximide and puromycin produces a suppression of DNA synthesis (Baserga, 1968; Donnelly and Sisken, 1967; Estensen and Baserga, 1966; Kishimoto and Lieberman, 1964; Verbin and Farber, 1967; Verbin et al., 1969).

Of interest are cell cycle studies following the incorporation of the amino acid analog, p-fluorophenylalanine, which inhibits protein synthesis. Sisken and Iwasaki (1969) found that p-fluorophenylalanine blocked cultured human amnion cells in mitosis and caused marked prolongation of metaphase time. They also reported that p-fluorophenylalanine inhibited the manufacture of a protein that was essential for the passage of cells from G_2 to mitosis. These findings were in agreement with their previous experiments (Sisken and Wilkes, 1967). The effects were dose dependent.

Hydroxyurea (Hu) is a derivative of hydroxylamine, which has been known since 1869. However, it was not used as an antitumor agent until after 1960 (Yarbro, 1968). Its primary effect is upon the synthesis of DNA (Rosenkranz et al., 1967; Yarbro, 1968). The incorporation of ^{3}H-TdR is blocked (Sartorelli and Creasey, 1969) but there is no effect on either protein or RNA synthesis (Stock, 1966b). The proteins manufactured by Hu-treated E. coli are normal (Rosenkranz et al., 1967). Sinclair (1965) using synchronized Chinese hamster cells, has shown that this agent has different effects upon the several compartments of the cell cycle. When added to cells in S they are damaged lethally. The exposure of G_1 cells prevents their entry into S. Cells in G_2 survive and pass through M into G_1. Their progress is then blocked.

L-Asparaginase is an unique compound in that its anticancer effect is dependent upon the selective inability of particular neoplastic cells to manufacture asparagine (Oettgen et al., 1967; Sartorelli and Creasey, 1969; Weinberger, 1971). The cytotoxic capabilities of this enzyme were first noted in 1953 when it was found that guinea pig serum halted the growth of transplanted lymphomas in mice or rats. L-Asparaginase was identified as the active agent (Oettgen et al., 1967; Weinberger, 1971). This compound does not affect the coexistent normal tissues of the host. Sensitive neoplastic cells need an external source of asparagine since their levels of asparagine synthetase are low; normal cells with high levels of asparagine synthetase can elaborate their own asparagine. Resistant tumors have high synthetase activity. In regenerating rat liver cells after partial hepatectomy, Becker and Broome (1967) demonstrated that the first wave of mitosis was delayed for 20 hours by L-asparaginase. They ascribed this effect to reduced RNA synthesis.

III. General Discussion

There are many ways to block the cell in its progress through the cycle of replication (Table I). The therapeutic usefulness of a cytotoxin is dependent on the efficiency of its incorporation into specific cellular biochemical events. The tissue distribution, the transport across cell membranes, the rates of metabolism of these compounds and their quantitative levels of incorporation determine their lethality (Kessel and Hall, 1967). However, it is not possible to consider the action of these drugs as though the tumor were an isolated cell population. Their effects must be viewed in the setting of the tumor-bearing patient.

For many years it was believed that cancer cells proliferated more rapidly than did normal cells and that the pattern of their proliferation was chaotic and continual. It seemed logical to expect that exposure of a tumor-bearing subject to a cytotoxin would result in the selective destruction of the more rapidly replicating tumor cells. Clinical experience has shown that, in general, the major limitation to the use of cytotoxins is the tolerance of the host organism to the destruction of the replicating normal cells of its gastrointestinal and hematopoietic systems, these being the chief proliferating adult populations. Indeed, whether or not tumor cells are adversely affected, some of these replicating normal cells usually are. During the past 10 years the study of the kinetics of normal and of tumor cell proliferation has shown that the wavelike character of the mitotic labeling curves of many cancer cells is consistent with a proliferation pattern that is under precise controls, not unlike those among normal cells. Autogenous tumors in animals (Hoffman and Post, 1966; Post and Hoffman, 1964, 1968) and in human subjects (Clarkson et al., 1965) have relatively long cycle and S times in relation to coexistent gut and blood cells.

In addition, the percentages of the populations of normal cells engaged in replication exceed those of tumor cells. Hence, if a cytotoxin is administered, it would be expected that more normal cells would be exposed to the drug than would coexistent tumor cells. Further, the frequency of exposure would be greater among the normal cells. In studies on the incorporation of ^{3}H-IUdR by breast tumor (Hoffman and Post, 1966) and sarcoma (Post and Hoffman, 1968) cells in relation to coexistent ileal crypt cells, it was found that the normal cells incorporated more cytotoxin and in far greater population percentages (Table III). These findings have provided an explanation for the clinically observed predilection of toxicity of many compounds for normal cells. Specific and exploitable metabolic attributes of cancer cells vis-à-vis coexistent normal cells have not yet been found.

In the experimental testing of drugs, the models used are often transplanted tumor cells, and these frequently are rapidly replicating with relatively

TABLE III

Summary of Data on Replication and Antimetabolite Incorporation by Normal and Tumor Cells[a]

Cell type	Generation time (hour)	DNA synthesis time (hour)	³H-TdR			³H-IUdR	
			Mean grain count (±probability error)	Interphase labeling (%)	Mitoses (%)	Mean grain count (±probability error)	Interphase labeling (%)
Sarcoma	40.0	24.0	7.2 ± 0.09	4.8	0.6	11.1 ± 0.20	11.4
Ileum	11.0–12.0	7.0	14.8 ± 0.15	37.2	6.7	20.2 ± 0.70	51.6
Breast tumor	45.0	10.0	14.6 ± 0.10	7.4	0.9	10.9 ± 0.30	2.7
Ileum	11.0–12.0	6.0	22.2 ± 0.80	41.6	6.3	15.5 ± 0.40	49.1
Hepatoma	31.0	17.0	12.5 ± 0.10	7.5	2.7	—	—
Ileum	11.0–12.0	6.0	11.2 ± 0.17	37.9	5.1	—	—

[a] Reprinted from British Journal of Cancer with permission of the publisher. From the work of Post and Hoffman (1968).

large proliferating pools in relation to the gut and blood cells of the host animal. In such a setting the tumor cells may compete successfully for the cytotoxic effects of the drug. When tried in human tumor-bearing subjects whose tumor cells may have long cycle and S times and small proliferating pools in relation to normal cells, the drug may be too toxic for therapeutic benefit. In considering cell replication and its control, several factors are important. These include the cycle time and its component intervals, the size of the proliferating and nonproliferating pools, and the transit times of the cells in these compartments. Recent studies in this laboratory have shown that in the tumor model of the autogenous rodent sarcoma, cells may be entering and leaving the proliferating pool from a nonproliferative state (Post and Hoffman, 1969b). While nonproliferative tumor cells have usually been considered as G_1O cells, this sarcoma contains a significant pool of G_2O cells which act as the source of tumor mitoses. The bidirectional flow of tumor cells between proliferation and nonproliferation is an added dimension to the problem of drug treatment in relation to drug exposure. Large numbers of cells may be sequestered in a nonproliferative stage beyond the reach of drug. This could be a major factor in the choice of drugs that affect only proliferating cells. The antitumor agent may even play a key role in determining the sizes of the proliferative and nonproliferative pools.

One phase of the study of the effects of drugs upon the cell cycle deals with the application of this information to the scheduling and the choice of therapy. If the tumor cells are replicating synchronously and if the time compartments and proliferating pool sizes are defined precisely, the drug might be selected for its site or sites of action. The timing and the frequency of administration could be determined with a high degree of accuracy and effectiveness. However, tumor cells divide asynchronously. At any given moment cells are randomly distributed throughout the cycle compartments. The achievement of the synchronization of human cells *in vivo* is unlikely and unrealistic. Furthermore, the effects of a synchronization technique upon cancer cells cannot be considered without effects upon the coexistent normal cells. The administration of a drug for the period occupied by $G_2 + M + G_1$ in each cell population should "catch" the members of the proliferative pool of these asynchronously dividing cells. An additional complexity to the subject is the wide range of variation of the time compartments of the cell cycle among the cells that make up a tumor (Barrett, 1966; Steel and Hanes, 1971). The data from mitotic labeling curves are average values. Finally, nonproliferative cells may become proliferative after days or weeks, and unless these characteristics are known, these cells would escape the therapy. At this stage in our knowledge therapy with antitumor agents remains largely empirical.

IV. Summary

The effects of several classes of antitumor drugs on the cell cycle have been discussed along with their application to the chemotherapy of cancer. Many of the studies on the effects of cytotoxins on cell cycle kinetics have been conducted with *in vitro* systems or with tumor transplants (Table I). Relating the results in such experimental models to the experience with the human cancer patient has frequently been nonproductive. However, successful results have been seen in certain tumors. Not enough is known yet about these neoplasms to help explain the success or failure of therapy. More information is needed regarding the replication kinetics of human cancer cells, especially those of the solid tumors of lung, breast, or gut, as well as about the sizes of the proliferating and nonproliferating pools. Studies should include the effects of antitumor agents upon these kinetic parameters, in responsive as well as in unresponsive tumors. Finally, new insights should be found from the continuing search for therapeutically exploitable differences between proliferating normal and cancer cells.

Acknowledgments

The authors wish to thank Dr. Robert Sklarew, Mrs. M. Miheyev, and Miss Cyrilla Ho for their technical assistance and Mrs. Ernestine Williams for her secretarial help. This work was supported by grants from the National Cancer Institute of the United States Public Health Service, the United States Atomic Energy Commission and the National Aeronautics and Space Administration.

References

Achenbach, C., Süss, R., Kinzel, V., Wieser, O., and Sturm, H. A. (1970). Time and dose dependent inhibition and enhancement of thymidine incorporation into fibroblasts by prednisolone. *Experientia* **26**, 405–406.

Barrett, J. C. (1966). A mathematical model of the mitotic cycle and its application to the interpretation of percentage labeled mitoses data. *J. Nat. Cancer Inst.* **37**, 443–450.

Baserga, R. (1968). Biochemistry of the cell cycle: A review. *Cell Tissue Kinet.* **1**, 167–191.

Baserga, R., Estensen, R. D., Petersen, R. O., and Layde, J. P. (1965). Inhibition of DNA synthesis in Ehrlich ascites cells by actinomycin D. I. Delayed inhibition by low doses. *Proc. Nat. Acad. Sci. U.S.* **54**, 745–751.

Becker, F. F., and Broome, J. D. (1967). L-asparaginase: Inhibition of early mitosis in regenerating rat liver. *Science* **156**, 1602–1603.

Bertalanffy, F. D., and Gibson, M. H. L. (1971). The *in vivo* effects of arabinosylcytosine on the cell proliferation of murine B16 melanoma and Ehrlich ascites tumor. *Cancer Res.* **31**, 66–71.

Borsa, J., and Whitmore, G. F. (1969). Cell killing studies on the mode of action of methotrexate on L-cells *in vitro*. *Cancer Res.* **29**, 737–744.

Brehaut, L. A. (1969). A delay in the G_2 period of cultured human leucocytes after treatment with rubidomycin (daunomycin). *Cell Tissue Kinet.* **2**, 311–318.

Brewen, J. G. (1965). The induction of chromatid lesions by cytosine arabinoside in post-DNA-synthetic human leucocytes. *Cytogenetics (Basel)* **4**, 28–36.

Bruce, W. R., Meeker, B. E., and Valeriote, F. A. (1966). Comparison of the sensitivity of normal hematopoietic and transplanted lymphoma colony-forming cells to chemotherapeutic agents administered *in vivo*. *J. Nat. Cancer Inst.* **37**, 233–245.

Cardinale, G., Cardinale, G., DeCaro, B. M., Handler, A. H., and Aboul-Enein, M. (1964). Effect of high doses of cortisone on bone marrow cell proliferation in the Syrian hamster. *Cancer Res.* **6**, 969–972.

Chu, M. Y., and Fischer, G. A. (1962). A proposed mechanism of action of 1-β-D-arabinofuranosyl cytosine as an inhibitor of the growth of leukemic cells. *Biochem. Pharmacol.* **11**, 423–430.

Clarkson, B., Ota, K., Ohkita, T., and O'Connor, A. (1965). Kinetics of proliferation of cancer cells in neoplastic effusion in man. *Cancer* **18**, 1189–1213.

Clifton, K. H., Szybalski, W., Heidelberger, C., Gollin, F. F., Ansfield, F. J., and Vermund, H. (1963). Incorporation of I^{125} labeled iododeoxyuridine into the deoxyribonucleic acid of murine and human tissues following therapeutic doses. *Cancer Res.* **23**, 1715–1723.

Cramer, J. W., and Morris, N. R. (1966). Absence of DNA synthesis in one-half of a population of mammalian tumor cells inhibited in culture by 5-iodo-2'-deoxyuridine. *Mol. Pharmacol.* **2**, 363–367.

Dalen, H., Oftebro, R., and Engeset, A. (1965). The effect of methotrexate and leucovorin on cell division in Chang cells. *Cancer* **18**, 41–48.

Delamore, I. W., and Prusoff, W. H. (1962). Effect of 5-iodo-2'-deoxyuridine on the biosynthesis of phosphorylated derivatives of thymidine. *Biochem. Pharmacol.* **11**, 101–112.

DeVita, V. T. (1971). Cell kinetics and the chemotherapy of cancer. *Cancer Chemother. Rep.* **2**, 23–33.

DeVita, V. T., Bray, D. A., Bostick, F., and Bagley, C. M. (1972). The effect of chemotherapy on the growth of leukemia L1210. II. Persistence of a nitrosourea-induced change in the growth characteristics of transplant generations. Personal communication.

Doniach, I., and Pelc, S. R. (1950). Autoradiograph technique. *Brit. J. Radiol.* **23**, 184–192.

Donnelly, G. M., and Sisken, J. E. (1967). RNA and protein synthesis required for entry of cells into mitosis and during the mitotic cycle. *Exp. Cell Res.* **46**, 93–105.

Erikson, R. L., and Szybalski, W. (1963). Molecular radiobiology of cell lines. III. Radiation-sensitizing properties of 5-iododeoxyuridine. *Cancer Res.* **23**, 122–130.

Estensen, R. D., and Baserga, R. (1966). Puromycin induced necrosis of crypt cells of the small intestine of mouse. *J. Cell Biol.* **30**, 13–22.

Farber, J., Rovera, G., and Baserga, R. (1971). Template activity of chromatin during stimulation of cellular proliferation in human diploid fibroblasts. *Biochem. J.* **122**, 189–195.

Farber, S., Diamond, L. K., Mercer, R. D., Sylvester, R. F., and Wolff, J. A. (1948). Temporary remission in acute leukemia in children produced by folic antagonist, 4-aminopteroyl glutamic acid (aminopterin). *N. Engl. J. Med.* **238**, 789–793.

Frankfurt, O. S. (1968). Effects of hydrocortisone adrenalin and actinomycin D on transition of cells to the DNA synthesis phase. *Exp. Cell Res.* **52**, 220–232.

Graham, F. L., and Whitmore, G. F. (1970). The effect of 1-β-D-arabinofuranosyl cytosine on growth viability and DNA synthesis of mouse L-cells. *Cancer Res.* **30**, 2627–2635.

Hampton, J. C. (1967). The effects of nitrogen mustard on the intestinal epithelium of the mouse. *Radiat. Res.* **30**, 576–589.

Henderson, E. S. (1969). Treatment of acute leukemia. *Semin. Hematol.* **6**, 271–319.

Henderson, I. C., Fischel, R. E., and Loeb, J. N. (1971). Suppression of liver DNA synthesis by cortisone. *Endocrinology* **88**, 1471–1476.

Hoffman, J., and Post, J. (1966). Replication and 5-iodo-2′-deoxyuridine-^{3}H incorporation by tumor and normal cells. *Cancer Res.* **26**, 1313–1318.

Hoffman, J., and Post, J. (1972). Effects of nitrogen mustard, methotrexate and hydrocortisone on cell replication in rat ileum and spleen. *Proc. Soc. Exp. Biol. Med.* **140**, 444–448.

Hoffman, J., Himes, M. B., Lapan, S., and Post, J. (1955). Responses of the liver to injury: Effects of cortisone upon acute carbon tetrachloride poisoning. *AMA Arch. Pathol.* **60**, 10–18.

Howard, A., and Pelc, S. R. (1953). Synthesis of desoxyribonucleic acid in normal and irradiated cells and its relation to chromosome breakage. *Heredity*, **6**, Suppl., 261–273.

Hryniuk, W. M., and Bertino, J. R. (1971). Growth rate and cell kill. *Ann N. Y. Acad. Sci.* **186**, 330–342.

Ioachim, H. L. (1971). The cortisone-induced wasting disease of newborn rats: Histopathological and autoradiographic studies. *J. Pathol.* **104**, 201–205.

Jackson, L. G., and Studzinski, G. P. (1968). Autoradiographic studies of the effects of inhibitors of protein synthesis on RNA synthesis in HeLa cells. *Exp. Cell Res.* **52**, 408–418.

Karon, M., and Benedict, W. (1970). Arabinosylcytosine-induced chromosome breakage and the cell cycle. *Clin. Res.* **18**, 219.

Karon, M., and Shirakawa, S. (1969). The locus of action of 1-β-D-arabinofuranosyl cytosine in the cell cycle. *Cancer Res.* **29**, 687–696.

Karon, M., and Shirakawa, S. (1970). Effects of 1-β-D-arabinofuranosyl-cytosine on cell cycle passage time. *J. Nat. Cancer Inst.* **45**, 861–887.

Kessel, D., and Hall, T. C. (1967). Amethopterin transport in Ehrlich ascites carcinoma and L1210 cells. *Cancer Res.* **27**, 1539–1543.

Kim, J. H., Gelbard, A. S., and Perez, A. G. (1968). Inhibition of DNA synthesis by actinomycin D and cycloheximide in synchronized HeLa cells. *Exp. Cell Res.* **53**, 478–487.

Kimball, A. P., and Wilson, M. J. (1968). Inhibition of DNA polymerase by β-D-arabinosylcytosine and reversal of inhibition by deoxycytidine-5′-triphosphate. *Proc. Soc. Exp. Biol. Med.* **127**, 429–432.

Kishimoto, S., and Lieberman, I. (1964). Synthesis of RNA and protein required for the mitosis of mammalian cells. *Exp. Cell Res.* **36**, 92–100.

Klevecz, R. R., and Stubblefield, E. (1967). RNA synthesis in relation to DNA replication in synchronized Chinese hamster cell cultures. *J. Exp. Zool.* **165**, 259–268.

Kollmorgen, G. M., and Griffin, M. J. (1969). The effect of hydrocortisone in HeLa cell growth. *Cell Tissue Kinet.* **2**, 111–122.

Layde, J. P., and Baserga, R. (1964). The effect of nitrogen mustard on the life cycle of Ehrlish ascites tumor cells *in vivo. Brit. J. Cancer* **18**, 150–158.

Levis, A. G., Marin, G., and Danieli, G. A. (1964). Differential inhibition of RNA and DNA synthesis by nitrogen mustard in cultured mammalian cells. *Caryologia* **17**, 427–431.

Lin, H., and Bruce, W. R. (1971). Effect of amethopterin on cells of experimental tumors. *Ann. N. Y. Acad. Sci.* **186**, 325–329.

Livingston, R. B., and Carter, S. R. (1968). Cytosine arabinoside (NSC-63878)—Clinical Brochure. *Cancer Chemother. Rep.* **1**, 179–205.

Margolis, S., Philips, F. S., and Steinberg, S. S. (1971). The cytotoxicity of methotrexate in mouse small intestine in relation to inhibition of folic acid reductase and of DNA synthesis. *Cancer Res.* **31**, 2037–2046.

Morris, N. R., and Cramer, J. W. (1966). DNA synthesis by mammalian cells inhibited in culture by 5-iodo-2′-deoxyuridine. *Mol. Pharmacol.* **2**, 1–9.

Ochoa, M., Jr. (1969). Alkylating agents in clinical cancer chemotherapy. *Ann. N. Y. Acad. Sci.* **163**, 921–930.

Ochoa, M., Jr., and Hirschberg, E. (1967). *Exp. Chemother.* **5**, 1–132.

Oettgen, H. F., Old, L. J., Boyse, E. A., Campbell, H. A., Philips, F. S., Clarkson, B. D., Tallal, L., Lieper, R. D., Schwartz, M. K., and Kim, J. H. (1967). Inhibition of leukemias in man by L-asparaginase. *Cancer Res.* **27**, 2619–2631.

Oliverio, V. T., and Zubrod, C. G. (1965). Clinical pharmacology of the effective antitumor drugs. *Annu. Rev. Pharmacol.* **5**, 335–356.

Post, J., and Hoffman, J. (1964). The replication time and pattern of carcinogen-induced hepatoma cells. *J. Cell Biol.* **22**, 341–350.

Post, J., and Hoffman, J. (1968). *In vivo* replication and anti-metabolite incorporation by coexistent normal and autogenous tumor cells. *Brit. J. Cancer* **22**, 149–154.

Post, J., and Hoffman, J. (1969a). Effects of 5-iodo-2′-deoxyuridine upon the replication of ileal and spleen cells *in vivo. Cancer Res.* **29**, 1859–1865.

Post, J., and Hoffman, J. (1969b). A G_2 population of cells in autogenous rodent sarcoma. *Exp. Cell Res.* **57**, 111–113.

Prusoff, W. H. (1960). Incorporation of iododeoxyuridine into the deoxyribonucleic acid of mouse Ehrlich ascites tumor cells *in vivo. Biochim. Biophys. Acta* **39**, 327–331.

Prusoff, W. H. (1963). A review of some aspects of 5-iododeoxyuridine and azauridine. *Cancer Res.* **23**, 1246–1259.

Prusoff, W. H. (1967). Recent advances in chemotherapy of viral diseases. *Pharmacol. Rev.* **19**, 209–250.

Puck, T. T., and Steffen, J. (1963). Life cycle analysis of mammalian cells. I. A method for localizing metabolic events within the life cycle, and its application to the action of colcemide and sublethal doses of x-irradiation. *Biophys. J.* **3**, 379–397.

Quastler, H., and Sherman, F. G. (1959). Cell population kinetics in intestinal epithelium of mouse. *Exp. Cell Res.* **17**, 420–438.

Rizzo, A. J., Heilpern, P., and Webb, T. E. (1971). Temporal changes in DNA and RNA synthesis in the regenerating liver of hydrocortisone-treated rats. *Cancer Res.* **31**, 876–881.

Rosenkranz, H. S., Winshell, E. B., Mednis, A., Carr, H. S., and Ellner, C. J. (1967). Studies with hydroxyurea. VII. Hydroxyurea and the synthesis of functional proteins. *J. Bacteriol.* **94**, 1025–1033.

Rovera, G., and Baserga, R. (1971). Early changes in the synthesis of acidic nuclear protein in human diploid fibroblasts stimulated to synthesize DNA by changing the medium. *J. Cell Physiol.* **77**, 201–212.

Sartorelli, A. C., and Creasey, W. A. (1969). Cancer chemotherapy. *Annu. Rev. Pharmacol.* **9**, 51–72.

Scott, R. B. (1970). Cancer chemotherapy—the first twenty-five years. *Brit. Med. J.* **4**, 259–265.

Sinclair, W. K. (1965). Hydroxyurea: Differential lethal effects on cultured mammalian cells during the cell cycle. *Science* **150**, 1729–1731.

Sisken, J. E., and Iwasaki, T. (1969). The effect of some amino acid analogs on mitosis and the cell cycle. *Exp. Cell Res.* **55**, 161–167.

Sisken, J. E., and Wilkes, E. (1967). The time of synthesis and the conservation of mitosis-related proteins in cultured human amnion cells. *J. Cell Biol.* **34**, 97–110.

Sisken, J. E., Kovacs, E., and Kinosita, R. (1960). Effects of cytoxan on DNA synthesis and the mitotic cycle in L4946 mouse leukemia. *Fed. Proc., Fed. Amer. Soc. Exp. Biol.* **19**, 134.

Steel, G. G., and Hanes, S. (1971). The technique of labeled mitoses: Analyses by automatic curve-fitting. *Cell Tissue Kinet.* **4**, 93–105.

Stein, G. S., and Brown, T. W. (1972). The synthesis of acidic chromosomal proteins during the cell cycle of HeLa S_3 cells. *J. Cell Biol.* **52**, 292–307.

Stock, J. A. (1966a). *Exp. Chemother.* **4**, 79–377.

Stock, J. A. (1966b). *Exp. Chemother.* **5**, 334–416.

Thomson, J. L., and Biggers, J. D. (1966). Effect of inhibition of protein synthesis on the development of pre-implantation mouse embryos. *Exp. Cell Res.* **41**, 411–427.

Tobey, R. A., and Ley, K. D. (1971). Isoleucine-mediated regulation of genome replication in various mammalian cell lines. *Cancer Res.* **31**, 46–51.

Tobey, R. A., Petersen, D. F., Anderson, E. C., and Puck, T. T. (1966). Life cycle analysis of mammalian cells. III. The inhibition of division in Chinese hamster cells by puromycin and actinomycin. *Biophys. J.* **6**, 567–581.

Todaro, G. J., Lazar, G. K., and Green, H. (1965). The initiation of cell division in a contact-inhibited mammalian cell line. *J. Cell. Comp. Physiol.* **66**, 325–334.

Verbin, R. S., and Farber, E. (1967). Effect of cycloheximide on the cell cycle of the crypts of the small intestine of the rat. *J. Cell Biol.* **35**, 649–658.

Verbin, R. S., Sullivan, R. J., and Farber, E. (1969). The effects of cycloheximide on the cell cycle of the regenerating rat liver. *Lab. Invest.* **21**, 179–182.

Vermund, H., and Gollin, F. F. (1968). Mechanisms of action of radiotherapy and chemotherapeutic adjuvants. *Cancer* **21**, 58–76.

Weinberger S. (1971). L-Asparaginase as an antitumor agent. *Enzymes* **12**, 143–160.

Wheeler, G. P. (1962). Studies related to the mechanisms of action of cytotoxic alkylating agents: A review. *Cancer Res.* **22**, 651–688.

Wheeler, G. P. (1967). Some biochemical effects of alkylating agents. *Fed. Proc., Fed. Amer. Soc. Exp. Biol.* **26**, 885–892.

Wheeler, G. P., and Alexander, J. A. (1969). Effects of nitrogen mustard and cyclophosphamide upon the synthesis of DNA in vivo and in cell-free preparation. *Cancer Res.* **29**, 98–109.

Wheeler, G. P., Bowdon, B. J., Adamson, D. J., and Vail, M. H. (1970a). Effects of certain nitrogen mustards upon the progression of cultured H. Ep. no. 2 cells through the cell cycle. *Cancer Res.* **30**, 100–111.

Wheeler, G. P., Bowdon, B. J., Adamson, D. J., and Vail, M. H. (1970b). Effect of 1, 3-bis(2-chloroethyl)-1-nitrosourea and some chemically related compounds upon the progression of cultured H. Ep. no. 2 cells through the cell cycle. *Cancer Res.* **30**, 1817–1827.

Wilkoff, L. J., Lloyd, H. H., and Dulmadge, E. A. (1971). Kinetic evaluation of the effect of actinomycin D, Daunomycin and mitomycin C on proliferating cultured leukemia L1210 cells. *Chemotherapy* **16**, 44–60.

Yarbro, J. W. (1968). Further studies on the mechanism of action of hydroxyurea. *Cancer Res.* **28**, 1082–1087.

Young, C. W. (1969). Actinomycin and antitumor antibiotics. *Amer. J. Clin. Pathol.* **52**, 130–137.

Young, C. W., and Burchenal, J. H. (1971). Cancer chemotherapy. *Annu. Rev. Pharmacol.* **11**, 369–386.

10

Effects of Purines, Pyrimidines, Nucleosides, and Chemically Related Compounds on the Cell Cycle

GLYNN P. WHEELER
and LINDA SIMPSON-HERREN

I.	Introduction	250
II.	Biochemical Mechanisms of Action	252
	A. General Discussion of Mechanisms	252
	B. Metabolic Pathways and Loci of Action of Agents	254
	C. Tabulation of Agents, Biochemical Effects on the Anabolism of Nucleic Acids, and Loci of Action on the Metabolic Pathways	255
III.	Effects of Agents on Cells	255
	A. General Discussion of Effects	255
	B. Tabulation of Effects of Agents	262
IV.	Correlation of Biochemical Effects and Effects on the Cell Cycle	263
	A. Agents Interfering with *de Novo* Synthesis of Purine Ribonucleotides and Pyrimidine Ribonucleotides	263
	B. Agents Interfering with Reduction of Ribonucleoside Diphosphates	282
	C. Agents Interfering with the Action of Thymidylate Synthetase	283
	D. Agents Interfering with the Action of DNA Polymerase	283
	E. Agents Incorporated into DNA	284
	F. Agents Incorporated into RNA	285
	G. Miscellaneous Agents	286
V.	Examples of the Logic and Use of These Agents for Cancer Chemotherapy	287
	A. Relationship of Tumor Population Kinetics to Chemotherapy	287
	B. Single versus Multiple Treatments	289
	C. Multiple Treatments versus Continuous Infusion	289
	D. Combinations for Therapy	290
	E. Conclusions	292
VI.	Concluding Comments	292
	References	293

I. Introduction

For many years, studies of cell division were directed primarily toward the events of mitosis, since visual recognition of the events made it easy to identify cells in that stage of the reproduction cycle, or cell cycle. The nonmitotic portion of the cycle was simply called interphase. Although it was known that some biosynthesis must occur during interphase, precisely what was synthesized and at which parts of interphase the synthesis occurred were largely unknown. Howard and Pelc (1953) presented evidence that the synthesis of DNA did not occur continuously throughout interphase, and these investigators defined the M, G_1, S, and G_2 phases of the cell cycle. Although the use of radioactive precursors and this concept of the cell cycle stimulated studies of the events occurring during interphase, much emphasis continued to be directed toward mitosis and the effects of chemicals on mitosis. This emphasis was made evident in the book written by J. J. Biesele in 1958 and entitled *Mitotic Poisons and the Cancer Problem*. In the opening pages of this book Biesele reviewed the definitions and classifications of mitotic poisons that a number of investigators had used, and then for use in his book he defined a mitotic poison as "any agent affecting mitosis, whether during or following treatment." He reviewed the evidence showing that a wide variety of compounds acted as mitotic poisons and the currently available evidence of the biochemical mechanisms of action of these compounds, but he could relate these mechanisms to the various parts of the cycle to only a limited degree, because little was known about the biochemistry of cells in specific parts of the cycle. He cataloged the compounds that caused "poisoning in prophase and earlier stages," "poisoning of chromosomes," and "poisoning in metaphase and later stages," and he pointed out that some compounds might poison in two or all of these ways, depending on the conditions and concentrations. The catalog included a number of purines and pyrimidines and related compounds including those listed in Table I. Biesele obviously recognized the possible importance of the effects of treatments of cells during interphase when he made the following statements.

> In summary, it is evident that inhibitors of various aspects of intermediary metabolism, nucleic acid metabolism, or protein metabolism may in some cell or other reduce the incidence of mitosis and in that sense act as preprophasic mitotic poisons. The great diversity of possible initial points of attack by mitotic poisons in the broad sense results from the complicated interlocking of the metabolic features of cell life.

The dearth of information prior to 1960 concerning the biochemistry of cells in the various parts of the cell cycle is reflected in the dates of the references in the reviews of this biochemistry by Baserga (1968), Mueller

TABLE I

CLASSIFICATION OF PURINES, PYRIMIDINES, AND RELATED COMPOUNDS AS MITOTIC POISONS[a]

Poisoning in prophase and earlier stages	
Caffeine	Adenosine
8-Ethoxycaffeine	2-Methyladenosine
Other methylxanthines	Crotonoside
8, 8'-Ethylenediaminoditheophylline	2-Hydroxyadenosine
Benzimidazole	2-Aminoadenosine
8-Azaguanine	Purine ribonucleoside
2, 6-Diaminopurine	6-Chloropurine ribonucleoside
2-Chloroadenine	6-Furfurylamino-9-β-D-ribofuranosylpurine
2-Methyladenine	
9-Methyladenine	6-Mercaptopurine ribonucleoside
2-Azaadenine	5-Bromocytidine
Purine	
6-Mercaptopurine	Purine ribonucleotide
6-Chloropurine	Adenylic acid
6-Methylpurine	Guanylic acid
6-Thioguanine	Cytidylic acid
6-(Methylthio)purine	
6-β-Indolylethylaminopurine	RNA
4-Aminopyrazolo[3, 4-d] pyrimidine	DNA

Poisoning of metaphase and later

Caffeine
Theophylline
Barbital
Adenine
2, 6-Diaminopurine
6-β-Indolylethylaminopurine
4-Aminopyrazolo[3, 4-d] pyrimidine
2, 6-Diaminopurine ribonucleoside
Guanosine

Poisoning of chromosomes

Stickiness	*Abnormal mitosis (bridging and fragmentation)*
8-Ethers of caffeine	2, 6-Diaminopurine
8-Thioethers of caffeine	Adenine (high concentrations)
Adenine	7-Methyladenine
6-Hydroxylaminopurine	6-Thioguanine
6-Thioguanine	6-Mercaptopurine ribonucleoside
6-Mercaptopurine	Purine
6-Mercaptopurine ribonucleoside	Caffeine
	8-Ethoxycaffeine
	1, 3, 7, 9-Tetramethyluric acid
	Barbital
	5-Aminouracil
	2-Aminouracil

[a] According to Biesele (1958).

(1969, 1971), Petersen *et al.* (1969), and Tobey *et al.* (1971). Since 1960 much information has been obtained. Studies of this biochemistry initially proceeded independent of the use of drugs but with the use of radioisotopes. Later, certain drugs were used to accomplish synchronization of cultures to better define the positions of cells in the cycle and to compare the sensitivities of cells to various other agents during specific portions of the cycle. In the meantime, the biochemical mechanisms of action of many agents were being worked out in a variety of biological systems without any direct concern for the cell cycle. As knowledge of the biochemistry of the cell cycle and knowledge of the mechanisms of action accumulated, it became possible to use each of these areas of knowledge to expand the other area. For example, an agent known to specifically inhibit the synthesis of DNA could be used to study the effects of cessation of the synthesis of DNA upon the progression of cells through G_2, M, and G_1. On the other hand, the relative sensitivities to a particular agent of cells in various phases of the cycle might yield information relating to the mechanism of action of the agent. It has also become possible to use the combined knowledge of these two areas to design drug regimens for use in experimental and clinical chemotherapy.

This chapter (1) presents mechanisms of action of certain naturally occurring purines and pyrimidines and their nucleosides and analogs of such compounds, (2) presents known effects of these bases, nucleosides, and analogs upon progression of cells through the cycle and upon viability, (3) relates biochemical mechanisms of action and effects upon progression through the cycle and upon viability, and (4) presents examples and some of the logic of the use of the above-mentioned agents, singly or in combination with each other or with agents of other types, for cancer chemotherapy. A number of published reviews have been used as secondary literature sources in preparing this chapter, and the reader can refer to the indicated reviews if he wishes to see the primary source material.

II. Biochemical Mechanisms of Agents

A. GENERAL DISCUSSION OF MECHANISMS

Most of the compounds that are considered in this review affect the synthesis and/or functioning of nucleic acids. Most of them do not elicit maximum metabolic inhibition per se but must be converted to the nucleotides or higher phosphates, which are the real inhibitors. However, since the cell membranes of mammalian cells are very slightly permeable to nucleotides and higher phosphates (because of the ionic nature of these compounds) but are permeable to many purines, pyrimidines, and related bases and nucleosides, the nonphosphorylated compounds are administered to the biologic system, and conversion to the nucleotides occurs intracellularly. Although the analog bases and nucleosides might compete with naturally occurring bases and nucleosides for certain anabolic enzymes, these enzymes are usually among those that are involved in "salvage" pathways and are not essential for cell survival. Therefore, competitive or noncompetitive inhibition of these enzymes might not affect the viability and functioning of cells. On the other hand, it is essential that the analogs successfully compete with these natural substrates to the extent that inhibitory concentrations of the analog nucleotides or higher phosphates are formed.

Once the nucleotides or higher phosphates are formed, there are three major ways in which they might function. (1) They may act as feedback (if the administered compounds are normally occurring compounds such as adenine, deoxyadenosine, or thymidine), pseudofeedback (if analogs are administered), or allosteric inhibitors of metabolic processes in the synthesis of nucleic acids. (2) They may compete with normal nucleotides for enzymes that catalyze the normal anabolic steps or interconversions of nucleotides. (3) They may be converted to the nucleoside triphosphates and then be incorporated into the nucleic acids. It would be expected that agents that function by ways (1) and (2) would affect the cells adversely soon after exposure to the agents, as there would soon be a deficit of essential intermediates in the synthesis of the nucleic acids; the time required for these effects to become evident would depend partially on the sizes of the pools of the intermediates at the time that exposure to the agent commenced and on the turnover times of these pools. If the agents are incorporated into the DNA or RNA, and this incorporation causes an immediate cessation of chain growth, the effects might be evident quickly; if incorporation permits continuation of chain growth but causes subsequent errors in base pairing and coding in replication, transcription, or translation, the detectable effects might be delayed. In some instances incorporation of analogs into the DNA increases the sensitivity of the cells to subsequent irradiation.

Catabolism of the agents or of the nucleotides derived from them can be a factor in limiting the cytotoxic effectiveness of an agent, if the product of the catabolism is nontoxic. The decreased concentration of the agent resulting from catabolism might also permit the dissociation of an analog–enzyme complex that prevents the synthesis of an essential intermediate. Therefore unless the concentration of the agent is maintained above a minimum level, the effect of the agent will probably be temporary. In living systems, transient effects of agents are commonly encountered, unless the agents are administered in multiple doses at suitable time intervals or are administered by continuous infusion over an extended period of time.

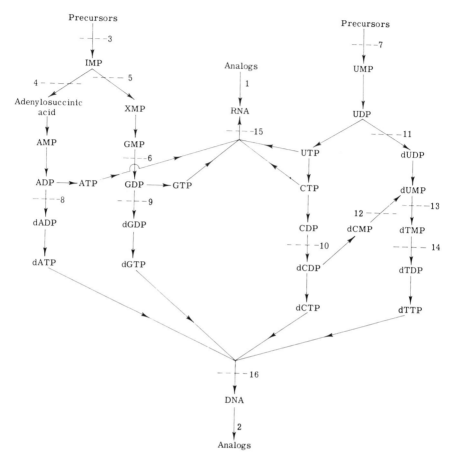

Fig. 1. Metabolic map of the synthesis of nucleic acids showing the sites of inhibition by agents included in this chapter.

B. Metabolic Pathways and Loci of Action of Agents

As stated above, the evidence presently available indicates that most of the compounds included in this review affect primarily the synthesis and functioning of nucleic acids. Figure 1 gives a simplified metabolic map for the biosynthesis of DNA and RNA. The early steps in the synthesis of inosine monophosphate (IMP) and uridine monophosphate (UMP) are not shown, the necessary cofactors and reactants are not given, and reversible reactions are not indicated. This map serves simply for the location of the steps at which the agents act. The loci of inhibition are indicated by dashed lines across the arrows, and each locus is given a number. These numbers are used in Table II to show the points of action of the individual agents.

C. Tabulation of Agents, Biochemical Effects on the Anabolism of Nucleic Acids, and Loci of Action on the Metabolic Pathways

Table II gives the biochemical effects on the anabolism of nucleic acids of the individual agents and the points of inhibition on the metabolic map. A number of the agents have multiple biochemical effects, and the relative importance of them will be discussed in Section IV. Some of the agents affect bacterial, botanical, and mammalian systems differently, and for inclusion in the table mammalian systems have been given priority. The agents included in Tables II and III are those for which there is reported evidence relating them to the cell cycle.

III. Effect of Agents on Cells

A. General Discussion of Effects

1. *Effects on Progression through the Cell Cycle*

As was pointed out in Section I, early observations of the effects of drugs on the progression of cells through the cell cycle were based on enumeration and examination of mitotic cells. An accumulation of cells at some stage of mitosis indicated an inhibition of progression through mitosis, whereas a decrease in mitotic index indicated a preprophase, or interphase, inhibition. These techniques are still valid and useful, but supplementary techniques have made it possible to gain more specific information about the point in the cycle at which inhibition of progression occurs. Newer techniques include the determination of (1) the labeling indexes after pulse or extended exposure to ^{3}H-thymidine and pulse or extended exposure to an agent; (2) the effects of an agent upon the rate of labeling cells and upon the rate of accumulation of the labeled and unlabeled cells at metaphase in the presence of colcemid; (3) effects of agents on the percent labeled mitosis curves; and (4) the effects of the agents on the DNA content of cells as determined by microspectrophotometry. These techniques are simply modifications of techniques used to determine the duration of the cell cycle and its phases in the absence of the test agent (Cleaver, 1967).

A number of the agents included in this review inhibit specifically the synthesis of DNA and, therefore, primarily affect cells in S phase; this effect is a retardation or prevention of progression of cells through the S phase. Since cells in the other phases of the cycle already have DNA complements characteristic of cells in those phases, they will proceed through the cycle until they reach the G_1–S transition point. If cells are exposed to such an agent for a sufficient period of time, the cells initially in G_1, M, and G_2 will proceed to a point near the G_1–S transition point, where they will

TABLE II
BIOCHEMICAL MECHANISMS OF ACTION OF AGENTS

Agent	Biochemical effects on the anabolism of nucleic acids	Sites of action on anabolic pathways to nucleic acids
Caffeine (NSC 5036), theophylline (NSC 2066), or theobromine (NSC 5039)	Inhibits 3',5'-cyclic nucleotide diesterase (Butcher and Sutherland, 1962); binds to DNA, particularly denatured DNA (Ts'o and Lu, 1964); inhibits purine nucleoside phosphorylases (Koch and Lamont, 1956)	
6-Mercaptopurine (NSC 755)	Incorporated into nucleic acids as 6-thioguanine; feedback inhibition of formation of phosphoribosylamine; competitive inhibitor of guanylic acid pyrophosphorylase; inhibits conversion of IMP to AMP (Balis, 1968; Heidelberger, 1967); converted to 6-methylthiopurine ribonucleotide, which, in turn, inhibits the formation of phosphoribosylamine (Bennett and Allan, 1971)	1–4
6-Thioguanine (NSC 752)	Converted to the ribonucleotide, which inhibits formation of phosphoribosylamine, conversion of IMP to XMP, and phosphorylation of GMP, and is incorporated into RNA and DNA (Montgomery et al., 1970; Agarwal, et al., 1971; LePage and Jones, 1961)	1–3, 5, 6
4-Aminopyrazolo [3, 4-d]pyrimidine (NSC 1393)	Converted to ribonucleoside mono-, di-, and triphosphates; inhibits formation of phosphoribosylamine (Montgomery et al., 1970)	3
2'-Deoxyadenosine (NSC 83258)	Inhibits dihydroorotase; converted to mono-, di-, and triphosphate; dAMP inhibits aspartate transcarbamylase and orotidylate decarboxylase; dATP inhibits reduction of ADP, GDP, UDP, and CDP to the corresponding deoxyribonucleoside diphosphates (Henderson, 1965; Reichard, 1968)	7–11
3'-Deoxyadenosine (Cordycepin) (NSC 401022)	The monophosphate inhibits the formation of phosphoribosylamine (Rottman and Guarino, 1964). The triphosphate inhibits the reduction of ADP and GDP (Chassy and Suhadolnik, 1968) and the formation of 5-phosphoribosyl-1-pyrophosphate (Ov-	1, 3, 7–9

10. PURINES, PYRIMIDINES, AND RELATED COMPOUNDS

Table II Cont'd.

Agent	Biochemical effects on the anabolism of nucleic acids	Sites of action on anabolic pathways to nucleic acids
	ergaard-Hansen, 1964). Incorporation into RNA prevents further elongation (Siev et al., 1969)	
9-β-D-Arabinofuranosyladenine (NSC 404241)	Converted to the triphosphate which inhibits DNA polymerase (York and LePage, 1966; Furth and Cohen, 1967, 1968) and ribonucleotide reductase (Moore and Cohen, 1967); is incorporated into DNA and acts as a chain terminator (Waqar et al., 1971)	2, 8–11, 16
6-(Methylthio) purine ribonucleoside (NSC 40774)	Inhibits synthesis of phosphoribosylamine (Heidelberger, 1967)	3
2'-Deoxyguanosine (NSC 22837)	Converted to the mono-, di-, and triphosphates; the monophosphate inhibits aspartate transcarbamylase (Henderson, 1965); the triphosphate inhibits the reduction of CDP, UDP, and GDP (Moore and Hurlburt, 1966)	7, 9, 10, 11
6-Thiodeoxyguanosine (NSC 71261)	Converted to the triphosphate and incorporated into DNA (Montgomery et al., 1970)	2
Puromycin (NSC 3055)	Interferes with protein synthesis, perhaps by binding to peptidyl RNA, thus interrupting extension of the protein chain and causing separation from the ribosomes (Montgomery et al., 1970)	
3'-Amino-3'-deoxy-N,N-dimethyladenosine (Aminonucleoside of puromycin) (NSC 3056)	May be demethylated to 3'-amino-3'-deoxyadenosine, which inhibits synthesis of RNA (Montgomery et al., 1970)	(15)[a]
7-Deazaadenosine (Tubercidin) (NSC 56408)	Incorporated into RNA and DNA and inhibits the synthesis of RNA, DNA, and protein (Heidelberger, 1967); inhibits formation of phosphoribosylamine (Hill and Bennett, 1969)	1–3
3',5'-Cyclic adenosine monophosphate (NSC 94017)	Activates protein kinases (Greengard, 1971)	

Table II Cont'd.

Agent	Biochemical effects on the anabolism of nucleic acids	Sites of action on anabolic pathways to nucleic acids
5-Aminouracil (NSC 22474)	Inhibits the synthesis of DNA in Chinese hamster cells (Chu, 1965) and slows the rate of synthesis of DNA and RNA in *Vicia faba* (Jakob and Trosko, 1965; Jakob, 1968)	(15, 16)
5-Fluorouracil (NSC 19893)	Is incorporated into RNA but not into DNA; is converted to the ribonucleoside triphosphate, which inhibits aspartate transcarbamylase and uridine kinase; is converted to 5-fluorodeoxyuridine monophosphate, which inhibits thymidylate synthetase (Heidelberger, 1967)	1, 7, 13
2-Thiouridine (no NSC No.) or 4-thiouridine (NSC 518132) [perhaps also 2-thiouracil (NSC 1905) or 2-thiocytosine (NSC 45755)]	Inhibits the synthesis of RNA at a concentration that does not inhibit the synthesis of DNA (Lozzio and Wigler, 1971). 2-Thiouracil is incorporated into the RNA of tobacco mosaic virus (Balis, 1968)	1, (15)
Thymidine (NSC 21548)	Inhibits synthesis of DNA prior to that of RNA and does not inhibit synthesis of protein (Kim et al., 1965). Inhibits aspartate transcarbamylase; the triphosphate inhibits thymidine kinase, deoxyuridine kinase, deoxycytidylate deaminase (Henderson, 1965), and the reduction of UDP and CDP (Moore and Hurlburt, 1966; Larsson and Reichard, 1966; Fujioka and Silber, 1970)	7, 10–12
5-Fluorodeoxyuridine (NSC 27640)	Is converted to the monophosphate, which inhibits thymidylate synthetase (Heidelberger, 1967)	13
5-Bromodeoxyuridine (NSC 38297)	Inhibits aspartate transcarbamylase and dihydroorotase (Henderson, 1965); inhibits synthesis of polysaccharides (Bischoff, 1971); incorporated into DNA (Eidinoff et al., 1959; Hakala, 1959; Littlefield and Gould, 1960; Djordjevic and Szybalski, 1960)	2, 7

10. PURINES, PYRIMIDINES, AND RELATED COMPOUNDS

Table II Cont'd.

Agent	Biochemical effects on the anabolism of nucleic acids	Sites of action on anabolic pathways to nucleic acids
5-Iododeoxyuridine (NSC 39661)	Inhibits aspartate transcarbamylase (Henderson, 1965); inhibits incorporation of formate and orotic acid into DNA; inhibits thymidine kinase; the monophosphate inhibits thymidylate kinase; the triphosphate may inhibit DNA polymerase; is incorporated into DNA (Heidelberger, 1967)	2, 7, 14, (16)
5-Bromodeoxycytidine (NSC 61765)	Inhibits aspartate transcarbamylase (Henderson, 1965); is deaminated either before or after phosphorylation and therefore has effects similar to 5-bromodeoxyuridine (Cramer et al., 1961)	2, 7
1-β-D-Arabinofuranosylcytosine (NSC 63878)	Inhibits DNA polymerase and is incorporated into DNA (Furth and Cohen, 1968; Graham and Whitmore, 1970b; Momparler, 1969; Inagaki et al., 1969); acts as a chain terminator in DNA (Waqar et al., 1971)	2, 16
1-β-D-Arabinofuranosyl-2-thiocytosine (no NSC No.)	Inhibits the synthesis of DNA more than the synthesis of RNA (Bremerskov et al., 1970)	(15, 16)
1-β-D-Arabinofuranosyl-5-fluorocytosine (NSC 529180)	Inhibits the synthesis of DNA with little, or no, effect upon the synthesis of RNA (Kim et al., 1966; Lenaz et al., 1969)	(16)
5-Azacytidine (NSC 102816)	Converted to the mono-, di-, and triphosphates and incorporated into RNA and DNA and inhibits synthesis of RNA and DNA in L1210 cells (Li et al., 1970a,b; Heidelberger, 1967); inhibits synthesis of RNA and indirectly inhibits synthesis of protein (Montgomery et al., 1970).	1, 2, (15, 16)
6-Azauridine (NSC 32074)	Is converted to the nucleotide which inhibits orotidylate decarboxylase (Montgomery et al., 1970)	7

^{a} The presence of a parentheses around numbers 15 or 16 indicates that the agent inhibited the synthesis of RNA or DNA but not necessarily at the polymerization step.

accumulate. If the agent is then removed from the cells, the cells initially in S and those accumulated near the G_1–S transition point will proceed through the cycle, and partial synchronization of the population will occur. Since DNA synthesis is prevented in cells in the S phase while the synthesis of RNA and of protein continue, a condition of unbalanced growth (Cohen and Studzinski, 1967) will result, and the cells will become larger than normal. Therefore, the cells of the synchronized culture will be abnormal and not characteristic of normal cells in any stage of the cycle (Studzinski and Lambert, 1969b; Petersen *et al.*, 1969; Whitmore, 1971). If the extent of unbalanced growth is excessive, the cells will die. Caution must be exercised in preparing synchronized cultures by using inhibitors of the synthesis of DNA and in interpreting data obtained with such cultures. Nevertheless, if one properly takes into account the undesirable features of these cultures, he can use them to obtain valuable information about the biochemistry of cells in the various phases of the cycle and about the effects of miscellaneous agents upon cells during the various phases (Sinclair, 1970; Whitmore, 1971).

Agents that interfere with the *de Novo* synthesis of ribonucleotides should inhibit the synthesis of both DNA and RNA and stop the progression of cells through S and through other phases in which the synthesis of RNA is necessary for progression to occur. Since cells in the latter part of G_2 and in M can proceed to divide in the absence of synthesis of RNA (Tobey *et al.*, 1971), cells might accumulate in the early part of G_1, but the extent of this accumulation would be too small to be of significant value for producing synchronized cultures. Unbalanced growth and formation of giant cells should not occur under these conditions, and cell viability should remain higher than under conditions of unbalanced growth (Whitmore *et al.*, 1969).

Agents that have multiple sites of action might inhibit the synthesis and function of both DNA and RNA, but the effects might be greater with respect to DNA or RNA. If the inhibition is greater for DNA, some unbalanced growth might occur, and cells might accumulate to some extent in the S phase or near the beginning of the S phase. If the inhibition is greater for RNA, cells in G_1, S, and, perhaps, early G_2 would progress through the cycle to an extent dependent on the sizes and the rates of turnover of the preexisting pools of RNA (Mueller, 1971; Tobey *et al.*, 1971).

2. *Cytotoxic Effects*

Determinations of the extent of cell survival and multiplication following exposure to specified agents are informative concerning the relations of drugs to the cell cycle. Colony formation by cultured cells (Puck *et al.*,

1956), spleen colony formation by normal hematopoietic cells (Till and McCulloch, 1961) and transplanted murine lymphoma cells (Bruce and van der Gaag, 1963), and bioassay of neoplastic cells (Skipper et al., 1961) to determine the viability of the cells following exposure to various drugs have been quite useful. The shape of the curve obtained, when the logarithm of the surviving fraction (when the period of exposure to the agent is short compared to the length of the cycle) is plotted against the concentration of the agent, is indicative of the cycle-stage specificity of the agent. When exponential curves are obtained, the usual interpretation is that the agent kills cells in all phases of the cell cycle; when the curve falls rapidly and then flattens for the higher concentrations, the usual interpretation is that the agent kills only cells that are in certain phases of the cell cycle and that increasing the concentration of the agent does not increase the cell kill (Bruce et al., 1966). These interpretations are based upon the idea that if all the cells in an asynchronous population are equally sensitive to an agent and the population is exposed to various concentrations of the agent for a short fixed time, cell kill will follow first-order kinetics (Wilcox, 1966); departure from first-order kinetics is indicative that all cells in the population are not equally sensitive to the agent. Axiomatically, if all the cells are equally sensitive to the agent and the population is exposed to a designated concentration of the agent for various periods of time, the cell kill will follow first-order kinetics; again departure from first-order kinetics indicates that all of the cells in the population are not equally sensitive to the agent (Wilcox, 1966; Wilkoff et al., 1967a,b).

Exposure of synchronized populations of cells to an agent when the populations are known to be in specific parts of the cell cycle and determination of the surviving fractions following such exposures make it possible to determine at what part of the cycle the cells are most sensitive to the agent (Madoc-Jones and Mauro, 1968; Mauro and Madoc-Jones, 1970).

There is evidence that cells that are not actively proliferating, but have the potential for beginning to proliferate under certain conditions, are less sensitive to certain types of agents than cells that are progressing through the cycle at the time of exposure. This fact is important in cancer chemotherapy, because the growth fractions of many solid tumors are quite low (Simpson-Herren and Lloyd, 1970; Skipper, 1971b). Differential cytotoxicities of drugs for proliferating and nonproliferating cells have been demonstrated in several experimental systems. Bruce and co-workers (1966; Bruce and Valeriote, 1968) have used normal hematopoietic stem cells and transplanted lymphoma cells. Schabel et al. (1965) treated suspensions of murine leukemia L1210 ascites cells in ascitic fluid [under these conditions the cells do not divide (Skipper et al., 1964)] with various agents and determined the extent of cell kill by bioassay; the data could be compared with results obtained by treating L1210 cells with the same agents *in vivo*. Madoc-Jones and Bruce (1967) treated suspension cultures of L cells during exponential and asymptotic stages of growth, and Thatcher and Walker (1969) treated proliferating glass-attached hamster cells and glass-attached cells that were part of a confluent sheet of cells and were not proliferating. The experimental results show that some agents have toxic effects on nonproliferating cells, while others do not; generally speaking, antimetabolites have low cytotoxicity for nonproliferating cells.

Under certain circumstances, some of the agents considered in this chapter cause cytotoxicity in somewhat indirect ways. Although the incorporation of 5-bromodeoxyuridine (after phosphorylation) into DNA may have cytotoxic effects per se, the presence of the 5-bromouracil moiety in the DNA labilizes the DNA to subsequent irradiation and causes increased cytotoxicity for a given dose of ultraviolet radiation (Djordjevic and Szybalski, 1960; Cleaver, 1968; Scaife, 1970). This increased lability and cytotoxicity has been used in a method of assaying the repair of damaged DNA (Regan et al., 1971). Although the initial incorporation of the 5-bromodeoxyuridine would occur primarily during the S phase, the cells then containing the 5-bromouracil would be more sensitive subsequently to radiation throughout the cycle. The presence of 5-iodouracil in the DNA has less effect than 5-bromouracil upon sensitivity to ultraviolet light, and the presence of 5-bromouracil in the DNA has less effect upon sensitivity to X-rays than to ultraviolet radiation (Scaife, 1970). The cytotoxicity of ultraviolet irradiation is also increased if cells are incubated in the presence of caffeine (at concentrations that cause little cytotoxicity per se) following irradiation; it is believed that the caffeine interferes with the repair of radiation damage (Cleaver, 1969; Domon and Rauth, 1969a). The incorporation of highly radioactive ^{3}H-thymidine (Cleaver, 1967), ^{125}I-5-iododeoxyuridine, and ^{131}I-5-iododeoxyuridine (Hofer and Hughes, 1971) into DNA results in death of the cells primarily as a result of the effects of radiation and perhaps partially as a result of transmutation of the atoms. By controlling the period of exposure of a proliferating asynchronous population of cells to one of these radioactive substrates, it is possible to allow only a specific fraction of the population of cells to incorporate the radioactive substrate and undergo radiation death; the cells that do not incorporate the radioactive substrate will remain viable, and a partially synchronized proliferating population results (Whitmore and Gulyas, 1966).

B. TABULATION OF EFFECTS OF AGENTS

Table III contains information about the effects of a number of agents on the cell cycle in a variety of biological systems. For some of the agents there is little information presently available, but for others many reports are in the literature. Although an effort has been made to cover the literature, no claim is made for complete coverage.

IV. Correlation of Biochemical Effects and Effects on the Cell Cycle

A. Agents Interfering with *de Novo* Synthesis of Purine Ribonucleotides and Pyrimidine Ribonucleotides

Since free purines do not occur on the *de Novo* pathway of the synthesis of purine ribonucleotides, the purine analogs per se do not compete with the normal substrates for an anabolic enzyme at any step of this synthesis, but they conceivably could compete with glutamine for 5-phosphoribosyl-1-pyrophosphate in the first step of the *de Novo* pathway. A number of ribonucleoside phosphates and deoxyribonucleoside phosphates of naturally occurring purines and purine analogs can inhibit the synthesis of phosphoribosylamine by feedback and pseudofeedback mechanisms (Hill and Bennett, 1969). However, Henderson (1965) concluded that there was no convincing evidence that the primary mechanism of action of any purine or pyrimidine analog was pseudofeedback inhibition of this *de Novo* synthesis. Subsequent to the time when Henderson wrote his review, Hill and Bennett (1969) observed that 6-(methylthio)purine ribonucleotide is a more potent inhibitor of 5-phosphoribosyl pyrophosphate amidotransferase than 6-mercaptopurine ribonucleotide and 6-thioguanine ribonucleotide. Since the inhibition of growth of cultured cells by 6-mercaptopurine is reversed or prevented by 4-amino-5-imidazolecarboxamide (Hakala and Nichol, 1964) and that of 6-(methylthio)purine ribonucleoside is reversed or prevented by 4-amino-5-imidazolecarboxamide or hypoxanthine (Bennett and Adamson, 1970), it appears that pseudofeedback may be the primary mechanism of action of this agent. Since 6-mercaptopurine ribonucleotide is methylated in *in vitro* and *in vivo* systems (Bennett and Allan, 1971), it appears that a portion (perhaps all) of the toxicity observed upon administering 6-mercaptopurine might be due to the pseudofeedback inhibition caused by the generated 6-(methylthio)purine ribonucleotide. Although methylation of 6-thioguanylic acid also occurs, the resulting 6-methylthioguanylic acid is a poorer inhibitor of phosphoribosyl pyrophosphate amidotransferase than 6-thioguanylic acid (Allan and Bennett, 1971). The phosphate of 7-deazaadenosine (tubercidin) inhibits phosphoribosyl pyrophosphate amidotransferase, but it is a relatively weak inhibitor. Following the exposure of cells to the ribonucleoside of 4-aminopyrazolo [3, 4-*d*] pyrimidine there is evidence of strong pseudofeedback inhibition of the *de Novo* synthesis of purine ribonucleotides, but it is still uncertain whether this is the primary mechanism

TABLE III
Effects of Agents on Cells[a]

Agent	Dose or concentration	Biological system	Effects on cells	Reference
Caffeine (NSC 5036)	$2 \times 10^{-3} M$	Mouse L cells	Small delays of $G_1 \to S$ and $G_2 \to M$; doubling time increased from 16 to 18–20 hours	Domon and Rauth (1969b)
	$10^{-3} M$	Human kidney T cells	Stimulation of mitoses in cells exposed in G_2	Scaife (1970)
	0.05%–1.0%	Allium cepa roots	Increases binucleated cells, decreases mitoses	Giménez-Martín et al. (1969)
	Various concentrations	2 Ultraviolet-irradiated Chinese hamsters cell lines	No effect on survival	Arlett (1967)
	$2 \times 10^{-3} M$	Ultraviolet-irradiated mouse L cells	Blocks progression through S, no effect on $G_1 \to S$ or $G_2 \to M$	Domon and Rauth (1969b)
	10^{-4}–$10^{-2} M$	Asynchronous UV-irradiated Chinese hamster cells	Inhibits first post-irradiation S phase, kills cells	Cleaver and Thomas (1969)
	10^{-3}–$10^{-2} M$	Asynchronous rat thymocytes in culture, X-irradiated	Protects cells against initial effects of high radiation doses on nucleic acid synthesis and retards degeneration	Myers (1971)
	$2 \times 10^{-3} M$	Mouse L cells treated with mitomycin	Cell kill increased in cells incubated with caffeine during first post-treatment cycle	Rauth et al. (1970)
6-Mercaptopurine (NSC 755)	$6.7 \times 10^{-5} M$	H. Ep. No. 2 Colcemid blocked	G_1/S blocked or delayed with no accumulation, S-phase cells unaffected; $G_2 \to M$ delayed, possibly killed G_1 cells; nonexponential cell kill	Wheeler et al. (1972)

			G$_1$ arrest	
—	J-128 cell culture			Vandevoorde and Hansen (1970)
5-30 mg/mouse given in six doses × 4 hours	Lymphoma colony forming and normal hematopoietic colony cells *in vivo*		Little effect on survival of either; may be either cycle specific or cycle-stage specific	Bruce et al. (1966)
6-Thioguanine (NSC 752)				
$6 \times 10^{-5} M$	H. Ep. No. 2 cells, Colcemid blocked		G$_1$/S blocked or delayed, S phase delayed, G$_2 \to$ M unaffected; nonexponential cell kill	Wheeler et al. (1972)
10^{-6}–$10^{-2} M$ for 72 hours	PHA-stimulated human lymphocytes		Decreases mitoses; above $10^{-3} M$ prevents mitoses and kills some cells	Price and Timson (1971)
4-Aminopyrazolo[3,4-d]pyrimidine (NSC 1393) plus guanine ($2 \times 10^{-4} M$)				
$5 \times 10^{-6} M$	Asynchronous Ehrlich cells *in vitro*		Accumulates cells in G$_1$ or early S; inhibits DNA synthesis exponentially, RNA very slightly, and protein about 50%; RNA and protein continuously degraded during treatment, cells remain viable	Schachtschabel et al. (1968)
2'-Deoxyadenosine (NSC 83258)				
$2.5 \times 10^{-3} M$	Chang appendix cells		Blocked mitoses, caused cell synchronization	Xeros (1962)
4×10^{-4}–$4 \times 10^{-3} M$	HeLa-S3		Exponential cell kill, non-cycle-stage specific	Kim et al. (1965)
3×10^{-5}–$5 \times 10^{-6} M$	Human leukocytes		Chromosome breaks in G$_2$, no effects in G$_1$ and S	Cohen and Shaw (1967)
$2 \times 10^{-3} M$	Human kidney cells		Slows DNA synthesis, does not block entry into S, synchronizes population; no chromosome aberrations	Galavazi and Bootsma (1966); Galavazi et al. (1966)

Table III Cont'd.

Agent	Dose or concentration	Biological system	Effects on cells	Reference
	—	*Vicia faba* root tips	Strong inhibition of DNA synthesis, little effect on RNA synthesis; chromosome breaks	Kihlman and Odmark (1966); Odmark and Kihlman (1965)
	1×10^{-3} M	X-irradiated HeLa S3 cells mitotically synchronized	Postirradiation treatment slightly enhances cell kill	Weiss and Tolmach (1967)
3'-Deoxyadenosine (Cordycepin) (NSC 401022)	10^{-4} M	*Allium cepa* root tips	Exposure in late interphase inhibits prophase→metaphase, increases mitotic index	González-Fernández et al. (1970)
	5×10^{-5} M	*Vicia faba* root tips	Inhibits mitotic spindle and produces extreme chromosome contraction; inhibits both RNA and DNA	Kihlman and Odmark (1966)
9-β-D-Arabinofuranosyladenine (NSC 404241)	1×10^{-3}–1×10^{-5} M	Asynchronous human leukocytes *in vitro*	Decrease in mitotic index correlates with chromosome breaks; S phase probably most sensitive, possibly also G_2	Nichols (1964)
	2×10^{-3} M	*Vicia faba*	Chromosome damage similar to that of other DNA inhibitors; little inhibition of mitoses after 6 hours exposure	Rao and Natarajan (1965)
	0.5–1.0×10^{-3} M	*Vicia faba*, colchicine blocked	Chromosome aberrations	Kihlman and Odmark (1966)

Compound	Concentration	System	Effects	Reference
6-(Methylthio)-purine ribonucleoside (NSC 40774)	3×10^{-5} M	H. Ep. No. 2 Colcemid blocked	$G_1 \to S$ blocked or delayed, S delayed, $G_2 \to M$ unaffected; nonexponential cell kill	Wheeler et al. (1972)
9-Alkyl-6-mercapto-purines	—	J-128 cell culture	Arrested growth in late S or G_2 only	Vandevoorde and Hansen (1970)
2'-Deoxy-guanosine (NSC 22837)	1.5×10^{-3} M	Chang appendix cells	Blocks mitoses and synchronizes cells	Xeros (1962)
	2×10^{-4} M	HeLa	Synchronization of cells; blocks DNA synthesis, but not synthesis of protein, RNA, or phospholipids.	Mueller (1963)
	2×10^{-3} M	Human kidney	Causes synchronization, no chromosome aberrations	Galavazi et al. (1966)
6-Thiodeoxy-guanosine (NSC 71261)	3.5×10^{-4} M	Chinese hamster ovary cells synchronized by double dThd block	Most sensitive in early and mid-S, with little effect on synthesis of DNA; M cells unaffected and G_1 cells only slightly sensitive; no progression or division delay even when treated in S and when survival was 11%.	Barranco and Humphrey (1971)
Puromycin (NSC 3055)	1.06×10^{-5} M	Rabbit kidney cortex cells cultured directly from rabbit	Prevents cells in most of G_2 from entering mitosis	Kishimoto and Lieberman (1964)
	1.06×10^{-4} M	Chinese hamster ovary cells synchronized with dThd	Blocks initiation of mitosis, G_2/M may be double block	Tobey et al. (1966)
	2.5 mg/mouse injected for 1-4 hours	Mouse small intestine	Blocks S-phase cells, slight delay of G_2 and M; G_1 possibly blocked; necrosis of dividing crypt cells in 3.5 hours	Estensen and Baserga (1966)

Table III Cont'd.

Agent	Dose or concentration	Biological system	Effects on cells	Reference
	$2.12 \times 10^{-5} M$	Asynchronous murine plasmacytoma cells	Marked inhibition of protein synthesis in 1 hour, cell viability decreases in 6 hours	Dehner et al. (1969)
	$10^{-3} M$	Pea root meristems synchronized with 5-FdUrd	Delayed transit of S but exposure at G_1/S boundary did not prevent onset of DNA synthesis.	Kovacs and Van't Hof (1971)
	$4.24 \times 10^{-5} M$	Chinese hamster cells synchronized with 5-FdUrd	Blocks at G_1/S, early S stopped, late S completes DNA synthesis	Epifanova et al. (1969)
	$2.12 \times 10^{-4} M$	HeLa S3, mitotically synchronized	G_1/S block, also blocks mitoses	Mauro and Madoc-Jones (1970)
	—	Cultured human amnion cells	Inhibits protein synthesis, major effect due to incorporation into protein rather than block	Sisken and Wilkes (1967)
	$1.06 \times 10^{-5} M$	X-irradiated HeLa cells, mitotically synchronized	Decreases survival of cells irradiated in S-phase	Djordjevic and Kim (1969)
	2.5 mg/mouse injected hourly for 1–4 hours	Mouse crypt cells	Inhibits DNA and RNA synthesis, stimulates protein synthesis, G_2 plus M cells slightly delayed, S cells blocked, possibly also G_1	Estensen and Baserga (1966)
3'-Amino-3'-deoxy-N,N-dimethyla-denosine (amino nucleoside of puromycin) (NSC 3056)	Range of concentrations	Exponentially growing HeLa cells	Low concentrations selectively inhibit ribosomal and transfer RNA with no effect on division; higher doses inhibit synthesis of nucleic acid and protein	Studzinski and Ellem (1966)

Compound	Concentration	System	Effect	Reference
	$3.4 \times 10^{-4}\ M$	X-irradiated HeLa cells mitotically synchronized	End of S phase most sensitive, G_2 intermediate, G_1 least sensitive	Djordjevic et al. (1968)
7-Deazaadenosine (Tubercidin) (NSC 56408)	—	Asynchronous Don cells	Not cycle-stage specific, exponential cell kill	Bhuyan et al. (1970)
3′,5′-Cyclic adenosine monophosphate (NSC 94017)	4–6 mg/kg	Rat bone marrow and thymus in vivo	Stimulates mitosis	Rixon et al. (1970)
	$10^{-3}\ M$	Human kidney cells in vitro synchronized by double dThd block	Inhibits mitosis	Scaife (1971)
	10^{-6}–$10^{-3}\ M$	Human kidney cells in vitro, asynchronous	No effect	Scaife (1971)
5-Aminouracil (NSC 22474)	5.9×10^{-4}–$1.18 \times 10^{-2}\ M$	Vicia faba roots	Block at S/G_2, decrease in synthesis of DNA and RNA; decrease in mitoses; synthesis of DNA continues at decreased level for 24-hours treatment	Socher and Davidson (1971)
	$2 \times 10^{-3}\ M$	Human kidney cells	Causes synchronization; no chromosome aberrations	Galavazi et al. (1966)
	10^{-3}–$10^{-2}\ M$	Vicia faba root tips, Haplopappus gracilis, Crepis capillaris	Mitotic synchrony and chromosome aberrations; possibly blocks G_2	Jakob and Trosko (1965); Eriksson (1966); Lavrovskii (1969); Resch and Schroeter (1969)

Table III Cont'd.

Agent	Dose or concentration	Biological system	Effects on cells	Reference
	—	L-strain fibroblasts	Causes synchronization	Cone and Tongier (1969)
	$3 \times 10^{-3} M$	HCAAT derived from HeLa	Blocks synthesis of DNA and RNA, causes synchronization; after release, DNA synthesized during first S phase but not RNA	Regan and Chu (1966)
5-Fluorouracil (NSC 19893)	5–18 mg/kg/day for 7 days	Ehrlich ascites *in vivo*	Decrease in DNA, increase in RNA and protein per cell, unbalanced growth; chromosome damage	Lindner (1959); Lindner et al. (1963)
	3.85×10^{-6}–$1.16 \times 10^{-5} M$	Asynchronous HeLa S3	Probably S phase specific, 6% survivors in 24 hours at any dose	Berry (1964)
	$3.85 \times 10^{-5} M$	HeLa cells synchronized by dThd	Maximum cytotoxicity to cells exposed in S phase	Adams et al. (1967)
	2.82×10^{-4}–$1.16 \times 10^{-3} M$	Asynchronous cultured L1210	Gompertzian cell kill, cycle-stage specific	Wilkoff et al. (1967b)
	0–1 mg/mouse (given as 6 doses × 4 hours) *in vivo*; 3.08×10^{-4}–$1.85 \times 10^{-3} M$ *in vitro*	Asynchronous mouse L cells; lymphoma colony forming, normal hematopoietic *in vivo*; asynchronous Don-C	Essentially all phases sensitive (except M possibly), exponential cell kill	Madoc-Jones and Bruce (1967); Bruce et al. (1967); Bhuyan et al. (1970)
	12.5 mg/kg	Clonogenic carcinoma in hamster cheek pouch	Synergism with X-ray	Goldenberg and Ammersdörfer (1970)
	Up to $1.16 \times 10^{-4} M$	H. Ep. No. 2	Inhibition of S-phase cells, partial synchrony at G_1/S, no delay of $G_2 \to M$, exponential cell kill at higher doses	Wheeler et al. (1972)

10. PURINES, PYRIMIDINES, AND RELATED COMPOUNDS

Compound	Concentration	Cell system	Effect	Reference
2-Thiocytosine (NSC 45755) 2-Thiouridine (no NSC No.) 4-Thiouridine (NSC 518132) 2-Thiouracil (NSC 1905)	4×10^{-4} M for 2 hours	Chinese hamster cells	Maximum sensitivity in late G_1 and G_2 phases corresponding to peaks of RNA synthesis	Lozzio and Wigler (1971)
Thymidine (NSC 21548)	10 μg per mouse	Mouse duodenal epithelium	Increased mitotic index apparently due to shortening of S phase	Greulich et al. (1961)
	2×10^{-3} M	Chang appendix cells	Inhibits DNA, permits synthesis of RNA and protein to continue, causes synchronization	Xeros (1962)
	2×10^{-3} M	Asynchronous HeLa cells	Increases mitotic index due to metaphase arrest	Barr (1963)
	5 to 20×10^{-3} M	Chinese hamster cells	Increases mitotic index and chromosome aberrations; induces synchrony	Yang et al. (1966)
	2.9×10^{-6} M	*Haplopappus gracilis* root meristem cells	Increased mitotic index due to increase in length of metaphase	Ames and Mitra (1967)
	2×10^{-3} M	HeLa S3	Block of DNA synthesis leads to unbalanced growth and cell death if time exceeds one cycle	Kim et al. (1965)
	2×10^{-3}–10^{-4} M	Asynchronous HeLa and HeLa S3	Block of DNA synthesis but viability retained up to 90 hours	Lambert and Studzinski (1967, 1969)
	10^{-2} M	BHK-21 hamster cells	Blocks DNA synthesis and accumulates cells at G_1/S, causes synchronization	Tobey et al. (1967b)

Table III Cont'd.

Agent	Dose or concentration	Biological system	Effects on cells	Reference
	$10^{-2} M$	Polyoma transformed P-183	Blocks all phases, usual replication when dThd removed; no synchrony	Tobey et al. (1967b)
	10^{-6}–$10^{-2} M$	HeLa S3	G_1 lengthened in dThd synchronized cells	Tobey et al. (1967a)
	$10^{-3} M$	Many cultured cells	Double dThd block causes synchronization.	Cleaver (1967)
	$7.5 \times 10^{-4} M$	Human kidney cells	Double dThd block causes synchronization, generation time shortened including G_1, S, and G_2	Galavazi and Bootsma (1966)
	2–$5 \times 10^{-3} M$	HeLa, HeLa S3 mitotically synchronized	Inhibits cell division by >90%, allows some DNA synthesis, does not effect rate of synthesis of RNA and protein when applied to G_1 cells; cells not arrested at G_2/S boundary but proceed slowly through S in regard to DNA synthesis and near normal through G_2 with regard to RNA and protein accumulation and cell enlargement	Studzinski and Lambert (1969a)
	$2 \times 10^{-3} M$	L cells, CHOP cells	Double dThd block accumulates cells in S phase due to decreased rate of DNA synthesis and increased length of S phase; no accumulation at G_1/S border	Bostock et al. (1971)

Compound	Dose/Concentration	System	Effect	Reference
	$7.5 \times 10^{-3} M$	Human kidney cells (synchronized by double dThd block)	No effect on $G_2 \to M$	Scaife (1970)
	$2 \times 10^{-3} M$	P 815 Y	Blocks DNA synthesis, phospholipid and protein synthesis continues at least 15 hours, cells synchronized by dThd remain abnormal for at least one generation	Bergeron (1971)
	$4 \times 10^{-5} – 2 \times 10^{-3} M$	X-irradiated HeLa cells, mitotically synchronized	Increases survival of cells irradiated at G_1/S or S; no effect on cells irradiated early G_1	Djordjevic and Kim (1969)
^{14}C-Thymidine and ^{3}H-Thymidine	—	Seedlings	^{14}C in methyl group produced twice as many chromosome aberrations as ^{14}C in pyrimidine ring; ^{3}H and ^{14}C at same dose and concentration produced similar number of aberrations	McQuade and Friedkin (1960)
^{3}H-Thymidine	1.9 Ci/mmole, $4.1 \times 10^{-5} M$	Asynchronous HeLa S3	Growth inhibition	Painter et al. (1964)
	0.02–2.5 μCi/ml for 30 minutes	Asynchronous HeLa S3	No effect on colony production	Painter et al. (1964)
	6 mCi/mouse, 6 doses × 4 hours	Normal hematopoietic and transplanted lymphoma colony forming lymphoma in vivo	80% cell kill of normal hematopoietic, 99.9% kill of lymphoma; cell-kill curve of cycle-stage specific type	Bruce and Meeker (1965); Bruce et al. (1966)
	1 μCi/ml (6.7 Ci/mmole)	L cells, L60T	Cell survival dropped to 55% in 1.5 hours and to 3% after 14 hours	Whitmore and Gulyas (1966); Cleaver (1967)

Table III Cont'd.

Agent	Dose or concentration	Biological system	Effects on cells	Reference
	3.3×10^{-6} M, 1, 2, or 3 Ci/mmole	Mouse L cells clone 929-E (synchronized by double dThd block)	Stimulates rate of precursor incorporation into DNA and histones in S, followed in G_2 by stimulation of precursor incorporation into nuclear RNA and nuclear acid protein	Krause (1967)
	—	Asynchronous Don cells	Cell kill reaches a saturation value: cycle-stage specific	Bhuyan et al. (1970)
	Up to 8.4×10^{-9} M	Asynchronous P 388 F lymphoma cells	Cell survival limited not by allosteric effect of dThd, but by tritium in DNA	Peterson et al. (1971)
	41.5 μCi/ml every 3 hours	L1210 cells in culture	Gompertzian cell kill (cycle-stage specific)	Wilkoff et al. (1972)
	4×10^{-7} M	Asynchronous H. Ep. No. 1	Unbalanced growth, irreversible after 24 hours	Eidinoff and Rich (1959); Rueckert and Mueller (1960)
5-Fluoro-deoxyuridine (NSC 27640)	10^{-8}–10^{-4} M	Asynchronous Vicia faba	S-phase cells halted, G_2 cells unaffected, chromosome damage in late S-phase cells but not in G_2 cells	Taylor et al. (1962)
	10^{-7} M	Asynchronous L cells	Cells in S fail to proceed, other cells proceed to S where synthesis of DNA does not initiate resulting in unbalanced growth. If block released, S-phase is shorter in next cycle	Till et al. (1963)

10. PURINES, PYRIMIDINES, AND RELATED COMPOUNDS

Concentration	Cell type	Effect	Reference
30 mg/kg	Ehrlich ascites *in vivo*	Unbalanced growth	Lindner et al. (1963)
4.1×10^{-7}–2.05×10^{-6} M	Strain L-M, Strain G_3, Don (Colcemid blocked)	Chromosome damage to early S-phase cells, little damage to those blocked at G_1/S	Hsu et al. (1964)
—	HeLa S3 mitotically synchronized	Selective disruption of small polysomes in S phase; no effect on polysomes in G_1	Robbins and Borun (1967)
2.05×10^{-6} M	Asynchronous HeLa	Inhibits DNA synthesis leading to unbalanced growth	Cohen and Studzinski (1967)
4.1×10^{-7} M	Human fetal diploid cells	Partial synchronization of cells; after release of block by dThd, S phase shortened	Priest et al. (1967)
4.1×10^{-7}–4.1×10^{-5} M	Chinese hamster fibroblasts, Don and Don-C, Colcemid synchronized	Early S phase: chromosome damage, late S cells not blocked (dThd pool sufficient), G_2 cells insensitive	Ockey et al. (1968)
10^{-7}–10^{-4} M	Chinese hamster cells, Colcemid synchronized	G_1, early and middle S most sensitive, M most resistant	Lozzio (1969)
10^{-6} M	Chinese hamster cells	Accumulation of cells in S (not G_1/S)	Epifanova et al (1969)
	Asynchronous Don cells	Cell kill reached saturation value: cycle-stage specific	Bhuyan et al. (1970)
1×10^{-6} M	P 815 Y	Prevents cell division, inhibits DNA synthesis, stops protein and phospholipid synthesis in 10 hours; decrease in cell volume after 30 hours correlates with loss of cell viability	Bergeron (1971)

Table III Cont'd.

Agent	Dose or concentration	Biological system	Effects on cells	Reference
	$4 \times 10^{-7} M$	H. Ep. No. 1 and HeLa	Causes synchronization and unbalanced growth	Nias and Fox (1971); Cohen and Studzinski (1967)
	$4.1 \times 10^{-5} M$	H. Ep. No. 2, Colcemid blocked	Partial block at or near G_1/S; delays or blocks S-phase cells, but not G_2 cells; nonexponential cell kill	Wheeler et al. (1972)
5-Bromo-deoxyuridine (NSC 38297)	$3 \times 10^{-5} M$	HeLa cells	Incorporated into DNA, permits only one S phase, yields nonviable giant cells	Hakala (1959)
	$3 \times 10^{-5} M$	Human D 98 S cells	Nontoxic but increases sensitivity to radiation	Djordjevic and Szybalski (1960)
	Up to $3.3 \times 10^{-5} M$	HeLa cells, mitotically synchronized	Maximum cell kill during maximum DNA synthesis; after 24 hours, DNA inhibited in subsequent cycle, RNA and protein only slightly reduced	Kim et al. (1967)
	$0.26 \times 10^{-6} M$ (continuous exposure)	P 388 F mouse lymphoma cells	Nontoxic; incorporated into DNA, increases doubling time and radiation sensitivity	Fox (1968)
	$1.3–1.7 \times 10^{-5} M$	Irradiated human kidney T cells (synchronized by double dThd block)	Increases unscheduled DNA synthesis, decreases cell survival; small fraction of UV irradiated 5-BrdUrd treated cells permanently blocked in G_2	Scaife (1970)

10. PURINES, PYRIMIDINES, AND RELATED COMPOUNDS

Compound	Concentration	Cell type	Effect	Reference
	10^{-5} M	Chinese hamster ovary cells mitotically synchronized, treated with 5-FdUrd prior to 5-BrdUrd	S phase delayed 3 hours, then normal progression	Taylor et al. (1971)
	1.3×10^{-5} M	Chinese hamster cells mitotically synchronized	Cells exposed at least one cell cycle prior to X-ray: M, most sensitive; G_1, intermediate; S, least sensitive	Dewey et al. (1971)
	10^{-8}–10^{-5} M	Mouse neuroblastoma C 1300 cells	Induces differentiation in absence of DNA synthesis	Schubert and Jacob (1970)
5-Bromo-deoxyuridine (NSC 38297) plus 5-fluoro-deoxyuridine (8.2×10^{-7} M)	3.3×10^{-5} M	Asynchronous HeLa cells	Cells divide only once, but DNA replicates at least two, and possibly four cycles.	Berkovitz et al. (1968)
5-Iodo-deoxyuridine (NSC 39661)	—	Human D 98 S cells	Short exposure increases ultraviolet sensitivity; long exposure causes cell death	Djordjevic and Szybalski (1960)
	High concentration (2×10^{-4} M)	Asynchronous P 815 Y murine mast cells	S phase most sensitive; about one-half cells fail to carry out subsequent DNA synthesis	Cramer and Morris (1966); Morris and Cramer (1966, 1968)
	0.28×10^{-6} M	P 388 F mouse lymphoma cells	Increases doubling time from 13 to 29 hours after treating culture for three doublings, also increases X-ray sensitivity	Fox (1968)
	Sub-lethal, dose dependent	Rat ileal cells and spleen lymphocytes in vivo	Delay of S and G_2 cells, decrease of DNA synthesis	Post and Hoffman (1969)

Table III Cont'd.

Agent	Dose or concentration	Biological system	Effects on cells	Reference
^{125}I-5-Iodo-deoxyuridine	1–50 μCi per mouse	L1210 in vivo	Cell survival curves almost identical with those of ^{3}H-dThd	Hofer and Hughes (1971)
5-Bromo-deoxycytidine (NSC 61765)	5×10^{-5} M (72 hours)	Chinese hamster HA-2 cells	Incorporation of 5-BrdCyd increases radiation sensitivity of exponentially growing cells and to a lesser extent, plateaued cells	Stewart et al. (1968)
1-β-D-Arabino-furanosylcytosine (NSC 63878)	2×10^{-6}–4×10^{-5} M	Asynchronous HeLa S3 and HeLa cells; asynchronous mouse fibroblasts	Inhibits DNA synthesis, leading to unbalanced growth; exposure in all phases; chromosome aberrations; minimum exposure for unbalanced growth to be lethal: 2 hours to length of S phase	Kim and Eidinoff (1965); Kim et al. (1968); Silagi (1965)
	8×10^{-5} M	HeLa S3 mitotically synchronized	Selective disruption of polysomes during S phase, no effect during G_1	Robbins and Borun (1967)
	4×10^{-6} M	X-irradiated HeLa S3 mitotically synchronized	Postirradiation treatment of S-phase cells with AraC enhances cell kill	Weiss and Tolmach (1967)
	4×10^{-6}–4×10^{-4} M	Asynchronous Don C	Sensitive in S-phase, possibly late G_1, cell kill related to length of exposure; after $2 \times T_c$, 5% do not label and a few survive	Karon and Shirakawa (1969)
	4×10^{-6}–4×10^{-5} M	Asynchronous Don C	Effects rate of S→G_2 more than G_1→S, inhibition of synthesis of DNA may occur without disturbances of transit time at lower dose	Karon and Shirakawa (1970)
	4×10^{-6}–4×10^{-4} M	Asynchronous Don C	DNA inhibition; chromosome breaks in G_1, S, G_2, but not M	Benedict et al. (1970)

Dose	Cell system	Effect	Reference
4×10^{-6}–4×10^{-4} M	Ultraviolet irradiated Don C	Exposure to ultraviolet prior to AraC decreased incidence of chromatid breaks	Benedict (1971)
—	Asynchronous Don C	Cycle-stage specific, cell survival reached a saturation value	Bhuyan et al. (1970)
1.3×10^{-5}–4×10^{-4} M	Asynchronous L1210 in vitro	Gompertzian cell kill, cycle-stage specific	Wilkoff et al. (1972)
10^{-5}–10^{-4} M	Don C. Colcemid synchronized	Unalterable cell kill of S-phase cells, unbalanced growth of G_1 cells may be reversed by deoxycytidine	Young and Fischer (1968)
5–40 mg/24 hours divided doses	Normal hematopoietic and transplanted lymphoma in vivo	Cycle-stage specific, probably S, determined by CFU; dose-survival curves reached saturation levels 24 and 48 hours, a few cells survived	Bruce et al. (1969)
5×10^{-6}–3×10^{-5} M	Human leukocytes	Chromosome breaks in G_2, no effect in G_1 or S	Cohen and Shaw (1967); Brewen (1965)
5×10^{-5} M	PHA-stimulated human leukocytes	Chromosome breaks in all phases of stimulated, but not unstimulated	Brewen and Christie (1967)
5×10^{-9}–5×10^{-8} M	PHA-stimulated leukocytes	Slows passage through S, reduces mitotic index, accumulates cells in early S phase	Brehaut and Fitzgerald (1968)
5×10^{-4} M	PHA-stimulated leukocytes	Cell kill in S, accumulation of labeled cells indicating cells enter S, no evidence for G_1/S block, low grain count of labeled cells	Brehaut and Fitzgerald (1968)

Table III Cont'd.

Agent	Dose or concentration	Biological system	Effects on cells	Reference
	$4 \times 10^{-5} M$	L cells	Selective killing of S-phase cells and giant cell formation, accumulation of G_1 cells and small accumulation of G_2 cells	Bremerskov et al. (1970)
	10–100 mg/kg	Mouse small intestine, spleen, and thymus	Lethal to S phase only in 2 hours; possible G_1 block	Lenaz et al. (1969)
	$4.1 \times 10^{-3} M$	Escherichia coli	Inhibits DNA synthesis, thymineless death	Atkinson and Stacey (1968); Cummings (1969)
	4 and 12 mg/kg	Human colonic carcinoma in hamster cheek pouch	Radiopotentiating effect with X-rays	Goldenberg (1970)
	$>7.2 \times 10^{-6} M$	Asynchronous mouse L cells	S-phase cells killed, G_1/S blocked	Graham and Whitmore (1970a)
	$<3.6 \times 10^{-7} M$	Asynchronous mouse L cells	Inhibits cell division during exposure up to 14 hours; giant cell formation, little loss of viability	Graham and Whitmore (1970a)
	25–50 mg/kg	B-16 Melanoma, Ehrlich ascites in vivo	S-phase cells sensitive except those close to end of DNA synthesis; ^{3}H-dThd uptake blocked within 15 min, partial synchrony of cells; S phase of subsequent cycle shortened, overall cell cycle lengthened	Bertalanffy and Gibson (1971)
1-β-D-Arabino-furanosyl-2-thiocytosine (no NSC No.)	$3.9 \times 10^{-5} M$	L cells, mitotically synchronized	S phase most sensitive, blocks G_1/S and possibly G_2/M, synthesis of DNA inhibited, but synthesis of RNA and protein increased—giant cell formation	Bremerskov et al. (1970)

10. PURINES, PYRIMIDINES, AND RELATED COMPOUNDS

Compound	Dose/Concentration	System	Effects	Reference
1-β-D-Arabinofuranosyl-5-fluorocytosine (NSC 529180)	50 μg/mouse on 3 consecutive days	AKR mice	Decrease of RNA synthesis in myeloid cells and DNA synthesis only in mature myeloid elements	Vesely and Šorm (1965)
	$4 \times 10^{-6} M$	Asynchronous HeLa S3 cells	Inhibits synthesis of DNA but not synthesis of RNA and protein; suggests unbalanced growth; cell viability declines after exposure equal to one generation	Kim et al. (1966)
	10 to 100 mg/kg	Mouse small intestine, spleen, and thymus	Lethal damage to cells synthesizing DNA, others escape damage. Inhibition of DNA synthesis must be sustained for several hours for irreversible damage.	Lenaz et al. (1969)
5-Azacytidine (NSC 102816)	1×10^{-5}–$1.18 \times 10^{-4} M$	Asynchronous Don C and L1210 in vitro	Primarily S-phase cell kill, minimal during G_1 and M; cell-kill curve of cycle-stage specific agent; chromosome damage	Li et al. (1970b)
	—	Don C cells	Cycle-stage specific from cell-kill curve	Bhuyan et al. (1970)
6-Azauridine (NSC 32074) plus deoxycytidine (NSC 83251) 10^{-6}–$10^{-4} M$ plus thymidine (NSC 21548) 10^{-6}–$10^{-4} M$	10^{-6}–$10^{-4} M$	Asynchronous murine mast cells—P-815-X2	Blocks in G_1, no delay of S or G_2 cells; inhibits RNA synthesis.	Burki (1971)

[a] The following abbreviations are used in this table: dThd, thymidine; 5-FdUrd, 5-fluorodeoxyuridine; 5-BrdUrd, 5-bromodeoxyuridine; 5-BrdCyd, 5-bromodeoxycytidine; AraC, 1-β-D-arabinofuranosylcytosine; CFU, colony-forming units; PHA, phytohemagglutinin.

of action of this purine analog (Montgomery et al., 1970). Thus, of the six agents listed in Table II as acting at Site 3 of the pathway (Fig. 1), only 6-mercaptopurine and 6-(methylthio)purine ribonucleoside are likely to have this as their major site of action.

Several of the agents can inhibit the *de Novo* synthesis of pyrimidine ribonucleotides by interfering with the functioning of aspartate transcarbamylase, dihydroorotase, or orotidylate decarboxylase. Of the ten agents shown in Table II to act at Site 7 (Fig. 1), only 6-azauridine, which specifically inhibits orotidylate decarboxylase, is likely to have this as a major site.

Agents that inhibit the *de Novo* synthesis of either purine ribonucleotides or pyrimidine ribonucleotides should prevent the synthesis of both DNA and RNA and affect cells in all phases of the cycle where the synthesis of RNA is required. 6-Mercaptopurine inhibits the *de Novo* synthesis of DNA and RNA equally (Bennett et al., 1963), and it is probable that 6-(methylthio)-purine ribonucleoside inhibits the synthesis of DNA and RNA equally. Therefore, these agents would be expected to halt the progression through the cycle of cells in G_1 and S (and perhaps the first part of G_2) without any large accumulation of cells at any one place in the cycle and without unbalanced growth and the formation of large cells. The observations given in Table III are consistent with these expectations. The fact that these agents affect cells in two phases of the cycle instead of only one but not the complete cycle might account for the difficulty encountered by Bruce et al. (1966) in classifying 6-mercaptopurine as a cycle specific or cycle-stage specific agent on the basis of the survival curve.

Realizing that 6-azauridine should prevent the synthesis of both DNA and RNA, Burki (1971) grew cells in the presence of 6-azauridine, deoxycytidine, and thymidine. Under these conditions, the synthesis of RNA decreased quickly, and there was a slower decrease in the synthesis of DNA. The cells proceeded through S, G_2, and M and accumulated in G_1. These effects are what one would expect. No report of the effects of 6-azauridine alone on the cell cycle has been found.

B. AGENTS INTERFERING WITH THE REDUCTION OF RIBONUCLEOSIDE DIPHOSPHATES

Thymidine, 2'-deoxyadenosine, 3'-deoxyadenosine, 9-β-D-arabinofuranosyladenine, and 2'-deoxyguanosine are listed in Table II as inhibitors of the reduction of ribonucleoside diphosphates. The evidence indicates that this probably is the major cite of action of 2'-deoxyadenosine, 2'-deoxyguanosine, and thymidine, but that it probably is not the major cite for the other two compounds. The observed effects (Table III) of these three agents upon the progression of cells through the cycle help in reaching this decision. All three of them cause greater inhibition of the synthesis of DNA than of RNA and synchronize the population of cells in or near the beginning of the S phase. The importance of the role of thymidine triphosphate in inhibiting the deamination of deoxycytidylate is not presently known.

C. Agents Interfering with the Action of Thymidylate Synthetase

Inhibition of thymidylate synthetase by the monophosphate of 5-fluorodeoxyuridine is the major site of action of this agent, and this is also an active site for administered 5-fluorouracil to the extent that the free base is converted to the monophosphate of 5-fluorodeoxyuridine. 5-Fluorouracil is also active by other mechanisms, and this increases the complexity of its effects upon the cell cycle. Whereas the cell kill curve indicated that 5-fluorodeoxyuridine is a cycle-stage specific agent (Bhuyan et al., 1970), the curves for 5-fluorouracil indicate that it is not cycle-stage specific (Bhuyan et al., 1970; Bruce et al., 1966). The cessation of progression of cells through the S phase, the resulting accumulation of cells in S or near the beginning of S, and the cell enlargement and unbalanced growth reflect the selective effect of preventing the synthesis of thymidylate and DNA. The specificity of the effect is also evident in the fact that added thymidine will overcome the inhibition and the cells will proceed through the cycle as a synchronized population. Evidently, 5-fluorodeoxyuridylate is bound to thymidylate synthetase rather tightly, as washing the cells does not relieve the inhibition; it is necessary to reverse the inhibition by adding thymidine. If reversal by thymidine is delayed too long, the cells die.

D. Agents Interfering with the Action of DNA Polymerase

1-β-D-Arabinofuranosylcytosine and 9-β-D-arabinofuranosyladenine inhibit DNA polymerase, and the latter can inhibit the reduction of ribonucleoside diphosphates to deoxyribonucleoside diphosphates; therefore they specifically inhibit the synthesis of DNA. Since 1-β-D-arabinofuranosyl-2-thiocytosine and 1-β-D-arabinofuranosylfluorocytosine inhibit the synthesis of DNA more than the synthesis of RNA, and have other effects similar to those of 1-β-D-arabinofuranosylcytosine, one might guess that these two agents also inhibit DNA polymerase, but apparently these compounds have not been tested in a cell-free system.

Many studies of the effects of 1-β-D-arabinofuranosylcytosine upon the progression of cells through the cycle and of the cell kill by this agent are reported in the literature. The results show that this agent specifically affects cells in the S phase, prevents the progression of cells through S but not through the other phases, and causes unbalanced growth and giant cell formation. These are the same effects caused by 5-fluorodeoxyuridine, which is not surprising, since both agents inhibit specifically the synthesis of DNA, although they do so by different mechanisms. Although much less experimental evidence is available for 9-β-D-arabinofuranosyladenine, 1-β-D-arabinofuranosyl-2-thiocytosine, and 1-β-D-arabinofuranosyl-5-fluorocytosine, the biochemical data and the effects upon cells are consistent with the possibility that they act by the same mechanism as 1-β-D-arabinofuranosylcytosine.

E. Agents Incorporated into DNA

Ten of the agents listed in Table II may be incorporated into DNA following conversion of them to deoxyribonucleoside triphosphates; some of these agents undergo metabolic alterations at some stage prior to incorporation into DNA. For some of them, incorporation into DNA is the most important site of action, but in others it is a secondary or less important site. Four effects of the incorporation of analogs into the DNA have been observed: (1) it may terminate the growth of the molecule and thus prevent replication or repair of DNA; (2) it may permit the synthesis of the DNA molecule to continue, but it may alter subsequent base pairing in replication and transcription; (3) it may permit the synthesis to continue, but the resulting DNA might be more sensitive to ultraviolet light and ionizing radiation; and (4) if the analog is radioactive, it may cause damage to the DNA by the effects of radiation and transmutation.

As was stated in Section IV,D, the chief site of action of 1-β-D-arabinofuranosylcytosine and of 9-β-D-arabinofuranosyladenine is probably inhibition of DNA polymerase, but Waqar et al. (1971) obtained evidence that these agents can be incorporated into nuclear DNA and terminate the growth of the chain. Since this inhibition of the synthesis would supplement the inhibition of the polymerase, both mechanisms would be consistent with the observed effects upon the cell cycle. Since there are conflicting reports concerning whether or not these agents serve as substrates of DNA polymerase and are incorporated into the DNA, the universality of this mechanism and its importance in various systems are uncertain at present.

It is probable that the main mechanism of action of 6-thiodeoxyguanosine is its incorporation into DNA. Some 6-thioguanine is also found in the DNA following exposure of cells to 6-thioguanine and much less following exposure to 6-mercaptopurine, although the ribonucleotides of these agents also act as pseudofeedback inhibitors. Cells are most sensitive to exposure to 6-thiodeoxyguanosine during the S phase, but its incorporation into the DNA does not cause any immediate interference with progression of the cells through the cycle and mitosis, and the toxicity becomes evident in subsequent cycles.

5-Bromodeoxyuridine and 5-iododeoxyuridine are extensively incorporated into DNA, and since 5-bromodeoxycytidine is easily deaminated in cells before or after phosphorylation, it also serves to introduce 5-bromouracil moieties into the DNA. This is probably the major mechanism of action of these agents, although they can also inhibit certain enzymes involved in the *de Novo* synthesis of pyrimidines. There is some evidence that cells in the S phase are most sensitive to these agents, and inhibition of thymidylate kinase by the monophosphate of 5-iododeoxyuridine may contribute toward S-phase specificity by this agent. The presence of 5-bromouracil or 5-iodouracil in the DNA sometimes slows the rate of synthesis of DNA, but the cells frequently can go through one or more cell cycles before dying. The analog-containing DNA is more sensitive to irradiation than normal DNA, and this difference has been useful in certain types of experiments.

Following exposure of cells to 7-deazaadenosine (tubercidin), 7-deazaadenine is present in both DNA and RNA. This agent is not cell cycle specific but rather gives a first-order cell kill.

5-Azacytidine is incorporated into DNA to a much smaller extent than into RNA, but it causes a greater inhibition of the synthesis of DNA than of RNA. Cell kill kinetics indicate that the agent is cycle-stage specific, and cells in the S phase are most sensitive to the agent. Cells treated in S are delayed in progressing to mitosis (Li *et al.*, 1970b). It has been suggested that incorporation into DNA is the chief mechanism of action of this agent, but, since it is incorporated into RNA to about four times the extent it is incorporated into DNA (Li *et al.*, 1970a), it appears likely that multiple mechanisms are involved. There is no experimental evidence to indicate whether or not incorporation into either DNA or RNA causes termination of chain growth.

The incorporation of ^{3}H-thymidine, ^{125}I-5-iododeoxyuridine, and ^{131}I-5-iododeoxyuridine and the subsequent toxicity were discussed in Section III,A,2.

F. Agents Incorporated into RNA

6-Thioguanine is incorporated into DNA and RNA, and to a small extent 6-mercaptopurine is converted into 6-thioguanylic acid, which is then incorporated into DNA. The relative importance of the roles of incorporation into DNA, incorporation into RNA, pseudofeedback inhibition of *de Novo* synthesis of purine ribonucleotides, and inhibition of guanylate kinase for 6-thioguanine is not clearly established. Since 6-thioguanylic acid is a considerably weaker pseudofeedback inhibitor than the ribonucleotides of 6-mercaptopurine and 6-(methylthio) purine, the greater toxicity of 6-thioguanine (Wheeler et al., 1972) must be due to other mechanisms. Since 6-thioguanine interferes with the progression of cells through S and probably through G_1 (Wheeler et al., 1972) while 6-thiodeoxyguanosine does not (Barranco and Humphrey, 1971), this additional effect of 6-thioguanine may be due to these other mechanisms. To some extent 6-thiodeoxyguanosine is cleaved to yield 6-thioguanine which then participates in the mechanisms mentioned above.

3′-Deoxyadenosine (cordycepin) is incorporated into RNA, and, since there is no hydroxyl group on carbon-3, termination of the growth of the molecule of RNA occurs. Therefore the inhibition of synthesis of RNA precedes the inhibition of synthesis of DNA and protein. Perhaps the observed interference with progression through mitosis is due to the prevention of synthesis of RNA during G_2.

As was discussed in Section IV,C, a major site of action of 5-fluorouracil is the inhibition of thymidylate synthetase by the 5-fluorodeoxyuridylate derived from it. Since it has toxic effects beyond those of 5-fluorodeoxyuridine, whose major site of action is the inhibition of thymidylate, there is evidently a supplementary mechanism of action. Perhaps this supplementary mechanism is the incorporation into RNA. Although 5-fluorouracil causes synchronization and unbalanced growth indicative of S-phase specificity in some systems, in others it gives exponential cell kill indicative of nonphase specificity. Perhaps the observed effects of this agent depend on the particular concentrations and biological system used.

The incorporation of 7-deazaadenosine and of 5-azacytidine into RNA and DNA was discussed in Section IV,E.

G. MISCELLANEOUS AGENTS

5-Aminouracil inhibits the synthesis of DNA more than the synthesis of RNA, but its site of action is not clearly defined. It causes synchronization.

9-Alkyl derivatives of 6-mercaptopurine stopped the growth of cultured cells in late S or G_2, but the mechanism of action of these agents is not known, and therefore they are omitted from Table II.

Puromycin interferes specifically with the synthesis of protein. Therefore it should inhibit cells in any phase of the cycle in which the synthesis of protein is required, and, in fact, inhibition in all phases of the cycle has been observed. This agent has been used to study the necessity for protein synthesis for progression of cells through certain parts of the cycle.

3′-Amino-3′-deoxy-N,N-dimethyladenosine, the aminonucleoside of puromycin, may be demethylated to 3′-amino-3′-deoxyadenosine, which can inhibit the synthesis of RNA. This inhibition could cause the observed delays in all phases of the cycle.

The important roles of 3′,5′-cyclic adenosine monophosphate are not yet clearly understood, but it is evident that this agent is intimately involved in processes controlling cell multiplication and function. Its mechanism of action is not absolutely established, but activation of protein kinases appears to be one of its major functions. Little is known about its effects on the progression of cells through the cycle, and the conflicting effects listed in Table III, where it stimulates mitosis in one system and inhibits mitosis in another system, make it evident that additional investigations with this agent are required.

The methylated xanthines, caffeine, theophylline, and theobromine, can inhibit 3′,5′-cyclic nucleotide diesterase and thus influence the concentration, and hence the magnitude of the effects of 3′,5′-cyclic adenosine monophosphate in cells. Therefore the effects of these two types of agents are closely linked, and again there are conflicting results given in Table III. The binding of the methylated xanthines to DNA might also interfere with the synthesis of DNA and delay progression through S. There is also evidence that caffeine inhibits the repair of DNA damaged by ultraviolet light (Cleaver, 1969; Domon and Rauth, 1969a), X-rays, and alkylating agents (Gaudin and Yielding, 1969).

V. Examples of the Logic and Use of These Agents for Cancer Chemotherapy

A. Relationship of Tumor Population Kinetics to Chemotherapy

A better understanding of the biological phenomena associated with tumor growth and drug sensitivity has developed since the discovery of the cell cycle by Howard and Pelc (1953) and introduction of the concept of growth fraction by Mendelsohn (1960), and this knowledge has provided a basis for a more rational approach to cancer chemotherapy.

The evidence indicates that sensitivity to cycle specific or cycle-stage specific agents is related to the proliferative state of the cell population (Mendelsohn, 1965; Skipper et al., 1970; Skipper, 1971a,b); tumors become less sensitive to drugs with increasing tumor mass and with decreasing growth rate. A decrease in growth rate has been observed in embryonic growth (Laird, 1966a), in the postnatal growth of birds and mammals (Laird, 1966b), in weight of fishes (Silliman, 1969), and in a variety of normal tissues, transplanted, induced, and spontaneous experimental tumor systems (Laird, 1969; Simpson-Herren and Lloyd, 1970). This change may result from (1) an increase in cell cycle time, (2) a decrease in growth fraction, i.e., a decrease in the ratio of proliferating cells to the total population, and (3) an increase in cell loss. Lala and Patt (1966) were the first to show that the decrease in growth rate of Ehrlich ascites could largely be accounted for by an increase in length of the cell cycle. A similar increase in generation time has been demonstrated in other ascitic tumors (Frindel et al., 1969; Tannock, 1969; Wiebel and Baserga, 1968; Choquet et al., 1970). In many solid tumors the length of the cell cycle either does not change with increasing mass or the change is not sufficient to account for the decrease in growth rate (Frindel et al., 1969; Hermens and Barendsen, 1969; Tannock, 1969; Simpson-Herren and Lloyd, 1970). Although changes in cell loss (Steel, 1967, 1968) may account for the decreasing growth rate of a few solid tumors (Frindel et al., 1968; Simpson-Herren and Griswold, 1971), a combination of changes in growth fraction, cell cycle time, and cell loss probably occurs in most tumors. Variations in the growth fraction may also be observed within a single tumor. Hermens and Barendsen (1969) have found that the percentage of ^{3}H-thymidine labeled cells in rhabdomyosarcoma of the rat decreases with distance from the periphery of the tumor. A similar observation was made by Shirakawa et al. (1970) in human melanoma. Tannock (1968, 1970), working with a transplantable tumor which grows in cords surrounding the capillaries, found a decrease in labeling index with distance from the capillary.

Lajtha (1963) first suggested the possibility that nonproliferative cells might be in a true resting state which would be consistant with lack of drug response of most solid tumors to some classes of drugs. Lala and Patt (1968) attempted to characterize the boundary between cycling and resting cells and Epifanova and Terskikh (1969) reviewed the current knowledge of nonproliferating cells. The question concerning the existence of a true G_0 or resting state, or alternatively, a prolonged G_1 or G_2 phase, remains unanswered. There, however, does appear to be agreement on the fact that some of the nonproliferative cells retain regenerative capacity (Laster et al., 1969; Epifanova and Terskikh, 1969). These studies also indicate that in solid tumors, particularly relatively large ones, the presence of large populations of nonproliferative cells contributes to the lack of drug sensitivity—nonmetabolizing cells are insensitive to analogs. It appears that two of the major factors which must be considered in matching drug schedules to the tumor are the length of the cell cycle and the proliferative fraction and any changes which may occur with increase or decrease in mass or perturbation of the population kinetics by therapy. For more detailed discussions of the relationship of cell kinetics to chemotherapy and radiotherapy see Steel and Lamerton (1969), Tubiana (1971), Schabel (1968, 1969a–c, 1970), DeVita (1971), Bergsagel (1970), Hahn (1967), and Andrews (1969).

Bruce et al. (1966) have classified drugs into three categories: (1) non-cycle specific—agents that kill resting cells and cells in all phases of the cell cycle at a similar rate (i.e., X-radiation, nitrogen mustard); (2) cycle-stage specific—agents that kill cells in only one stage of the cell cycle and spare cells in other stages and resting cells (i.e., 1-β-D-arabinofuranosylcytosine); (3) cycle specific—agents which kill cells in all stages of the cell cycle but spare resting cells (i.e., 5-fluorouracil). The different characteristics may be used to advantage in combination therapy to achieve maximum tumor cell kill within the limits of acceptable toxicity to the host.

B. Single versus Multiple Treatments

Skipper et al. (1964) found that a given dose of an effective drug would kill a similar fraction of the cell population, regardless of the size of population. Skipper et al. (1967) then investigated the scheduling of 1-β-D-arabinofuranosylcytosine to take advantage of the S-phase specificity against murine leukemic cells. They found that a single high dose killed about 90% of the viable cells if given 2 days after implant of 10^5 L1210 cells. The observed percent cell kill is very close to the average percent of the cell cycle occupied by synthesis of DNA. However, when the drug was administered in divided doses over a 24-hour period to allow cells in G_2, M, and G_1 to progress to the S phase and the courses were repeated at suitable intervals, up to 10^6 cells could be eradicated. The increase in life span over the control with the multiple dose schedule was 450% compared to 200% for the single daily dose schedule, with no increase in toxicity. Freireich et al. (1970) adapted this schedule to the kinetic parameters of human leukemia, giving continuous infusion for 5-day periods repeated every 2 weeks and concluded that intermittent courses of treatment designed to cover at least one generation time were superior to daily continuous treatment.

Mathé (1969) has shown that a single massive dose of 6-mercaptopurine given 2 days after implant of L1210 cells is more effective than a similar dose given on the sixth day postimplant and both schedules are more effective than the same total dose given in small repeated daily doses for 20 days. Mathé et al. (1970a) later concluded that large, intermittent doses of phase- or cycle-specific agents should replace daily therapy. Large intermittent doses also have less immunosuppressive effect on the host, and there is less opportunity for development of drug resistance.

C. Multiple Treatment versus Continuous Infusion

Blenkinsopp (1969) compared multiple injections with continuous infusion of ^{3}H-thymidine in stratified squamous epithelium of mice. The labeling index following a 24-hour infusion of thymidine-^{3}H was 41% compared to 77.6% following five injections in 24 hours. However, he found that if the 24-hour infusion was preceded by a 24-hour infusion with saline, the percentage of cells labeled was 66%; the difference between this value and 77.6% was not considered significant. He attributed the low value found after a single 24-hour infusion to trauma related to initiation of the infusion. In a related study, Blenkinsopp (1970) concluded that multiple intraperitonial injections did not disturb normal cell proliferation. Consistent with these observations is the finding of Neil et al. (1970) that 1-β-D-arabinofuranosylcytosine 5'-adamantoate, a sustained release form of 1-β-D-arabinofuranosylcytosine, was almost as effective given in a single dose as the latter given on an optimum schedule, where the drug is administered in multiple doses over one to two average cell cycles.

D. Combinations for Therapy

1. *Purines, Pyrimidines, and Related Agents*

Few instances of increased therapeutic effectiveness have been observed when antimetabolites are used in combination. However, Schmidt et al. (1970) reported that the LD_{10} and LD_{50} for 6-thioguanine was more than doubled when administered in combination with 1-β-D-arabinofuranosylcytosine at levels ranging from two-thirds to one-sixth of the LD_{10}. High proportions of cures were obtained with the combination compared to a twofold to fourfold increase in survival time when either agent was administered alone on optimal dose schedules against mice bearing an intravenous challenge of 10^6 L1210 cells. These investigators attributed the benefits of this combination to reduced toxicity of 6-thioguanine and the potential of administering larger doses of this agent.

2. *These Agents with Alkylating Agents, Radiation, or Surgery*

Tumors with low growth fractions (most solid tumors) respond poorly to therapy with cycle- or cycle-stage-specific agents; however, when the tumor population is reduced by a cytotoxic agent the cells apparently respond to the "uncrowding" or perturbation by returning to a proliferative state and becoming more sensitive to therapy. In early work with methotrexate, Venditti and Goldin (1964) found that a priming dose of cyclophosphamide substantially increased the effectiveness of methotrexate used alone in treatment of L1210 leukemia. Sartorelli and Booth (1965) reported synergistic inhibition of solid Sarcoma 180 and L1210 leukemia when mitomycins were given in combination with either 6-thioguanine or 5-fluorouracil. Laster et al. (1969), using a knowledge of the tumor kinetics, achieved cell "cures" of moderately advanced Carcinoma 755 with a combination of cyclophosphamide and 6-mercaptopurine. Multiple doses of 1-β-D-arabinofuranosylcytosine following 1,3-bis(2-chloroethyl)-1-nitrosourea (Tyrer et al.,

1967) or cyclophosphamide (Hoffman et al., 1968) gave better chemotherapeutic response than either agent used alone. Griswold et al. (1970) reported an increase in the sensitivity of Plasmacytoma No. 1 to 1-β-D-arabinofuranosylcytosine following the administration of a noncurative dose of cyclophosphamide.

D. L. Bruce et al. (1970) found that light anesthesia of AKR mice with either halothane or nitrous oxide significantly reduced the destruction of normal murine hematopoietic stem cells by 1-β-D-arabinofuranosylcytosine, while neither agent affected the extent of damage to lymphoma cells.

Much greater difficulty has been encountered in demonstrating synergism of drug combinations in therapy of human cancers than of experimental animals, and these problems have been discussed by Clarkson et al. (1971), Freireich et al. (1970), Holland and Glidewell (1970), Mathé et al. (1970a,b), Mukherji et al. (1971), Nemoto and Dao (1971), Johnson and Wolberg (1971), and Costanzi and Coltman (1969). Good therapeutic response of human leukemia has been achieved using cyclic therapy of combinations of four or more drugs and cyclic therapy has also improved response of some human solid tumors.

Combinations of purines and pyrimidines and related compounds with radiation have significantly improved the therapeutic response. Incorporation of 5-bromodeoxyuridine into the DNA of cultured cells prior to irradiation increases the sensitivity of cultured cells (Table III). For therapy of human brain tumors, Sano et al. (1968) used continuous infusion of this agent followed by irradiation with encouraging results.

Gaudin and Yeilding (1969) treated hamsters bearing cyclophosphamide-resistant plasmacytoma with X-rays or cyclophosphamide followed by caffeine or chloroquine with resultant tumor regression. X-ray or cyclophosphamide alone had little or no effect on the growth rate of the tumor. These investigators suggest that resistance to alkylating agents may result from an active repair mechanism and the resistance was overcome by inhibition of repair. Care must be used in selection of the agents for combination therapy, however; 6-azauridine (Grozdanovič et al., 1968) apparently protects cells against X-ray damage when present during irradiation. Myers (1971) found that caffeine protects rat thymocytes from the damaging effects of high doses of X-radiation and retards degeneration. The latter result is contrary to the effects of caffeine combined with ultraviolet irradiation.

Tubiana (1971) and Frindel and Tubiana (1971) reviewed the relationship of the kinetics of cell proliferation and radiotherapy. Although radiobiology of the cell cycle is a relatively new field of investigation, it is of practical interest. Cell killing has been significantly improved with fractionated dose schedules and lends hope that a quantitative understanding of repair of sublethal damage, return of proliferative capacity, and effects of radiation on the cell cycle may lead to more efficient combination of chemotherapeutic agents with radiation.

Although limited clinical studies where chemotherapy has been administered in a conventional manner either prior to or following surgery have not been encouraging, it does appear likely that surgical removal of the accessible tumor might result in "uncrowding" of the remaining population with a resultant increase in drug sensitivity. Simpson-Herren and Griswold (1972) have found that surgical removal of 70% to 90% of a subcutaneously implanted Plasmacytoma No. 1 results in a 50% increase in the labeling index of the residual tumor by the second postoperative day and the elevated labeling index persists for about one week. Using Lewis Lung Carcinoma, Laster et al. (1971) showed that surgical removal of the primary subcutaneous tumor followed by 1-(2-chloroethyl)-3-(4-methylcyclohexyl)-1-nitrosourea (NSC 95441) significantly increased the percentage of tumor-free long-term survivors. Based on the results with other combinations, there is ample reason to think that combining surgery with chemotherapy, taking into account the effects that therapy may have on wound healing and the immune system of the host, may give encouraging results.

E. CONCLUSIONS

Cain (1971) in discussing a strategy for cure of cancer said, "In the treatment of experimental animal tumors matching of drug schedules to host and tumor can make the difference between no apparent result or cure." Success in the treatment of choriocarcinoma in women with methotrexate and Burkitt's lymphoma with cyclophosphamide is, perhaps, largely due to unusual drug sensitivity associated with a high growth fraction. Slow-growing (low growth fraction) human tumors present a more difficult problem and successful chemotherapy of these tumors will probably require use of multiple drugs and careful matching of drug schedules to the population kinetics of tumor and host both before and during therapy.

VI. Concluding Comments

The generation of information of the biochemical events occurring during the cell cycle, of the biochemical mechanisms of action of purines, pyrimidines, and related compounds, and of the effects of these agents upon cells in various parts of the cell cycle continues. The interrelationships of these bits of information are of interest academically, aesthetically, and practically. They contribute to the basic understanding of events occurring during the growth and multiplication of cells and of the control mechanisms involved. The meshing of biological and biochemical events to give predictable results has a certain entrancing beauty to the natural scientist, and the predictability of the results makes it possible to use drugs, including those discussed in this chapter, for therapy. Already it is possible to use more effectively drugs that have been available for many years and to use an expanded sphere in the logical design of new drugs. Although an attempt has been made to present in this chapter the currently available knowledge of the effects of purines, pyrimidines, and related compounds upon the cell cycle, it is anticipated this review will quickly become obsolete, as research in these areas proceeds rapidly.

References

Adams, J. E., Breed, N. L., and Valenti, C. (1967). Enhancement of 5-fluorouracil cytotoxicity in synchronized human malignant cells in culture. *Tex. Rep. Biol. Med.* **25**, 342–349.

Agarwal, R. P., Scholar, E. M., Agarwal, K. C., and Parks, R. E., Jr. (1971). Identification and isolation on a large scale of guanylate kinase from human erythrocytes—effects of monophosphate nucleotides of purine analogs. *Biochem. Pharmacol.* **20**, 1341–1354.

Allan, P. W., and Bennett, L. L., Jr. (1971). 6-Methylthioguanylic acid, a metabolite of 6-thioguanine. *Biochem. Pharmacol.* **20**, 847–852.

Ames, I. H., and Mitra, J. (1967). An effect of exogenous thymidine on the cell cycle in *Haplopappus gracilis*. *J. Cell. Physiol.* **69**, 253–258.

Andrews, J. R. (1969). Combined cancer radiotherapy and chemotherapy: The relevance of cell population kinetics and pharmacodynamics. *Cancer Chemother. Rep.* **53**, 313–324.

Arlett, C. F. (1967). A failure of specific inhibitors to affect survival of Chinese hamster cells exposed to ultra violet light. *Int. J. Radiat. Biol.* **13**, 369–376.

Atkinson, C., and Stacey, K. A. (1968). Thymineless death induced by cytosine arabinoside. *Biochim. Biophys. Acta* **166**, 705–707.

Balis, M. E. (1968). "Antagonists and Nucleic Acids." North-Holland Publ., Amsterdam.

Barr, H. J. (1963). The effects of exogenous thymidine on the mitotic cycle. *J. Cell. Comp. Physiol.* **61**, 119–127.

Barranco, S. C., and Humphrey, R. M. (1971). The effects of β-2'-deoxythioguanosine on survival and progression in mammalian cells. *Cancer Res.* **31**, 583–586.

Baserga, R. (1968). Biochemistry of the cell cycle: A review. *Cell Tissue Kinet.* **1**, 167–191.

Benedict, W. F. (1971). Chromatid breakage: Cytosine arabinoside-induced lesions inhibited by ultraviolet irradiation. *Science* **171**, 680–682.

Benedict, W. F., Harris, N., and Karon, M. (1970). Kinetics of 1-β-D-arabinofuranosylcytosine-induced chromosome breaks. *Cancer Res.* **30**, 2477–2483.

Bennett, L. L., Jr., and Adamson, D. J. (1970). Reversal of the growth inhibitory effects of 6-methylthiopurine ribonucleoside. *Biochem. Pharmacol.* **19**, 2172–2176.
Bennett, L. L., Jr. and Allan, P. W. (1971). Formation and significance of 6-methylthiopurine ribonucleotide as a metabolite of 6-mercaptopurine. *Cancer Res.* **31**, 152–158.
Bennett, L. L., Jr., Simpson, L., Golden, J., and Barker, T. L. (1963). The primary site of inhibition by 6-mercaptopurine on the purine biosynthetic pathway in some tumors *in vivo. Cancer Res.* **23**, 1574–1580.
Bergeron, J. J. M. (1971). Different effects of thymidine and 5-fluorouracil 2'-deoxyriboside on biosynthetic events in cultured P815Y mast cells. *Biochem. J.* **123**, 385–390.
Bergsagel, D. E. (1970). Cell kinetics and solid tumor chemotherapy. *In* "Oncology 1970" (R. T. Clark *et al.*, eds.), Vol. II, pp. 174–183. Yearbook Publ., Chicago, Illinois.
Berkovitz, A., Simon, E. H., and Toliver, A. (1968). DNA replication in the absence of cell division in BUdR-FUdR treated HeLa cells. *Exp. Cell Res.* **53**, 497–505.
Berry, R. J. (1964). A comparison of effects of some chemotherapeutic agents and those of x-rays on the reproductive capacity of mammalian cells. *Nature (London)* **203**, 1150–1153.
Bertalanffy, F. D., and Gibson, M. H. L. (1971). The *in vivo* effects of arabinosylcytosine on the cell proliferation of murine B16 melanoma and Ehrlich ascites tumor. *Cancer Res.* **31**, 66–71.
Bhuyan, B. K., Scheidt, L. G., and Fraser, T. J. (1970). Cell cycle specificity of several antitumor agents. *Proc. Amer. Ass. Cancer Res.* **11**, 8.
Biesele, J. J. (1958). "Mitotic Poisons and the Cancer Problem." Elsevier, Amsterdam.
Bischoff, R. (1971). Acid mucopolysaccharide synthesis by chick amnion cell cultures—inhibition by 5-bromodeoxyuridine. *Exp. Cell Res.* **66**, 224–236.
Blenkinsopp, W. K. (1969). Comparison of multiple injections with continuous infusion of tritiated thymidine, and estimation of the cell cycle time. *J. Cell Sci.* **5**, 575–582.
Blenkinsopp, W. K. (1970). Absence of effect of multiple intraperitoneal injections on cell proliferation, and estimation of the cell cycle time in the epithelium of the oesophagus and forestomach in mice, hamsters and rats. *Cell Tissue Kinet.* **3**, 83–88.
Bostock, C. J., Prescott, D. M., and Kirkpatrick, J. B. (1971). An evaluation of the double thymidine block for synchronizing mammalian cells at the G_1-S border. *Exp. Cell Res.* **68**, 163–168.
Brehaut, L. A., and Fitzgerald, P. H. (1968). The effects of cytosine arabinoside on the cell cycle of cultured human leucocytes: A microdensitometric and autoradiographic study. *Cell Tissue Kinet.* **1**, 147–152.
Bremerskov, V., Kaden, P., and Mittermayer, C. (1970). DNA synthesis during the life cycle of L cells: Morphological, histochemical and biochemical investigations with arabinosylcytosine and thioarabinosylcytosine. *Eur. J. Cancer* **6**, 379–392.
Brewen, J. G. (1965). The induction of chromatid lesions by cytosine arabinoside in post-DNA-synthetic human leucocytes. *Cytogenetics (Basel)* **4**, 28–36.
Brewen, J. G., and Christie, N. T. (1967). Studies on the induction of chromosomal aberrations in human leukocytes by cytosine arabinoside. *Exp. Cell Res.* **46**, 276–291.

Bruce, D. L., Lin, H., and Bruce, W. R. (1970). Reduction of colony-forming cell sensitivity to arabinosylcytosine by halothane anesthesia. *Cancer Res.* **30**, 1803-1805.

Bruce, W. R., and Meeker, B. E. (1965). Comparison of the sensitivity of normal hematopoietic and transplanted lymphoma colony-forming cells to tritiated thymidine. *J. Nat. Cancer Inst.* **34**, 849-856.

Bruce, W. R., and Valeriote, F. A. (1968). Normal and malignant stem cells and chemotherapy. "The Proliferation and Spread of Neoplastic Cells," pp. 409-420. Williams & Wilkins, Baltimore, Maryland.

Bruce, W. R., and van der Gaag, H. (1963). A quantitative assay for the number of murine lymphoma cells capable of proliferation *in vivo*. *Nature (London)* **199**, 79-80.

Bruce, W. R., Meeker, B. E., and Valeriote, F. A. (1966). Comparison of the sensitivity of normal hematopoietic and transplanted lymphoma colony-forming cells to chemotherapeutic agents administered *in vivo*. *J. Nat. Cancer Inst.* **37**, 233-245.

Bruce, W. R., Valeriote, F. A., and Meeker, B. E. (1967). Survival of mice bearing a transplanted syngeneic lymphoma following treatment with cyclophosphamide, 5-fluorouracil, or 1, 3-bis(2-chloroethyl)-1-nitrosourea. *J. Nat. Cancer Inst.* **39**, 257-266.

Bruce, W. R., Meeker, B. E., Powers, W. E., and Valeriote, F. A. (1969). Comparison of the dose- and time-survival curves for normal hematopoietic and lymphoma colony-forming cells exposed to vinblastine, vincristine, arabinosylcytosine, and amethopterin. *J. Nat. Cancer Inst.* **42**, 1015-1025.

Burki, H. R. (1971). DNA content of mastocytoma cells in cell culture after inhibition of RNA synthesis with 6-azauridine. *Cancer Res.* **31**, 1188-1191.

Butcher, R. W., and Sutherland, E. W. (1962). Adenosine 3',5'-phosphate in biological materials. I. Purification and properties of cyclic 3',5'-nucleotide phosphodiesterase and use of this enzyme to characterize adenosine 3',5'-phosphate in human urine. *J. Biol. Chem.* **237**, 1244-1250.

Cain, B. F. (1971). Experimental cancer chemotherapy: Strategy for cure. *N. Z. Med. J.* **73**, 85-90.

Chassy, B. M., and Suhadolnik, R. J. (1968). Nucleoside antibiotics. II. Biochemical tools for studying the structural requirements for the interaction at the catalytic and regulatory sites of ribonucleotide reductase from *Escherichia coli*. *J. Biol. Chem.* **243**, 3538-3541.

Choquet, C., Chavaudre, N., and Malaise, E. P. (1970). The influence of allogeneic inhibition and tumour age on the kinetics of L1210 leukaemia *in vivo*. *Eur. J. Cancer* **6**, 373-378.

Chu, E. H. Y. (1965). Effects of ultraviolet radiation on mammalian cells. II. Differential uv and x-ray sensitivity of chromosomes to breakage in 5-aminouracil synchronized cell populations. *Genetics* **52**, 1279-1294.

Clarkson, B., Todo, A., Ogawa, M., Gee, T., and Fried, J. (1971). Consideration of the cell cycle in chemotherapy of acute leukemia. *Recent Results Cancer Res.* **36**, 88-118.

Cleaver, J. E. (1967). Thymidine metabolism and cell kinetics. *Front. Biol.* **6**.

Cleaver, J. E. (1968). Repair replication and degradation of bromouracil-substituted DNA in mammalian cells after irradiation with ultraviolet light. *Biophys. J.* **8**, 775-791.

Cleaver, J. E. (1969). Repair replication of mammalian cell DNA: Effects of compounds that inhibit DNA synthesis or dark repair. *Radiat. Res.* **37**, 334-348.

Cleaver, J. E., and Thomas, G. H. (1969). Single strand interruptions in DNA and the effects of caffeine in Chinese hamster cells irradiated with ultraviolet light. *Biochem. Biophys. Res. Commun.* **36**, 203–208.

Cohen, L. S., and Studzinski, G. P. (1967). Correlation between cell enlargement and nucleic acid and protein content of HeLa cells in unbalanced growth produced by inhibitors of DNA synthesis. *J. Cell. Physiol.* **69**, 331–340.

Cohen, M. M., and Shaw, M. W. (1967). Specific effects of viruses and antimetabolites on mammalian chromosomes. "The Chromosome—Structural and Functional Aspects," pp. 50–66. Williams & Wilkins, Baltimore, Maryland.

Cone, C. D., Jr., and Tongier, M., Jr. (1969). Mitotic synchronization of L-strain fibroblasts with 5-aminouracil as determined by time-lapse cinephotography. *Sci. Tech. Aerosp. Rep.* **7**, 1095.

Costanzi, J. J., and Coltman, C. A., Jr. (1969). Combination chemotherapy using cyclophosphamide, vincristine, methotrexate and 5-fluorouracil in solid tumors. *Cancer* **23**, 589–596.

Cramer, J. W., and Morris, N. R. (1966). Absence of DNA synthesis in one-half of a population of mammalian tumor cells inhibited in culture by 5-iodo-2'-deoxyuridine. *Mol. Pharmacol.* **2**, 363–367.

Cramer, J. W., Prusoff, W. H., and Welch, A. D. (1961). 5-Bromo-2'-deoxycytidine (BCDR). II. Studies with murine neoplastic cells in culture and *in vitro*. *Biochem. Pharmacol.* **8**, 331–335.

Cummings, D. J. (1969). Comments on "thymineless death induced by cytosine arabinoside." *Biochim. Biophys. Acta* **179**, 237–238.

Dehner, L. P., Pettengill, O. S., Sorenson, G. D., and Carroll, S. B. (1969). Effects of puromycin on murine myeloma cells *in vitro*. *Arch. Pathol.* **88**, 211–224.

DeVita, V. T. (1971). Cell kinetics and the chemotherapy of cancer. *Cancer Chemother. Rep.* **2**, 23–33.

Dewey, W. C., Stone, L. E., Miller, H. H., and Giblak, R. E. (1971). Radiosensitization with 5-bromodeoxyuridine of Chinese hamster cells x-irradiated during different phases of the cell cycle. *Radiat. Res.* **47**, 672–688.

Djordjevic, B., and Kim, J. H. (1969). Modification of radiation response in synchronized HeLa cells by metabolic inhibitors: Effects of inhibitors of DNA and protein synthesis. *Radiat. Res.* **37**, 435–450.

Djordjevic, B., and Szybalski, W. (1960). Genetics of human cell lines. III. Incorporation of 5-bromo- and 5-iododeoxyuridine into the deoxyribonucleic acid of human cells and its effect on radiation sensitivity. *J. Exp. Med.* **112**, 509–531.

Djordjevic, B., Kim, J. H., and Kim, S. H. (1968). Modification of radiation response in synchronized HeLa cells by metabolic inhibitors. I. Effect of an inhibitor of RNA-synthesis, puromycin aminonucleoside. *Int. J. Radiat. Biol.* **14**, 1–7.

Domon, M., and Rauth, A. M. (1969a). Effects of caffeine on ultraviolet-irradiated mouse L cells. *Radiat. Res.* **39**, 207–221.

Domon, M., and Rauth, A. M. (1969b). Ultraviolet-light irradiation of mouse L cells: Effects on cells in the DNA synthesis phase. *Radiat. Res.* **40**, 414–429.

Eidinoff, M. L., and Rich, M. A. (1959). Growth inhibition of a human tumor cell strain by 5-fluoro-2'-deoxyuridine: Time parameters for subsequent reversal by thymidine. *Cancer Res.* **19**, 521–524.

Eidinoff, M. L., Cheong, L., and Rich, M. A. (1959). Incorporation of unnatural pyrimidine bases into deoxyribonucleic acid of mammalian cells. *Science* **129**, 1550–1551.

Epifanova, O. I., and Terskikh, V. V. (1969). On the resting periods in the cell life cycle. *Cell Tissue Kinet.* **2**, 75–93.

Epifanova, O. I., Smolenskaya, I. N., Sevastyanova, M. V., and Kurdyumova, A. G. (1969). Effects of actinomycin D and puromycin on the mitotic cycle in synchronized cell culture. *Exp. Cell Res.* **58**, 401–410.

Eriksson, T. (1966). Partial synchronization of cell division in suspension cultures of *Haplopappus gracilis*. *Physiol. Plant.* **19**, 900–910.

Estensen, R. D., and Baserga, R. (1966). Puromycin-induced necrosis of crypt cells of the small intestine of mouse. *J. Cell Biol.* **30**, 13–22.

Fox, M. (1968). The effect of halogenated nucleosides on the sensitivity of a mouse lymphoma P388F to x-rays and methyl methanesulphonate. *Int. J. Cancer* **3**, 382–389.

Freireich, E. J., Bodey, G. P., Hart, S., Rodriguez, V., Whitecar, J. P., and Frei, E., III. (1970). Remission induction in adults with acute myelogenous leukemia. *Recent Results Cancer Res.* **30**, 85–91.

Frindel, E., and Tubiana, M. (1971). Radiobiology and the cell cycle. *In* "The Biochemistry of Disease" (R. Baserga, ed.), Vol. 1, pp. 391–447. Dekker, New York.

Frindel, E., Malaise, E., and Tubiana, M. (1968). Cell proliferation kinetics in five human solid tumors. *Cancer* **22**, 611–620.

Frindel, E., Valleron, A. J., Vassort, F., and Tubiana, M. (1969). Proliferation kinetics of an experimental ascites tumour of the mouse. *Cell Tissue Kinet.* **2**, 51–65.

Fujioka, S., and Silber, R. (1970). Purification and properties of ribonucleotide reductase from leukemic mouse spleen. *J. Biol. Chem.* **245**, 1688–1693.

Furth, J. J., and Cohen, S. S. (1967). Inhibition of mammalian DNA-polymerase by the 5′-triphosphate of 9-β-D-arabinofuranosyladenine. *Cancer Res.* **27**, 1528–1533.

Furth, J. J., and Cohen, S. S. (1968). Inhibition of mammalian DNA polymerase by the 5′-triphosphate of 1-β-D-arabinofuranosylcytosine and the 5′-triphosphate of 9-β-D-arabinofuranosyladenine. *Cancer Res.* **28**, 2061–2067.

Galavazi, G., and Bootsma, D. (1966). Synchronization of mammalian cells *in vitro* by inhibition of the DNA synthesis. II. Population dynamics. *Exp. Cell Res.* **41**, 438–451.

Galavazi, G., Schenk, H., and Bootsma, D. (1966). Synchronization of mammalian cells *in vitro* by inhibition of the DNA synthesis. I. Optimal conditions. *Exp. Cell Res.* **41**, 428–437.

Gaudin, D., and Yielding, K. L. (1969). Response of a "resistant" plasmacytoma to alkylating agents and x-ray in combination with the "excision repair inhibitors" caffeine and chloroquine. *Proc. Soc. Exp. Biol. Med.* **131**, 1413–1416.

Giménez-Martín, G., Meza, I., López-Sáez, J. F., and González-Fernández, A. (1969). Kinetics of binucleate cell production by caffeine. *Cytologia* **34**, 29–35.

Goldenberg, D. M. (1970). Radiopotentiating action of 1-β-D-arabinofuranosylcytosine in a xenografted human colonic carcinoma. *Proc. Soc. Exp. Biol. Med.* **134**, 510–512.

Goldenberg, D. M., and Ammersdörfer, E. (1970). Synergistic effects of x-rays and drugs on a human tumor xenograft, GW-39. *Eur. J. Cancer* **6**, 73–80.

González-Fernández, A., Fernández-Gómez, M. E., Stockert, J. C., and López-Sáez, J. F. (1970). Effect produced by inhibitors of RNA synthesis on mitosis. *Exp. Cell Res.* **60**, 320–326.

Graham, F. L., and Whitmore, G. F. (1970a). The effect of 1-β-D-arabinofuranosylcytosine on growth, viability, and DNA synthesis of mouse L-cells. *Cancer Res.* **30**, 2627–2635.

Graham, F. L., and Whitmore, G. F. (1970b). Studies in mouse L-cells on the incorporation of 1-β-D-arabinofuranosylcytosine into DNA and on inhibition of DNA polymerase by 1-β-D-arabinofuranosylcytosine 5'-triphosphate. *Cancer Res.* **30**, 2636–2644.

Greengard, P. (1971). On the reactivity and mechanism of action of cyclic nucleotides. *Ann. N. Y. Acad. Sci.* **185**, 18–26.

Greulich, R. C., Cameron, I. L., and Thrasher, J. D. (1961). Stimulation of mitosis in adult mice by administration of thymidine. *Proc. Nat. Acad. Sci. U.S.* **47**, 743–748.

Griswold, D. P., Jr., Simpson-Herren, L., and Schabel, F. M., Jr. (1970). Altered sensitivity of a hamster plasmacytoma to cytosine arabinoside (NSC-63878). *Cancer Chemother. Rep.* **54**, 337–346.

Grozdanovič, J., Vich, Z., and Truxová, G. (1968). The combined effect of 6-azauridine and irradiation *in vivo* and *in vitro*. I. The acute effect on the mouse Ehrlich ascitic carcinoma. *Neoplasma* **15**, 247–249.

Hahn, G. M. (1967). Cellular kinetics, cell cycles and cell killing. *Biophysik* **4**, 1–14.

Hakala, M. T. (1959). Mode of action of 5-bromodeoxyuridine on mammalian cells in culture. *J. Biol. Chem.* **234**, 3072–3076.

Hakala, M. T., and Nichol, C. A. (1964). Prevention of the growth-inhibitory effect of 6-mercaptopurine by 4-aminoimidazole-5-carboxamide. *Biochim. Biophys. Acta* **80**, 665–668.

Heidelberger, C. (1967). Cancer chemotherapy with purine and pyrimidine analogues. *Annu. Rev. Pharmacol.* **7**, 101–124.

Henderson, J. F. (1965). Effects of anticancer drugs on biochemical control mechanisms. *Progr. Exp. Tumor Res.* **6**, 84–125.

Hermens, A. F., and Barendsen, G. W. (1969). Changes of cell proliferation characteristics in a rat rhabdomyosarcoma before and after x-irradiation. *Eur. J. Cancer* **5**, 173–189.

Hill, D. L., and Bennett, L. L., Jr. (1969). Purification and properties of 5-phosphoribosyl pyrophosphate amidotransferase from adenocarcinoma 755 cells. *Biochemistry* **8**, 122–130.

Hofer, K. G., and Hughes, W. L. (1971). Radiotoxicity of intranuclear tritium, [125]iodine and [131]iodine. *Radiat. Res.* **47**, 94–109.

Hoffman, G. S., Waravdekar, V. S., Venditti, J. M., and Mantel, N. (1968). Sequential chemotherapy with cytoxan and 1-β-D-arabinofuranosylcytosine hydrochloride (Ara-C) in mice with advanced leukemia L1210. *Proc. Amer. Ass. Cancer Res.* **9**, 31.

Holland, J. F., and Glidewell, O. (1970). Complementary chemotherapy in acute leukemia. *Recent Results Cancer Res.* **30**, 95–108.

Howard, A., and Pelc, S. R. (1953). Synthesis of desoxyribonucleic acid in normal and irradiated cells and its relation to chromosome breakage. *Heredity* **6**, Suppl., 261–273.

Hsu, T. C., Humphrey, R. M., and Somers, C. E. (1964). Responses of Chinese hamster and L cells to 2'-deoxy-5-fluoro-uridine and thymidine. *J. Nat. Cancer Inst.* **32**, 839–851.

Inagaki, A., Nakamura, T., and Wakisaka, G. (1969). Studies on the mechanism of action of 1-β-D-arabinofuranosylcytosine as an inhibitor of DNA synthesis in human leukemic leukocytes. *Cancer Res.* **29**, 2169–2176.

Jakob, K. M. (1968). The inhibition of RNA synthesis in *Vicia faba* by 5-aminouracil. *Exp. Cell Res.* **52**, 499–506.

Jakob, K. M., and Trosko, J. E. (1965). The relation between 5-amino uracil-induced mitotic synchronization and DNA synthesis. *Exp. Cell Res.* **40**, 56–67.

Johnson, R. O., and Wolberg, W. H. (1971). Cellular kinetics and their implications for chemotherapy of solid tumors, especially cancer of the colon. *Cancer* **28**, 208–212.

Karon, M., and Shirakawa, S. (1969). The locus of action of 1-β-D-arabinofuranosylcytosine in the cell cycle. *Cancer Res.* **29**, 687–696.

Karon, M., and Shirakawa, S. (1970). Effect of 1-β-D-arabinofuranosylcytosine on cell cycle passage time. *J. Nat. Cancer Inst.* **45**, 861–867.

Kihlman, B. A., and Odmark, G. (1966). Effects of adenine nucleosides on chromosomes, cell division and nucleic acid synthesis in *Vicia faba*. *Hereditas* **56**, 71–82.

Kim, J. H., and Eidinoff, M. (1965). Action of 1-β-D-arabinofuranosylcytosine on the nucleic acid metabolism and viability of HeLa cells. *Cancer Res.* **25**, 698–702.

Kim, J. H., Kim, S. H., and Eidinoff, M. L. (1965). Cell viability and nucleic acid metabolism after exposure of HeLa cells to excess thymidine and deoxyadenosine. *Biochem. Pharmacol.* **14**, 1821–1829.

Kim, J. H., Eidinoff, M. L., and Fox, J. J. (1966). Action of 1-β-D-arabinofuranosyl-5-fluorocytosine on the nucleic acid metabolism and viability of HeLa cells. *Cancer Res.* **26**, 1661–1664.

Kim, J. H., Gelbard, A. S., Perez, A. G., and Eidinoff, M. L. (1967). Effect of 5-bromodeoxyuridine on nucleic acid and protein synthesis and viability in HeLa cells. *Biochim. Biophys. Acta* **134**, 388–394.

Kim, J. H., Perez, A. G., and Djordjevic, B. (1968). Studies on unbalanced growth in synchronized HeLa cells. *Cancer Res.* **28**, 2443–2447.

Kishimoto, S., and Lieberman, I. (1964). Synthesis of RNA and protein required for the mitosis of mammalian cells. *Exp. Cell Res.* **36**, 92–101.

Koch, A. L., and Lamont, W. A. (1956). The metabolism of methylpurines by *Escherichia coli*. II. Enzymatic studies. *J. Biol. Chem.* **219**, 189–201.

Kovacs, C. J., and Van't Hof, J. (1971). Mitotic delay and the regulating events of plant cell proliferation: DNA replication by a G1/S population. *Radiat. Res.* **48**, 95–106.

Krause, M. O. (1967). Tritiated thymidine effects on DNA, RNA, and protein synthetic rates in synchronized L-cells. *J. Cell. Physiol.* **70**, 141–154.

Laird, A. K. (1966a). Dynamics of embryonic growth. *Growth* **30**, 263–275.

Laird, A. K. (1966b). Postnatal growth of birds and mammals. *Growth* **30**, 349–363.

Laird, A. K. (1969). Dynamics of growth in tumors and in normal organisms. *Nat. Cancer Inst., Monogr.* **30**, 15–28.

Lajtha, L. G. (1963). Differential sensitivity of the cell life cycle: On the concept of the cell cycle. *J. Cell. Comp. Physiol.* **62**, Suppl., 143–145.

Lala, P. K., and Patt, H. M. (1966). Cytokinetic analysis of tumor growth. *Proc. Nat. Acad. Sci. U.S.* **56**, 1735–1742.

Lala, P. K., and Patt, H. M. (1968). A characterization of the boundary between the cycling and resting states in ascites tumor cells. *Cell Tissue Kinet.* **1**, 137–146.

Lambert, W. C., and Studzinski, G. P. (1967). Recovery from prolonged unbalanced growth induced in HeLa cells by high concentrations of thymidine. *Cancer Res.* **27**, 2364–2369.

Lambert, W. C., and Studzinski, G. P. (1969). Thymidine as a synchronizing agent. II. Partial recovery of HeLa cells from unbalanced growth. *J. Cell. Physiol.* **73**, 261–266.

Larsson, A., and Reichard, P. (1966). Enzymatic synthesis of deoxyribonucleotides. X. Reduction of purine ribonucleotides; allosteric behavior and substrate specificity of the enzyme system from *Escherchia coli* B. *J. Biol. Chem.* **241**, 2540–2549.

Laster, W. R., Jr., Mayo, J. G., Simpson-Herren, L., Griswold, D. P., Jr., Lloyd, H. H., Schabel, F. M., Jr., and Skipper, H. E. (1969). Success and failure in the treatment of solid tumors. II. Kinetic parameters and "cell cure" of moderately advanced carcinoma 755. *Cancer Chemother. Rep.* **53**, 169–188.

Laster W. R., Jr., Mayo, J. G., Andrews, C. M., and Schabel, F. M., Jr. (1971). Lewis lung carcinoma as an experimental model for studies with anticancer drugs as surgical adjuvants. *Proc. Amer. Ass. Cancer Res.* **12**, 7.

Lavrovskii, V. A. (1969). Synchronization of the cell division in *Crepis capillaris* root meristems treated with 5-aminouracil. *Genetika* **5**, 5–9.

Lenaz, L., Sternberg, S. S., and Philips, F. S. (1969). Cytotoxic effects of 1-β-D-arabinofuranosyl-5-fluorocytosine and of 1-β-D-arabinofuranosylcytosine in proliferating tissues in mice. *Cancer Res.* **29**, 1790–1798.

LePage, G. A., and Jones, M. (1961). Further studies on the mechanism of action of 6-thioguanine. *Cancer Res.* **21**, 1590–1594.

Li, L. H., Olin, E. J., Buskirk, H. H., and Reineke, L. M. (1970a). Cytotoxicity and mode of action of 5-azacytidine on L1210 leukemia. *Cancer Res.* **30**, 2760–2769.

Li, L. H., Olin, E. J., Fraser, T. J., and Bhuyan, B. K. (1970b). Phase specificity of 5-azacytidine against mammalian cells in tissue culture. *Cancer Res.* **30**, 2770–2775.

Lindner, A. (1959). 5-Fluorouracil and Ehrlich ascites cells. *Cancer Res.* **19**, 189–194.

Lindner, A., Kutkam, T., Sankaranarayanan, K., Rucker, R., and Arradondo, J. (1963). Inhibition of Ehrlich ascites tumor with 5-fluorouracil and other agents. *Exp. Cell Res., Suppl.* **9**, 485–508.

Littlefield, J. W., and Gould, E. A. (1960). The toxic effect of 5-bromodeoxyuridine on cultured epithelial cells. *J. Biol. Chem.* **235**, 1129–1133.

Lozzio, C. B. (1969). Lethal effects of fluorodeoxyuridine on cultured mammalian cells at various stages of the cell cycle. *J. Cell. Physiol.* **74**, 57–62.

Lozzio, C. B., and Wigler, P. W. (1971). Cytotoxic effects of thiopyrimidines. *J. Cell. Physiol.* **78**, 25–32.

McQuade, H. A., and Friedkin, M. (1960). Radiation effects of thymidine-^{3}H and thymidine-^{14}C. *Exp. Cell Res.* **21**, 118–125.

Madoc-Jones, H., and Bruce, W. R. (1967). Sensitivity of L cells in exponential and stationary phase of 5-fluorouracil. *Nature (London)* **215**, 302–303.

Madoc-Jones, H., and Mauro, F. (1968). Interphase action of vinblastine and vincristine: Differences in their lethal action through the mitotic cycle of cultured mammalian cells. *J. Cell Physiol.* **72**, 185–196.

Mathé, G. (1969). Operational research in cancer chemotherapy. Chemotherapy in the strategy of cancer treatment. *Recent Results Cancer Res.* **21**, 72–96.

Mathé, G., Schneider, M., and Schwarzenberg, L. (1970a). The time factor in cancer chemotherapy. *Eur. J. Cancer* **6**, 23–31.

Mathé, G., Amiel, J. L., Schwarzenberg, L., Schneider, M., Cattan, A., Hayat, M., DeVassal, F., and Schlumberger, J. R. (1970b). Methods and strategy for the treatment of acute lymphoblastic leukemia. *Recent Results Cancer Res.* **30**, 109–137.

Mauro, F., and Madoc-Jones, H. (1970). Age responses of cultured mammalian cells to cytotoxic drugs. *Cancer Res.* **30**, 1397–1408.

Mendelsohn, M. L. (1960). The growth fraction: A new concept applied to tumors. *Science* **132**, 1496.

Mendelsohn, M. L. (1965). The kinetics of tumor cell proliferation. "Cellular Radiation Biology," pp. 498–513. Williams & Wilkins, Baltimore, Maryland.

Momparler, R. L. (1969). Effect of cytosine arabinoside 5'-triphosphate on mammalian DNA polymerase. *Biochem. Biophys. Res. Commun.* **34**, 465–471.

Montgomery, J. A., Johnston, T. P., and Shealy, Y. F. (1970). Drugs for neoplastic diseases. *In* "Medicinal Chemistry" (A. Burger, ed.), 3rd ed., Part I, pp. 680–783. Wiley (Interscience), New York.

Moore, E. C., and Cohen, S. S. (1967). Effects of arabinonucleotides on ribonucleotide reduction by an enzyme system from rat tumor. *J. Biol. Chem.* **242**, 2116–2118.

Moore, E. C., and Hurlburt, R. B. (1966). Regulation of mammalian deoxyribonucleotide biosynthesis by nucleotides as activators and inhibitors. *J. Biol. Chem.* **241**, 4802–4809.

Morris, N. R., and Cramer, J. W. (1966). DNA synthesis by mammalian cells inhibited in culture by 5-iodo-2'-deoxyuridine. *Mol. Pharmacol.* **2**, 1–9.

Morris, N. R., and Cramer, J. W. (1968). 5-Iodo-2'-deoxyuridine and DNA synthesis in mammalian cells. *Exp. Cell Res.* **51**, 555–563.

Mueller, G. C. (1963). Molecular events in replication of nuclei. *Exp. Cell Res., Suppl.* **9**, 144–149.

Mueller, G. C. (1969). Biochemical events in the animal cell cycle. *Fed. Proc., Fed. Amer. Soc. Exp. Biol.* **28**, 1780–1789.

Mueller, G. C. (1971). Biochemical perspectives of the G_1 and S intervals in the replication cycle of animal cells: A study in the control of cell growth. *In* "The Biochemistry of Disease" (R. Baserga, ed.), Vol. 1, pp. 269–307. Dekker, New York.

Mukherji, B., Yagoda, A., Oettgen, H. F., and Krakoff, I. H. (1971). Cyclic chemotherapy in lymphoma. *Cancer* **28**, 886–893.

Myers, D. K. (1971). The initial effect of x-radiation on thymidine incorporation into DNA in thymus cells. *Radiat. Res.* **47**, 731–740.

Neil, G. L., Wiley, P. F., Manak, R. C., and Moxley, T. E. (1970). Antitumor effect of 1-β-D-arabinofuranosylcytosine 5'-adamantoate (NSC 117614) in L1210 leukemic mice. *Cancer Res.* **30**, 1047–1054.

Nemoto, T., and Dao, T. L. (1971). 5-Fluorouracil and cyclophosphamide in disseminated breast cancer. Relationship between chemotherapy and hormonal therapy. *N.Y. State J. Med.* **71**, 554–558.

Nias, A. H. W., and Fox, M. (1971). Synchronization of mammalian cells with respect to the mitotic cycle. *Cell Tissue Kinet.* **4**, 375–398.

Nichols, W. H. (1964). *In vitro* chromosome breakage induced by arabinosyladenine in human leukocytes. *Cancer Res.* **24**, 1502–1504.

Ockey, C. H., Hsu, T. C., and Richardson, L. C. (1968). Chromosome damage induced by 5-fluoro-2'-deoxyuridine in relation to the cell cycle of the Chinese hamster. *J. Nat. Cancer Inst.* **40**, 465–473.

Odmark, G., and Kihlman, B. A. (1965). Effects of chromosome-breaking purine derivatives on nucleic acid synthesis and on the levels of adenosine 5'-triphosphate and deoxyadenosine 5'-triphosphate in bean root tips. *Mutat. Res.* **2**, 274–286.

Overgaard-Hansen, K. (1964). The inhibition of 5-phosphoribosyl-1-pyrophosphate formation by cordycepin triphosphate in extracts of Ehrlich ascites tumor cells. *Biochim. Biophys. Acta* **80**, 504–507.

Painter, R. B., Drew, R. M., and Rasmussen, R. E. (1964). Limitations in the use of carbon-labeled and tritium-labeled thymidine in cell culture studies. *Radiat. Res.* **21**, 355–366.

Petersen, D. F., Tobey, R. A., and Anderson, E. C. (1969). Synchronously dividing mammalian cells. *Fed. Proc., Fed. Amer. Soc. Exp. Biol.* **28**, 1771–1779.

Peterson, A. R., Fox, B. W., and Fox, M. (1971). Toxicity of tritiated thymidine in P388F lymphoma cells. I. Incorporation into DNA and cell survival. *Int. J. Radiat. Biol.* **19**, 123–132.

Post, J., and Hoffman, J. (1969). The effects of 5-iodo-2'-deoxyuridine upon the replication of ileal and spleen cells *in vivo. Cancer Res.* **29**, 1859–1865.

Price, D. J., and Timson, J. (1971). 6-Thioguanine: Antimitotic effect on human lymphocytes *in vitro* prevented by adenine. *Brit. J. Cancer* **25**, 106–108.

Priest, J. H., Heady, J. E., and Priest, R. E. (1967). Synchronization of human diploid cells by fluorodeoxyuridine. The first ten minutes of synthesis in female cells. *J. Nat. Cancer Inst.* **38**, 61–72.

Puck, T. T., Marcus, P. I., and Cieciura, S. J. (1956). Clonal growth of mammalian cells *in vitro*—growth characteristics of colonies from single HeLa cells with and without a "feeder" layer. *J. Exp. Med.* **103**, 273–284.

Rao, R. N., and Natarajan, A. T. (1965). Effect of 9-β-D-arabinofuranosyladenine on *Vicia faba* chromosomes. *Cancer Res.* **25**, 1764–1769.

Rauth, A. M., Barton, B., and Lee, C. P. Y. (1970). Effects of caffeine on L-cells exposed to mitomycin C. *Cancer Res.* **30**, 2724–2729.

Regan, J. D., and Chu, E. H. Y. (1966). A convenient method for assay of DNA synthesis in synchronized human cell cultures. *J. Cell Biol.* **28**, 139–143.

Regan, J. D., Setlow, R. B., and Ley, R. D. (1971). Normal and defective repair of damaged DNA in human cells: A sensitive assay utilizing the photolysis of bromodeoxyuridine. *Proc. Nat. Acad. Sci. U.S.* **68**, 708–712.

Reichard, P. (1968). The biosynthesis of deoxyribonucleotides. *Eur. J. Biochem.* **3**, 259–266.

Resch, A., and Schroeter, D. (1969). Effect of 5-aminouracil on the interphase structure of the nucleus in *Vicia faba* root tips. *Planta* **89**, 342–351.

Rixon, R. H., Whitfield, J. F., and MacManus, J. P. (1970). Stimulation of mitotic activity in rat bone marrow and thymus by exogenous adenosine 3'5'-monophosphate (cyclic AMP). *Exp. Cell Res.* **63**, 110–116.

Robbins, E., and Borun, T. W. (1967). The cytoplasmic synthesis of histones in HeLa cells and its temporal relationship to DNA replication. *Biochemistry* **57**, 409–416.

Rottman, F., and Guarino, A. J. (1964). The inhibition of phosphoribosylpyrophosphate amidotransferase activity by cordycepin monophosphate. *Biochim. Biophys. Acta* **89**, 465–472.

Rueckert, R. R., and Mueller, G. C. (1960). Studies on unbalanced growth in tissue culture. I. Induction and consequences of thymidine deficiency. *Cancer Res.* **20**, 1584–1591.

Sano, K., Hoshino, T., and Nagai, M. (1968). Radiosensitization of brain tumor cells with a thymidine analogue (bromouridine). *J. Neurosurg.* **28**, 530–538.

Sartorelli, A. C., and Booth, B. A. (1965). The synergistic antineoplastic activity of combinations of mitomycins with either 6-thioguanine or 5-fluorouracil. *Cancer Res.* **25**, 1393–1400.

Scaife, J. F. (1970). Mitotic G_2 delay induced in synchronized human kidney cells by uv and x-irradiation and its relation to DNA strand breakage, repair and transcription. *Cell Tissue Kinet.* **3**, 229–242.

Scaife, J. F. (1971). Cyclic 3′-5′-adenosine monophosphate: Its possible role in mammalian cell mitosis and radiation-induced mitotic G_2-delay. *Int. J. Radiat. Biol.* **19**, 191–195.

Schabel, F. M., Jr. (1968). *In vivo* leukemic cell kill kinetics and "curability" in experimental systems. "The Proliferation and Spread of Neoplastic Cells," pp. 379–408. Williams & Wilkins, Baltimore, Maryland.

Schabel, F. M., Jr. (1969a). Drug treatment of malignant tumors of man and animals —a rational approach to cancer chemotherapy. *S. Med. Bull.* **57**, 40–46.

Schabel, F. M., Jr. (1969b). The use of tumor growth kinetics in planning "curative" chemotherapy of advanced solid tumors. *Cancer Res.* **29**, 2384–2389.

Schabel, F. M., Jr (1969c). Cellular kinetics and its implication in cancer chemotherapy. "Neoplasia in Childhood," pp. 61–78. Yearbook Publ., Chicago, Illinois.

Schabel, F. M., Jr. (1970). Concept and practice of total tumor cell kill. *In* "Oncology 1970" (R. L. Clark *et al.*, eds.), Vol. II, pp. 35–45. Yearbook Publ., Chicago, Illinois.

Schabel, F. M., Jr., Skipper, H. E., Trader, M. W., and Wilcox, W. S. (1965). Experimental evaluation of potential anticancer agents. XIX. Sensitivity of nondividing leukemic cell populations to certain classes of drugs *in vivo. Cancer Chemother. Rep.* **48**, 17–30.

Schachtschabel, D. O., Killander, D, Zetterberg, A., McCarthy, R. E., and Foley, G. E. (1968). Effects of 4-aminopyrazolo(3, 4-d)pyrimidine in combination with guanine on nucleic acid and protein synthesis by Ehrlich ascites cells in culture— biochemical and cytochemical analysis. *Exp. Cell Res.* **50**, 73–80.

Schmidt, L. H., Montgomery, J. A., Laster, W. R., Jr., and Schabel, F. M., Jr. (1970). Combination therapy with arabinosylcytosine and thioguanine. *Proc. Amer. Ass. Cancer Res.* **11**, 70.

Schubert, D., and Jacob, F. (1970). 5-Bromodeoxyuridine-induced differentiation of a neuroblastoma. *Proc. Nat. Acad. Sci. U.S.* **67**, 247–254.

Shirakawa, S., Luce, J. K., Tannock, I., and Frei, E., III. (1970). Cell proliferation in human melanoma. *J. Clin. Invest.* **49**, 1188–1199.

Siev, M., Weinberg, R., and Penman, S. (1969). The selective interruption of nucleolar RNA synthesis in HeLa cells by cordycepin. *J. Cell Biol.* **41**, 510–520.

Silagi, S. (1965). Metabolism of 1-β-D-arabinofuranosylcytosine in L cells. *Cancer Res.* **25**, 1446–1453.

Silliman, R. P. (1969). Comparison between Gompertz and von Bertalanffy curves for expressing growth in weight of fishes. *J. Fish Res. Bd. Can.* **26**, 161–165.

Simpson-Herren, L., and Griswold, D. P., Jr. (1971). Kinetic studies of 7, 2-dimenthylbenz(α)anthracene(DMBA)-induced mammary adenocarcinoma and a derived transplanted line. *Proc. Amer. Ass. Cancer Res.* **12**, 7.

Simpson-Herren, L., and Griswold, D. P., Jr. (1972). Unpublished results.

Simpson-Herren, L., and Lloyd, H. H. (1970). Kinetic parameters and growth curves for experimental tumor systems. *Cancer Chemother. Rep.* **54**, 143–174.

Sinclair, R. (1970). Synchrony in cultured animal cells. *In Vitro* **5**, 79–93.

Sisken, J. S., and Wilkes, E. (1967). The time of synthesis and the conservation of mitosis-related proteins in cultured human amnion cells. *J. Cell Biol.* **34**, 97–110.

Skipper, H. E. (1971a). Cancer chemotherapy is many things: G. H. A. Clowes memorial lecture. *Cancer Res.* **31**, 1173–1180.
Skipper, H. E. (1971b). The cell cycle and chemotherapy of cancer. *In* "The Biochemistry of Disease" (R. Baserga, ed.), Vol. 1, pp. 358–387. Dekker, New York.
Skipper, H. E., Schabel, F. M., Jr., Trader, M. W., and Thomson, J. R. (1961). Experimental evaluation of potential anticancer agents. VI. Anatomical distribution of leukemic cells and failure of chemotheapy. *Cancer Res.* **21**, 1154–1164.
Skipper, H. E., Schabel, F. M., Jr., and Wilcox, W. S. (1964). Experimental evaluation of potential anticancer agents. XIII. On the criteria and kinetics associated with "curability" of experimental leukemias. *Cancer Chemother. Rep.* **35**, 1–111.
Skipper, H. E., Schabel, F. M., Jr., and Wilcox, W. S. (1967). Experimental evaluation of potential anticancer agents. XXI. Scheduling of arabinosylcytosine to take advantage of its S-phase specificity against leukemia cells. *Cancer Chemother. Rep.* **51**, 125–165.
Skipper, H. E., Schabel, F. M., Jr., Mellett, L. B., Montgomery, J. A., Wilkoff, L. J., Lloyd, H. H., and Brockman, R. W. (1970). Implications of biochemical, cytokinetic, pharmacologic, and toxicologic relationships in the design of optimal therapeutic schedules. *Cancer Chemother. Rep.* **54**, 431–450.
Socher, S. H., and Davidson, D. (1971). 5-Aminouracil treatment. A method for estimating G_2. *J. Cell Biol.* **48**, 248–252.
Steel, G. G. (1967). Cell loss as a factor in the growth rate of human tumours. *Eur. J. Cancer* **3**, 381–387.
Steel, G. G. (1968). Cell loss from experimental tumours. *Cell Tissue Kinet.* **1**, 193–207.
Steel, G. G., and Lamerton, L. F. (1969). Cell population kinetics and chemotherapy. I. The kinetics of tumor cell populations. *Nat. Cancer Inst., Monogr.* **30**, 29–50.
Stewart, J. E., Hahn, G. M., Parker, V., and Bagshaw, M. A. (1968). Chinese hamster cell monolayer cultures. II. X-ray sensitivity and sensitization by 5-bromodeoxycytidine in the exponential and plateau periods of growth. *Exp. Cell Res.* **49**, 293–299.
Studzinski, G. P., and Ellem, K. A. O. (1966). Relationship between RNA synthesis, cell division, and morphology of mammalian cells. I. Puromycin aminonucleoside as an inhibitor of RNA synthesis and division in HeLa cells. *J. Cell Biol.* **29**, 411–421.
Studzinski, G. P., and Lambert, W. C. (1969a). Thymidine as a synchronizing agent. I. Nucleic acid and protein formation in synchronous HeLa cultures treated with excess thymidine. *J. Cell. Physiol.* **73**, 109–118.
Studzinski, G. P., and Lambert, W. C. (1969b). Fallacies in the induction of synchrony in HeLa cell cultures by inhibitors of DNA synthesis. *In Vitro* **4**, 139–140.
Tannock, I. F. (1968). The relation between cell proliferation and the vascular system in a transplanted mouse mammany tumour. *Brit. J. Cancer* **22**, 258–273.
Tannock, I. F. (1969). A comparison of cell proliferation parameters in solid and ascites Ehrlich tumors. *Cancer Res.* **29**, 1527–1534.
Tannock, I. F. (1970). Population kinetics of carcinoma cells, capillary endothelial cells, and fibroblasts in a transplanted mouse mammary tumor. *Cancer Res.* **30**, 2470–2476.
Taylor, J. H., Haut, W. F., and Tung, J. (1962). Effects of fluorodeoxyuridine on DNA replication, chromosome breakage, and reunion. *Proc. Nat. Acad. Sci. U.S.* **48**, 190–198.

Taylor, J. H., Myers, T. L., and Cunningham, H. L. (1971). Programmed synthesis of deoxyribonucleic acid during the cell cycle. *In Vitro* **6**, 309–321.

Thatcher, C. J., and Walker, I. G. (1969). Sensitivity of confluent and cycling embryonic hamster cells to sulfur mustard, 1,3-bis(2-chloroethyl)-1-nitrosourea, and actinomycin D. *J. Nat. Cancer Inst.* **42**, 363–368.

Till, J. E., and McCulloch, E. A. (1961). A direct measurement of the radiation sensitivity of normal mouse bone marrow cells. *Radiat. Res.* **14**, 213–222.

Till, J. E., Whitmore, G. F., and Gulyas, S. (1963). Deoxyribonucleic acid synthesis in individual L-strain mouse cells. *Biochim. Biophys. Acta* **72**, 277–289.

Tobey, R. A., Petersen, D. F., Anderson, E. C., and Puck, T. T. (1966). Life cycle analysis of mammalian cells. III. The inhibition of division in Chinese hamster cells by puromycin and actinomycin. *Biophys. J.* **6**, 567–581.

Tobey, R. A., Anderson, E. C., and Petersen, D. F. (1967a). The effect of thymidine on the duration of G_1 in Chinese hamster cells. *J. Cell Biol.* **35**, 53–59.

Tobey, R. A., Petersen, D. F., and Anderson, E. C. (1967b). Differential response of two derivatives of the BHK21 hamster cell to thymidine. *J. Cell. Physiol.* **69**, 341–344.

Tobey, R. A., Petersen, D. F., and Anderson, E. C. (1971). Biochemistry of G_2 and mitosis. *In* "The Biochemistry of Disease" (R. Baserga, ed.), Vol. 1, pp. 309–353. Dekker, New York.

Ts'o, P. O. P., and Lu, P. (1964). Interaction of nucleic acids. I. Physical binding of thymine, adenine, steriods, and aromatic hydrocarbons to nucleic acids. *Proc. Nat. Acad. Sci. U.S.* **51**, 17–24.

Tubiana, M. (1971). The kinetics of tumour cell proliferation and radiotherapy. *Brit. J. Radiol.* **44**, 325–347.

Tyrer, D. D., Kline, I., Venditti, J. M., and Goldin, A. (1967). Separate and sequential chemotherapy of mouse leukemia L1210 with 1-β-D-arabinofuranosylcytosine hydrochloride and 1,3-bis(2-chloroethyl)-1-nitrosourea. *Cancer Res.* **27**, 873–879.

Vandevoorde, J. P., and Hansen, H. J. (1970). Carcinostatic action of 6-mercaptopurine and derivatives. *Proc. Amer. Ass. Cancer. Res.* **11**, 80.

Venditti, J. M., and Goldin, A. (1964). Chemotherapy of advanced mouse leukemia L1210: Comparison of methotrexate alone and in sequential therapy. *Cancer Res.* **24**, 1457–1460.

Vesely, J., and Šorm, F. (1965). Cytological and metabolic effects of a new antileukemic analog 5-azacytidine in normal mice followed auto-radiographically with tritium. *Neoplasma* **12**, 3–9.

Waqar, M. A., Burgoyne, L. A., and Atkinson, M. R. (1971). Deoxyribonucleic acid synthesis in mammalian nuclei. Incorporation of deoxyribonucleotides and chain-terminating nucleotide analogues. *Biochem. J.* **121**, 803–809.

Weiss, B. G., and Tolmach, L. J. (1967). Modification of x-ray-induced killing of HeLa S3 cells by inhibitors of DNA synthesis. *Biophys. J.* **7**, 779–795.

Wheeler, G. P., Bowdon, B. J., Adamson, D. J., and Vail, M. H. (1972). Comparison of the effects of several inhibitors of the synthesis of nucleic acids upon the viability and progression through the cell cycle of cultured H. Ep. No. 2 cells. *Cancer Res.* **32**, 2661–2669.

Whitmore, G. F. (1971). Natural and induced synchronous cultures. *In Vitro* **6**, 276–285.

Whitmore, G. F., and Gulyas, S. (1966). Synchronization of mammalian cells with tritiated thymidine. *Science* **151**, 691–694.

Whitmore, G. F., Borsa, J., Bacchetti, S., and Graham, F. (1969). Discussion: Mammalian cell killing by inhibitors of DNA synthesis. *Recent Results Cancer Res.* **17**, 109–117.

Wiebel, F., and Baserga, R. (1968). Cell proliferation in newly transplanted Ehrlich ascites tumor cells. *Cell Tissue Kinet.* **1**, 273–280.

Wilcox, W. S. (1966). The last surviving cancer cell: The chances of killing it. *Cancer Chemother. Rep.* **50**, 541–542.

Wilkoff, L. J., Dixon, G. J., Dulmadge, E. A., and Schabel, F. M., Jr. (1967a). Effect of 1,3-bis(2-chloroethyl)-1-nitrosourea (NSC-409962) and nitrogen mustard (NSC-762) on kinetic behavior of cultured L1210 leukemic cells. *Cancer Chemother. Rep.* **51**, 7–18.

Wilkoff, L. J., Wilcox, W. S., Burdeshaw, J. A., Dixon, G. J., and Dulmadge, E. A. (1967b). Effect of antimetabolites on kinetic behavior of proliferating cultured L1210 leukemia cells. *J. Nat. Cancer Inst.* **39**, 965–975.

Wilkoff, L. J., Dulmadge, E. A., and Lloyd, H. H. (1972). Kinetics of effect of 1-β-D-arabinofuranosylcytosine, its 5'-palmitoyl ester, 1-β-D-arabinofuranosyluracil, and tritiated thymidine on the viability of cultured leukemia L1210 cells. *J. Nat. Cancer Inst.* **48**, 685–695.

Xeros, N. (1962). Deoxyriboside control and synchronization of mitosis. *Nature (London)* **194**, 682–683.

Yang, S-J., Hahn, G. M., and Bagshaw, M. A. (1966). Chromosome aberrations induced by thymidine. *Exp. Cell Res.* **42**, 130–135.

York, J. L., and LePage, G. A. (1966). A proposed mechanism for the action of 9-β-D-arabinofuranosyladenine as an inhibitor of the growth of some ascites cells. *Can. J. Biochem.* **44**, 19–26.

Young, R. S. K., and Fischer, G. A. (1968). The action of arabinosylcytosine on synchronously growing populations of mammalian cells. *Biochem. Biophys. Res. Commun.* **32**, 23–29.

Author Index

Numbers in italics refer to the pages on which the complete references are listed.

A

Abdou, N. I., 171, *181*
Abell, C. W., 154, 166, *181, 186*
Aboul-Enein, M., 231, *243*
Achenbach, C., 231, *242*
Adams, J. E., 270, *293*
Adams, J. F., 30, 31, 32, 33, 34, 35, 36, 37, 38, 39, 40, 41, *46, 48*
Adamson, D. J., 222, 223, 225, 227, *246, 247*, 263, 264, 265, 267, 270, 276, 286, *294, 300, 305*
Adelman, M. R., *21*
Adinolfi, M., 142, *182*
Agarwal, K. C., 256, *293*
Agarwal, R. P., 256, *293*
Agarwal, S. S., 148, 149, 168, 172, 173, *182, 189*
Ahern, T., 167, 168, *187*
Ahmad-Zadeh, C., 170, *194*
Albrecht, M., 79, *91*
Albright, M. L., 126, *130*
Aldrich, D. J., 20, *22*
Alexander, J. A., 222, 225, *246*
Alfert, M., 100, *127*
Alford, R. H., 140, *182*
Aliosso, M. D., 97, *129*
Allan, P. W., 256, 263, *293, 294*
Allfrey, V. G., 148, 154, 155, 157, 161, 167, *187, 189, 191*
Altmann, H. W., 100, 115, *135*
Amaldi, F., 162, *182*
Ames, I. H., 271, *293*

Amiel, J. L., 291, *300*
Ammersdörfer, E., 270, *297*
Anderson, E. C., 62, *65*, 223, 230, *246*, 252, 260, 267, 271, 272, *302, 305*
Andrews, C. M., 292, *300*
Andrews, J. R., 288, *293*
Ansfield, F. J., 235, *243*
Antopol, W., 81, *91*
Argyris, T. S., 124, 125, *127*
Arlett, C. F., 264, *293*
Armentrout, S. A., 12, *22*
Armstrong, J. J., 20, *22*
Arradondo, J., 270, 275, *300*
Ash, J. F., 20, *23*
Asofsky, R., 171, *182*
Atkins, M., 167, 168, *187*
Atkinson, C., 280, *293*
Atkinson, M. R., 257, 259, 284, *305*
Attardi, B., 160, *182*
Attardi, G., 160, 162, *182*

B

Bacchetti, S., 260, *306*
Bach, F., 142, 148, 167, 171, *182, 186, 192*
Bade, E. G., 108, *127*
Baes, C., 120, *127*
Bagley, C. M., 222, 223, 225, 227, *243*
Bagshaw, M. A., 271, 278, *294, 306*
Bain, B., 141, *182*
Baldia, L. B., 107, *135*
Balduzzi, R., 97, *129*

Balis, M. E., 256, 258, *293*
Banerjee, M. R., 114, *127*
Baney, R. N., 176, *188*
Banfield, W. G., 121, 122, *131*
Barbason, H., 101, *127*
Barbiroli, B., 97, 101, 102, *127*
Barendsen, G. W., 288, *298*
Barka, T., 124, 125, *127*
Barker, B. E., 141, 145, *182, 184*
Barker, T. L., 282, *294*
Barkham, P., 156, *183*
Barnum, C. P., 120, *136*
Barocke, E., 79, *91*
Barondes, S. H., 14, *22*
Barr, H. J., 271, *293*
Barranco, S. C., 267, 286, *293*
Barrett, J. C., 241, *242*
Barton, A. D., 120, *127*
Barton, B., 264, *302*
Bartuska, D., 120, *133*
Baserga, R., 96, 109, 121, 125, 126, *127, 135,* 205, 213, *215,* 220, 221, 222, 223, 225, 226, 230, 238, *242, 243, 245, 246,* 250, 267, 268, 288, *293, 297, 306*
Bauer, C., 27, *48*
Baugnet-Mahieu, L., 120, *127*
Bauman, J., 196, 212, *216*
Bauminger, S., 143, 149, 172, *185*
Beard, Jr., N. S., 12, *22*
Becker, F. F., 81, *93,* 111, 115, 121, *127, 134,* 222, 238, *242*
Behnke, O., 21, *22*
Benacerraf, B., 153, *185*
Bender, M. A., 147, *182*
Benedict, W. F., 223, 235, *244,* 278, 279, *293*
Bennett, B., 147, 173, *182*
Bennett, Jr., L. L., 256, 257, 263, 282, *293, 294, 298*
Bensch, K. G., 17, 18, *22*
Bergeron, J. J. M., 273, 275, *294*
Bergsagel, D. E., 288, *294*
Berkovitz, A., 277, *294*
Berlin, M., 28, *46, 47*
Bernard, G., 46, *48*
Bernengo, M. G., 173, *191*
Berry, R. J., 270, *294*
Bertalanffy, F. D., 233, *242,* 280, *294*
Bertino, J. R., 229, *244*
Bertrand, M., 68, *92*

Bezmalinovic, Z., 123, *136*
Bhoopalam, N., 171, *186*
Bhuyan, B. K., 118, *130,* 259, 269, 270, 274, 275, 279, 281, 283, 285, *294, 300*
Bianchi, N. O., 177, *182*
Biberfeld, G., 139, 142, *182*
Biberfeld, P., 139, 142, *182*
Bickle, T. A., 170, *194*
Biesele, J. J., 250, 251, *294*
Biggers, J. D., 222, 231, *246*
Binz, C., 68, *91*
Bischoff, R., 20, *22,* 258, *294*
Blake, J. T., 139, 171, *192*
Blenkinsopp, W. K., 289, 290, *294*
Bloom, B. R., 147, 148, 173, 174, *182*
Blum, J. J., 49, 51, *64*
Bochert, G., 142, *192*
Bodey, G. P., 289, 291, *297*
Boecker, W., 113, *135*
Böök, J. A., 140, *188*
Börjeson, J., 141, 172, *182, 183*
Boiron, M., 170, *193*
Booth, B. A., 290, *302*
Bootsma, D., 176, *182,* 265, 267, 269, 272, *297*
Boquet, P. L., 76, *92*
Borberg, H., 149, *182*
Borg, K., 26, *47*
Borisy, G. G., 14, 15, 16, 17, *21, 22, 23*
Borjeson, H., 141, 145, 172, *184*
Borsa, J., 228, *243,* 260, *306*
Borun, T. W., 275, 278, *302*
Bostick, F., 222, 223, 225, 227, *243*
Bostock, C. J., 272, *294*
Bowdon, B. J., 222, 223, 225, 227, *246, 247,* 264, 265, 267, 270, 276, 286, *305*
Boyse, E. A., 238, *245*
Bradley, M. O., 20, *23*
Brändle, H., 122, *133*
Brain, M. C., 126, *127*
Brawerman, G., 170, 180, *188*
Bray, D. A., 222, 223, 225, 227, *243*
Brdar, B., 81, *91*
Brecher, G., 147, 148, 149, *189*
Breed, N. L., 270, *293*
Brehaut, L. A., 223, 231, *243,* 279, *294*
Breitner, J., 147, *193*
Bremerskov, V., 259, 280, *294*
Brewen, J. G., 223, 235, *243,* 279, *294*

AUTHOR INDEX

Brightman, S. A., 107, *135*
Brittinger, G., 151, 155, 161, *186*
Brockman, R. W., 6, 7, *10*, 287, *304*
Broome, J. D., 121, *127*, 222, 238, *242*
Brostrom, C. O., 154, *192*
Brown, R. F., 119, *128*
Brown, S. S., 121, *128*
Brown, T. W., 221, *246*
Brownhill, L. E., 141, *184*
Bruce, D. L., 291, *295*
Bruce, W. R., *243, 245,* 261, 262, 265, 270, 273, 279, 282, 283, 289, 291, *295, 300*
Brues, A. M., 105, 121, *128*
Brunstetter, F. H., 142, *183*
Bryan, J., 15, 17, 18, *22*
Bryant, B. J., 107, *128*
Bryceson, A. D. M., 143, 147, *189*
Bucher, N. L. R., 99, 101, 102, 103, 104, 106, 109, 110, 122, 123, *128, 133*
Budke, L., 176, *182*
Büchner, F., 100, 101, 102, 116, *128, 133*
Bunyaratvej, S., 116, 117, *135*
Burchenal, J. H., 227, *247*
Burdeshaw, J. A., 261, 270, *306*
Burgoyne, L. A., 257, 259, 284, *305*
Burke, G. C., 140, *185*
Burki, H. R., 281, 282, *295*
Burki, K., 124, *134*
Burnet, F. M., 175, *182*
Busanny-Caspari, W., 100, *128*
Buskirk, H. H., 259, 285, *300*
Butcher, E., 153, *192*
Butcher, R. W., 154, *183*, 256, *295*
Butter, D., 97, *135*
Byrt, P., 139, 171, *193*

C

Cain, B. F., 292, *295*
Cairnie, A. B., 200, 213, *215*
Cameron, I. L., 36, 41, 43, 44, 45, *47, 48,* 50, 52, 53, 54, 59, 61, 62, *64, 65,* 100, 108, *128,* 271, *298*
Camiener, G. W., 118, *130*
Campbell, H. A., 238, *245*
Campbell, I. L., 68, *92*
Cantarow, A., 120, *133*
Cantelmo, F., 209, *216*
Cantor, C. R., 16, *22*

Cardinale, G., 231, *243*
Carlson, K., 15, 17, *23,* 37, 39, *48*
Caron, G. A., 142, *190*
Carr, H. S., 238, *245*
Carriere, R., 100, *128*
Carroll, S. B., 268, *296*
Carter, C. E., 118, 119, *130*
Carter, S. B., 20, *22*
Carter, S. R., 233, *245*
Cassir, R., 120, *134*
Castells, J., 121, *132*
Cater, D. B., 101, 105, 122, *128, 130*
Casy, A. F., 70, *91*
Cattan, A., 291, *300*
Cave, M., 177, *183*
Cederberg, A., 126, *134*
Celozzi, E., 126, *136*
Ceppelini, R., 171, *190*
Chalmers, D. G., 141, *183*
Chambon, P., 164, *187*
Chang, L. O., 125, *128*
Chari-Bitron, A., 82, *91*
Charvát, Z., 124, *135*
Chasey, D., 16, 18, *22*
Chassy, B. M., 256, *295*
Chavaudre, N., 288, *295*
Cheong, L., 258, *296*
Chernozemski, I. N., 108, 117, *128*
Chessin, L. N., 141, 145, 150, 172, *182, 183, 184, 188*
Chezzi, C., 159, *192*
Chiesara, E., 111, *128*
Chiga, M., 115, *134*
Chmelar, V., 124, *135*
Choquet, C., 288, *295*
Chordikian, F., 120, *134*
Christie, N. T., 279, *294*
Chu, E. H. Y., 258, 270, *295, 302*
Chu, M. Y., 224, 233, *243*
Cieciura, S. J., 260, 261, *302*
Ciovîrnache, M., 115, *132*
Claman, H. N., 142, *183*
Clarkson, B. D., 238, 239, *243, 245,* 291, *295*
Clarkson, T. W., 26, 28, *47*
Cleaver, J. E., 255, 262, 264, 272, 273, 287, *295*
Clifton, K. H., 235, *243*
Cline, M. J., 175, *184*
Clouet, D. H., 68, 81, *91*

Cohen, A., 121, *128*
Cohen, C., 16, 19, *22*
Cohen, G. M., 82, *91*
Cohen, L. S., 260, 275, 276, *296*
Cohen, M. M., 155, 161, *189,* 265, 279, *296*
Cohen, S. S., 71, 75, *92, 93,* 257, 259, *297, 301*
Colot, H. V., 170, *192*
Coltman, Jr., C. A., 291, *296*
Condie, R. M., 138, *190*
Cone, Jr., C. D., 270, *296*
Connett, R. J., 49, *64*
Conti, F., 111, *128*
Cooke, A., 164, 173, *183, 187*
Cooper, E. H., 156, *183*
Cooper, H. L., 139, 141, 147, 148, 149, 153, 157, 158, 159, 160, 161, 162, 163, 164, 165, 167, 172, 173, 175, 178, 181, *183, 185, 186, 187*
Cooper, M. D., 139, 142, *187*
Cooperband, S. R., 150, 156, *191*
Corbin, J. D., 154, *192*
Corssen, G., 79, *91*
Costanzi, J. J., 291, *296*
Costea, N., 171, *186*
Côté, J., 116, *128*
Cotten, M. de V., 68, *91*
Coulson, A. S., 141, *183*
Cox, B. M., 80, *91*
Craddock, V. M., 117, *129*
Craig, N. C., 60, *64,* 164, *183*
Cramer, J. W., 235, *243, 245,* 259, 277, *296, 301*
Creasey, W. A., 233, 234, 238, *246*
Cross, M. E., 155, 161, *184*
Crowley, G. M., 151, *185*
Cummings, D. J., 280, *296*
Cummins, J. E., 50, 62, *64*
Cunningham, H. L., 277, *305*
Curry, S. H., 68, 70, *92*

D

Dalen, H., 223, 228, *243*
Danieli, G. A., 223, 227, *245*
Daniels, M., 21, *22*
Daniller, A. I., 123, *136*
Dao, T. L., 291, *301*
Daoust, R., 113, 114, *129*

Darnell, J. E., 159, 160, 162, 163, 170, 178, 180, *184*
Datta, R. K., 81, *91*
Davidson, C. S., 113, *131*
Davidson, D., 269, *304*
Davidson, N., 36, *47*
Davie, J., 139, 171, *192*
Davis, J. C., 122, 123, *129*
de Bianchi, M., 177, *182*
De Caro, B. M., 231, *243*
DeDeken, R. H., 57, *64*
DeDeken-Grenson, M., 57, *64*
Dehner, L. P., 168, *296*
de La Chapelle, A. 142, *185*
DeLamore, I. W., 235, *243*
Delhanty, J., 49, 63, *65*
Delius, H., 170, *194*
Deml, F., 122, 123, 124, *135*
Deneau, G. A., 68, *93*
DeRecondo, A. M., 102, 117, *129*
Desai, L. S., 155, 161, *184*
DeVassal, F., 291, *300*
DeVita, V. T., 222, 223, 225, 227, *243,* 288, *296*
Devynck, M. A., 72, 74, *92*
Dewey, W. C., 277, *296*
Dewey, W. L., 70, *91*
Diamond, L. K., 228, *243*
Dietz, A., 173, *191*
Ditroia, J. F., 99, 101, 102, 103, 104, 106, *128*
Dixon, G. J., 261, 270, *306*
Dixon, R. L., 111, *129*
Djordjevic-Camba, V., 124, *131*
Djordjevic, B., 258, 262, 268, 269, 273, 276, 277, 278, *296, 299*
Domon, M., 262, 264, 287, *296*
Doniach, I., 220, *243*
Donnelly, G. M., 238, *243*
Douglas, S. D., 141, 143, 145, 146, 161, 172, *182, 183, 184*
Dow, D. S., 152, 153, 154, 179, *191*
Downing, D. T., 116, *129*
Dray, S., 173, *193*
Drew, R. M., 273, *302*
Dulmadge, E. A., 223, 231, *247,* 261, 270, 274, 279, *306*
Dumonde, D. C., 143, 147, *189*
Dunn, M. R., 107, *135*
Dustin, Jr., P., 13, *22*

AUTHOR INDEX

Dutton, G., 14, *22*
Dvořáčková, I., 122, 123, 124, *135*

E

Ebaugh, F. G., 148, 149, *189*
Echave Llanos, J. M., 97, *129*
Eddy, N. B., 68, *92*
Edery, H., 70, *92*
Edmonds, M., 170, 180, *184*
Edwards, J. L., 98, 101, 105, 108, 109, *129*
Eidinoff, M. L., 258, 259, 265, 271, 274, 276, 278, 281, *296, 299*
Eigsti, O. J., 13, *22*
Elkerbout, F., 2, 5, *10*
Elkind, M. M., 196, 199, 204, 209, *215, 216, 217*
Ellem, K. A. O., 268, *304*
Ellner, C. J., 238, *245*
Elrod, H., 59, 60, 61, *64*
Elsbach, P., 76, 81, *93, 94*
Elves, M. W., 141, 171, *184*
Emerson, J. D., 126, *130*
Emmerson, J., 21, *22*, 59, *64*
Engberg, J., 59, *64*
Engeset, A., 223, 228, *243*
Ennis, H. L., 165, *184*
Epifanova, O. I., 268, 275, 288, *297*
Epstein, C., 100, 109, *129*
Epstein, L. B., 175, *184*
Erbenová, Z., 122, 123, *135*
Erickson, R. L., 235, *243*
Eriksson, T., 269, *297*
Erlichman, J., 154, *184*
Erne, K., 26, *47*
Erslev, A. J., 107, *131*
Estensen, R. D., 205, *215,* 222, 223, 230, 238, *242, 243,* 267, 268, *297*
Everett, N. B., 138, *184*
Ewald, J. L., 148, 149, 168, 172, 173, *189*
Ezell, D., 77, *92*

F

Fabrikant, J. I., 99, 101, 102, 105, 106, 107, *129*
Fanger, H., 141, *184*
Farber, E., 119, *135,* 176, *188,* 222, 223, 231, 238, *246*
Farber, J., 221, *243*
Farber, S., 228, *243*
Farnes, P., 141, 145, *182, 184*
Fausto, N., 119, *130*
Fawcett, D. W., 145, *184*
Feinenclegen, L. E., 209, *216*
Feit, H., 14, 15, 17, *22*
Feldman, M., 147, *185*
Fernández-Gómez, M. E., 266, *297*
Findstad, J., 138, *190*
Finnin, B. C., 80, *93*
Firschein, I. L., 148, 167, *186*
Fischel, R. E., 122, 123, *130,* 231, *244*
Fischer, G. A., 224, 233, *243,* 279, *306*
Fisher, D. B., 151, *184*
Fiskesjö, G., 28, *47*
Fitschen, W., 122, *136*
Fitzgerald, P. H., 138, 141, *184, 190,* 279, *294*
Florey, H. W., 122, 123, *134*
Fogel, S., 10, *10*
Foley, G. E., 155, 161, *184,* 265, *303*
Forbes, I. J., 171, *185*
Forer, A., 21, *22*
Forsdyke, D. R., 157, *185*
Forster, J., 153, *185*
Forti, G., 171, *190*
Foster, J. M., 150, 156, 181, *191*
Fouts, J. R., 111, *129*
Fox, B. W., 274, *302*
Fox, J. J., 259, 281, *299*
Fox, M., 274, 276, 277, *297, 301, 302*
Franceschini, P., 171, *190*
Frankfurt, O. S., 222, 224, 230, 231, *244*
Franklin, T. J., 2, *10*
Fraser, T. J., 259, 269, 270, 274, 275, 279, 281, 283, 285, *294, 300*
Frayssinet, C., 102, 117, *129*
Frei, E., 166, *190*
Frei, III, E., 288, 289, 291, *297, 303*
Freireich, E. J., 289, 291, *297*
Frenster, J. H., 161, *189*
Fried, J., 291, *295*
Friedkin, M., 17, *23,* 273, *300*
Friedman, R. M., 147, 173, 175, *185*
Friedrich, G., 209, *216*
Friedrich-Freksa, H., 107, 108, 116, *134*
Frindel, E., 288, 291, *297*
Fromageot, P., 72, 74, 81, *91, 92*
Fry, R. J. M., 115, 124, *133,* 209, 213, *216*

Fudenberg, H. H., 145, *184*
Fujioka, M., 117, 118, 120, *129*
Fujioka, S., 258, *297*
Furth, J. J., 257, 259, *297*

G

Gabriel, M. L., 10, *10*
Gailibert, F., 170, *193*
Galavazi, G., 265, 267, 269, 272, *297*
Gale, E. F., 71, 72, 73, 76, *91*
Gallo, R. C., 173, *192*
Gardner, B., 142, *182*
Gardner, L. I., 77, *92*
Garofalo, M., 117, 118, 120, 122, *134*
Garwes, D. J., 74, *93*
Gaudin, D., 287, 291, *297*
Gee, T., 291, *295*
Gelbard, A. S., 222, 230, 231, *244*, 276, *299*
Gell, P. G. H., 142, *193*
Geschwind, I. I., 100, *127*
Gesner, B., 149, *182*
Gianelli, F., 142, *182*
Gibbons, I. R., 14, *23*, 39, *48*
Giblak, R. E., 277, *296*
Gibson, E. M., 165, *183*
Gibson, M. H. L., 233, *242*, 280, *294*
Gibson, S., 28, *46*
Gigon, P. L., 111, *130*
Gillette, E. L., 203, 209, *216*
Gillette, J. R., 111, *130*
Gilman, A., 69, *91*
Giménez-Martin, G., 264, *297*
Ginsburg, H., 147, *185*
Giudice, G., 118, 119, *129*
Glade, P. R., 147, 148, 174, *182*
Glidewell, O., 291, *298*
Gniazdowski, M., 164, *187*
Goddard, J. W., 120, *133*
Goldberg, M. L., 140, *185*
Golden, J., 282, *294*
Goldenberg, D. M., 270, 280, *297*
Goldfarb, S., 116, *134*
Goldin, A., 290, 291, *305*
Goldstein, I. J., 140, 149, *185*
Gollin, F. F., 235, *243*, *246*
Gonzalez, E. M., 120, *129*
González-Fernández, A., 264, 266, *297*
Good, R. A., 138, *190*
Goodman, L. S., 69, *91*

Gordon, J., 143, 147, *185*
Gordon, J. A., 120, *130*
Gorovsky, M. A., 60, *64*
Gottlieb, L. I., 119, *130*
Gould, E. A., 258, *300*
Gould, H. J., 170, *185*
Goutier, R., 120, *127*
Gowans, J. L., 138, *185*
Graef, J. W., 164, 173, *185*
Gräsbeck, R., 140, 142, *185*, *194*
Graham, F. L., 234, *244*, 259, 260, 280, *297*, *298*, *306*
Gram, E., 111, *130*
Granboulan, N., 159, *192*
Granchelli, F. E., 70, *91*
Granger, G. A., 147, 173, *187*
Gray, G. D., 118, *130*
Greaves, M. F., 143, 149, 172, *185*, *190*
Green, H., 220, 221, *246*
Green, I., 153, *185*
Green, R. C., 27, *47*
Greene, F. E., 111, *130*
Greene, R., 72, 74, 75, *91*
Greengard, P., 257, *298*
Greulich, R. C., 271, *298*
Grierson, D., 170, *185*
Griffin, E. E., 59, *64*
Griffin, M. J., 222, 231, *244*
Grimstone, A. V., 18, *22*
Grisham, J. W., 97, 98, 100, 101, 102, 104, 105, 106, 107, 109, 110, 126, *130*
Griswold, Jr., D. P., 288, 290, 291, 292, *298*, *300*, *303*
Grosclaude, J., 159, *192*
Gross, P. R., 170, *192*
Grossman, J., 154, 155, *186*
Grozdanovič, J., 291, *298*
Grunfeld, Y., 70, *92*
Gruskin, R., 140, 141, *186*
Guarino, A. J., 256, *302*
Guarino, A. M., 111, *130*
Günther, 101, *130*
Gulyas, S., 262, 273, 274, *305*
Gurd, F. R. N., 27, *47*
Guzek, J. W., 122, 123, *130*
Gyermek, L., 49, *64*

H

Hämmerling, W., 100, 101, 102, *133*
Härkönen, M., 126, *130*

AUTHOR INDEX

Hagemann, R. F., 195, 196, 197, 200, 201, 202, 203, 204, 207, 209, 210, 213, 214, *216*
Hahn, G. M., 271, 278, 288, *298, 304, 306,*
Hakala, M. T., *258,* 263, 276, *298*
Hakala, T. R., 126, *132*
Hale, A. J., 156, *183*
Hall, T. C., 121, *133,* 239, *244*
Hampton, J. C., 222, 226, *244*
Handler, A. H., 231, *243*
Handmaker, S. D., 147, 152, 154, 158, 159, 164, 173, *185, 187*
Hanes, S., 241, *246*
Hanko, E., 26, *47*
Hansen, H. J., 265, 267, *305*
Hansson, A., 28, *48*
Hardy, S. J. S., 170, *188*
Harkness, R. D., 102, *130*
Harris, L. S., 70, *91*
Harris, N., 278, *293*
Harrison, S. C., 16, 19, *22*
Hart, S., 289, 291, *297*
Hartenstein, R., 116, *133*
Hartmann, K.-U., 120, *130*
Hashem, N., 148, *167, 186*
Hausen, P., 152, 158, 159, 173, *185, 191*
Haut, W. F., 274, *304*
Havemann, K., 166, *185*
Hay, E. D., 161, *185*
Hayat, M., 291, *300*
Hayden, G. A., 151, *185*
Heady, J. E., 275, *302*
Hedeskov, C. J., 156, *186*
Heidelberger, C., 120, *130,* 235, *243,* 256, 257, 258, 259, *298*
Heilpern, P., 122, 123, *134,* 231, *245*
Heine, W. D., 98, 99, *135*
Heiniger, H. J., 209, *216*
Heller, P., 171, *186*
Hemingway, J. T., 122, *130*
Henderson, D. W., 171, *185*
Henderson, E. S., 225, *244*
Henderson, I. C., 122, 123, *130,* 231, *244*
Henderson, J. F., 256, 257, 258, 259, 263, *298*
Henderson, P. T., 111, *130*
Henney, C. S., 154, *186*
Hermens, A. F., 288, *298*
Henry, P. H., 159, 160, *189, 194*
Hersh, E. M., 141, *190*

Heubner, W., 79, *91*
Hill, D. L., 9, *10,* 257, 263, *298*
Hill, Jr., R. B., 120, *130*
Himes, M. B., 231, *244*
Hirsch, A. H., 154, *184*
Hirschberg, E., *245*
Hirschberg, S. E., 126, *131*
Hirschhorn, K., 140, 141, 142, 148, 151, 155, 167, 171, *182, 186, 192*
Hirschhorn, R., 151, 154, 155, 161, *186*
Hirose, K., 28, 33, *48*
Hofer, K. G., 262, 278, *298*
Hoffman, G. S., 291, *298*
Hoffman, J., 97, 98, 99, 100, 117, *133,* 222, 223, 224, 225, 226, 227, 228, 229, 231, 232, 234, 235, 236, 237, 239, 240, 241, *244, 245,* 277, *302*
Hoffman, P. F., 141, 145, 172, *184*
Holland, J. F., 291, *298*
Hollander, N., 147, *185*
Hollerman, C. E., 140, 149, *185*
Hollister, L. E., 67, *92*
Hollister, R. M., 113, *131*
Holm, G., 148, *186*
Holmes, B. E., 101, 105, 122, *128*
Holt, L. J., 142, *186*
Holt, P. J. L., 147, 149, 181, *188*
Holtzer, H., 20, *22*
Hopper, G. D. K., 165, *186*
Horváth, E., 122, *130*
Hoshino, T., 291, *302*
Howard, A., 220, *244,* 250, 287, *298*
Howes, J. F., 70, *91*
Hryniuk, W. M., 229, *244*
Hsu, T. C., 275, *298, 301*
Huang, C. Y., 98, 99, *133*
Hübner, K., 101, *130*
Hughes, W. L., 26, 27, 39, *47,* 262, 278, *298*
Hultin, T., 111, *135*
Humphrey, R. M., 267, 275, 286, *293, 298*
Hurlburt, R. B., 257, 258, *301*
Hurwitz, A., 118, 119, *130*
Husáková, A., 124, 125, *135*
Husband, E. M., 141, *188*
Hwang, M., 160, *182*
Hyde, T. A., 122, 123, *129*

I

Ihnen, M., 122, *133*
Inagaki, A., 259, *298*
Inoué, S., 13, 15, *22*
Insalaco, J. R., 49, *65*
Ioachim, H. L., 231, *244*
Ishikawa, H., 20, *22*
Israels, M. C. G., 141, 171, *184*
Iwasaki, T., 222, 238, *246*

J

Jackson, B., 97, *131*
Jackson, E. B., 121, *128*
Jackson, H., 139, 171, *193*
Jackson, L. G., 230, *244*
Jacob, F., 161, *186*, 277, *303*
Jaffe, J. J., 101, *131*
Jahiel, R. I., 126, *131*
Jakob, K. M., 258, 269, *298, 299*
Jakoi, E. R., 56, 57, 59, 60, 61, *64*
James, K., 141, *188*
Jamieson, G. A., 151, *185*
Janossy, G., 172, *190*
Jansen, M., 75, *92*
Japundzic, I., 124, *131*
Japundzic, M., 124, *131*
Jensen, R., 36, *47*
Jeon, K. W., 50, *64*
Jeter, J., 50, 52, 53, 54, 62, *64*
Johnson, L. D., 154, 166, *181, 186*
Johnson, R. O., 291, *299*
Johnston, T. P., 256, 257, 259, 282, *301*
Joklik, W. K., 122, 123, *134*
Jones, M., 256, *300*
Joyce, C. R. B., 68, 70, *92*
Juhn, S. K., 114, *131*

K

Kabat, E. A., 149, *189*
Kaden, P., 259, 280, *294*
Kamper, C., 199, 204, *216*
Kaplan, J. G., 151, 152, 153, 154, 179, *191*
Karon, M., 223, 224, 233, 234, 235, *244*, 278, *293, 299*
Kasakura, S., 143, 147, 173, *186*
Katchalski, E., 140, 142, 149, 150, 154, *190*
Kay, J. E., 140, 148, 150, 151, 152, 154, 156, 157, 158, 159, 161, 162, 164, 165, 167, 168, 173, *183, 186, 187*
Kedinger, C., 164, *187*
Kemp, C. W., 166, *181*
Kennedy, B. J., 118, *135*
Kennedy, J. S., 70, *91*
Kenney, F. T., 119, *129*
Kent, G., 126, *135*
Kersten, K. J., 111, *130*
Kessel, D., 239, *244*
Kewitz, H., 79, *91*
Kibrick, S., 181, *191*
Kihlman, B. A., 266, *299, 301*
Kikuchi, Y., 177, *187*
Killander, D., 155, 156, 161, *187, 192*, 265, *303*
Kim, J. H., 222, 230, 231, 238, *244, 245*, 258, 259, 265, 268, 269, 271, 273, 276, 278, 281, *296, 299*
Kim, S. H., 258, 265, 269, 271, *296, 299*
Kimball, A. P., 224, 233, 234, *244*
Kincade, P. W., 139, 142, *187*
King, J., 124, *136*
King, S., 77, *92*
Kinosita, R., 108, *131*, 226, *246*
Kinzel, V., 231, *242*
Kirchner, H., 142, *187, 192*
Kirkpatrick, J. B., 272, *294*
Kishimoto, S., 230, 238, *244*, 267, *299*
Kisieleski, W. E., 209, *216*
Kiviranta, A., 126, *130*
Klein, K. M., 115, *127*
Klein, R. E., 108, 109, *129*
Kleinfeld, R. G., 115, 121, *131*
Kleinsmith, L., 154, 155, *187*
Klevecz, R. R., 221, *244*
Kline, I., 290, 291, *305*
Klinge, W., 101, 103, *131*
Klinman, N. R., 107, *131*
Klug, A. J., 18, *22*
Knezevic, B., 124, *131*
Knight, J., 68, *93*
Koch, A., 98, 101, 105, *129*
Koch, A. L., 256, *299*
Koga, M., 117, 118, 120, *129*
Kohn, H., 213, *216*
Kolb, W. P., 147, 173, *187*
Kollmorgen, G. M., 222, 231, *244*
Kolodny, R. L., 148, 167, *186*

Koransky, W., 124, 125, *134*
Kornfeld, R., 149, *187*
Kornfeld, S., 149, 150, 152, 166, *187, 189*
Kovacs, C. J., 268, *299*
Kovacs, E., 226, *246*
Kovács, K., 122, *130*
Kraft, N., 173, *188*
Krakoff, I. H., 291, *301*
Krause, M. O., 274, *299*
Krejczy, K., 120, *129*
Kreke, C. W., 27, *47*
Kronman, R., 126, *131*
Krueger, H. N., 68, *92*
Kula, N. S., 117, *134*
Kumar, A., 60, *64*
Kurdyumova, A. G., 268, 275, *297*
Kurland, A. A., 78, *93*
Kurland, C. G., 170, *188*
Kuschinsky, K., 81, *92*
Kutkam, T., 270, 275, *300*
Kuzmich, M. J., 14, *22*

L

Lafarge, C., 117, *129*
Lahtiharju, A., 122, *131*
Laird, A. K., 120, *127*, 287, *299*
Lajtha, L. G., 288, *299*
Lala, P. K., 288, *299*
Lambert, W. C., 176, *193*, 260, 271, 272, *299, 304*
Lamelin, J. P., 153, *185*
Lamelin, J.-P., 147, *188*
Lamerton, L. F., 213, *215*, 288, *304*
Lamont, W. A., 256, *299*
Landau, J. V., 78, *92*
Landing, B. H., 121, 122, *131*
Landy, M., 141, 147, 150, 174, *188*
Lapan, S., 231, *244*
Larsson, A., 258, *300*
Laster, Jr., W. R., 288, 290, 292, *300, 303*
Lavrovskii, V. A., 269, *300*
Lawrence, H. S., 147, 174, *188*
Laws, J. O., 115, *131*
Lawton, A. R., 139, 142, *187*
Layde, J. P., 222, 226, 230, *242, 245*
Layman, D. L., 125, *127*
Lazar, G. K., 220, 221, *246*
Leake, C. D., 77, *92*
Leblond, C. P., 97, *132*, 195, *216*

LeBouton, A. V., 99, *131*
Lebow, S., 81, *93*
LeBreton, E., 117, *129*
LeDain, G., 68, *92*
Leduc, E. H., 125, *136*
Lea, M. A., 120, *131*
Lee, C. P. Y., 264, *302*
Lee, K. L., 123, 124, *131*
Lee, R. E., 176, *188*
Lee, S. Y., 170, 180, *188*
Lee, Y. C., 60, *64*
Leevy, C. M., 113, *131*
Lehmann, H. E., 68, *92*
Leick, V., 59, *64*
Leiken, S., 140, *188*
Lejsek, K. J., 125, *135*
LeLièvre, P., 101, *127*
Lenaz, L., 259, 280, 281, *300*
Leon, M. A., 140, 149, *191*
Leong, G. F., 126, *130*
LePage, G. A., 256, 257, *300, 306*
Lesher, J., 201, 204, *216*
Lesher, S., 195, 196, 197, 200, 201, 202, 203, 204, 209, 210, 212, 213, 214, *216*
Leventhal, B. G., 141, 147, 162, 165, 167, *185, 187, 190*
Levis, A. G., 223, 227, *245*
Levis, W. R., 141, *188*
Ley, K. D., 220, *246*
Ley, R. D., 262, *302*
Li, L. H., 259, 281, 285, *300*
Lichenstein, L., 154, *186*
Lieberman, I., 117, 118, 119, 120, *128, 129*, 230, 238, *244*, 267, *299*
Lieberman, M. W., 176, *188*
Liener, I. E., 140, *193*
Lieper, R. D., 238, *245*
Lin, H., 245, 291, *295*
Lindahl-Kiessling, K. M., 140, 149, 150, 156, *188*
Lindell, T. J., 164, *188*
Lindner, A., 270, 275, *300*
Ling, N. R., 141, 142, 147, 149, 171, 181, *186, 188*
Linstead, D., 49, *64*
Lindsten, J., 28, *48*
Lipkin, M., 177, *188*
Littau, V. C., 161, *189*
Littlefield, J. W., 258, *300*
Livingston, R. B., 233, *245*

Lloyd, H. H., 6, 7, *10*, 223, 231, *247*, 261, 274, 279, 287, 288, 290, *300, 303, 304, 306*
Lloyd, K. O., 149, *189*
Loeb, J. N., 122, 123, *130*, 231, *244*
Loeb, L. A., 148, 149, 168, 172, 173, *182, 189*
Loening, U. E., 170, *185*
Löser, R., 76, *92*
Looney, W. B., 105, 125, *128, 131*
López-Sáez, J. F., 264, 266, *297*
Lorch, I. J., 50, *64*
Lowenstein, L., 141, 143, 147, 173, *182, 186*
Lozzio, C. B., 258, 271, 275, *300*
Lu, P., 256, *305*
Lucas, D. O., 151, *189*
Lucas, Z. J., 152, 158, 159, *189*
Luce, J. K., 288, *303*
Luduena, M. A., 20, *23*
Lutter, L., 170, *188*
Lutzner, M., 145, *182*
Luyckx, A., 121, *132*
Lycette, R. R., 141, *190*

M

McBride, W. G., 63, *64*
McCarthy, R. E., 265, *303*
McLean, D. K., 83, 84, 85, 86, 87, 88, *92*
McCluskey, I. S., 173, *191*
McCulloch, E. A., 261, *305*
McDonald, J. W. D., 154, *189*
MacDonald, R. A., 100, 103, 108, 113, 126, *131, 132*
MacDonnell, D. M., 160, *190*
MacGillivray, A. J., 153, 156, *189*
MacGregor, D. D., 138, *185*
McGuire, M., 142, *182*
MacHaffie, R. A., 156, *189*
McIntyre, O. R., 148, 149, *189*
McKellar, M., 97, 99, 108, *132*
MacKinney, A. A., 147, 148, 149, *189*
MacLean, L. D., 143, 147, *185*
MacManus, J. P., 269, *302*
McQuade, H. A., 273, *300*
Madden, S. C., 161, *193*
Madera-Orsini, F., 126, *135*
Madoc-Jones, H., 197, *216*, 261, 262, 268, 270, *300*

Madsen, N. B., 26, 27, *47*
Magasanik, B., 72, 74, 75, *91*
Magnus, D. R., 124, *127*
Magour, S., 124, 125, *134*
Maini, M. M., 113, 114, *132*
Maini, R. N., 143, 147, *189*
Malaise, E., 288, *297*
Malaise, E. P., 288, *295*
Malawista, S. E., 17, *22*
Malinsky, J., 121, *134*
Mallory, G. K., 126, *132*
Malt, R. A., 109, 110, 120, *128, 129, 135*
Manak, R. C., 290, *301*
Mandal, R. K., 165, *189*
Mandel, J. L., 164, *187*
Mannering, G. J., 82, *91*
Mantel, N., 291, *298*
Marantz, R., 17, 18, *22*
Marble, B. B., 105, *128*
Marchand, R., 99, *131*
Marcus, P. I., 260, 261, *302*
Margolis, S., 223, 229, *245*
Marin, G., 223, 227, *245*
Marquardt, H., 114, *132*
Marshak, A., 121, *132*
Marsland, D., 15, *22*, 78, *92*
Masuko, K., 116, *134*
Mathé, G., 289, 291, *300*
Mathyl, J., 101, *131*
Mattson, A., 150, 156, *188*
Maurer, W., 108, *127*
Mauro, F., 197, *216*, 261, 268, *300*
Maxwell, D. S., 37, 39, 46, *48*
Maxwell, R. A., 49, *65*
Mayo, J. G., 288, 290, 292, *300*
Mazia, D., 17, 18, *23*
Mechoulam, R., 70, *92*
Mednis, A., 238, *245*
Mee, L. K., 101, 105, 122, *128*
Meeker, B. E., *243*, 261, 265, 270, 273, 279, 282, 283, 289, *295*
Meldolesi, J., 111, *128*
Melka, J., 124, *135*
Mellett, L. B., 6, 7, *10*, 287, *304*
Melvin, J. B., 113, 118, *132*
Mendecki, J., 170, 180, *188*
Mendelsohn, J., 150, 152, *189*
Mendelsohn, M. L., 287, *301*
Menyhárt, J., 103, *132*
Mercer, R. D., 228, *243*

AUTHOR INDEX

Merigan, T. C., 175, *184*
Merler, E., 151, *189*
Meschan, I., 121, *132*
Messier, B., 97, *132*
Meza, I., 264, *297*
Miller, H. H., 277, *296*
Miller, J. F. A. P., 175, *193*
Miller, O. N., 123, 124, *131*
Minick, T., 126, *135*
Mironescu, S., 115, *132*
Mirsky, A. E., 148, 154, 155, 157, 161, 167, *187, 189, 191*
Mitra, J., 271, *293*
Mittermayer, C., 259, 280, *294*
Mittwoch, U., 49, 63, *64*
Miura, Y., 81, *93*
Möller, G., 147, 149, 150, 153, *189*
Mohr, U., 116, *132*
Molnar, F., 113, 114, *129*
Momparler, R. L., 259, *301*
Monjardino, J. P. P. V., 153, 156, *189*
Monod, J., 161, *186*
Montgomery, J. A., 6, 7, *10*, 256, 257, 259, 282, 287, 290, *301, 303, 304*
Moolten, F. L., 122, 123, *133*
Moore, E. C., 257, 258, *301*
Morel, C., 159, *192*
Morris, H. P., 120, 125, *128, 131*
Morris, N. R., 235, *243, 245*, 277, *296, 301*
Morris, P. W., 164, *188*
Moses, W. B., 199, *216*
Moulder, J. E., 15, *23*
Moxley, T. E., 290, *301*
Mozai, T., 26, 36, *48*
Mueller, G. C., 77, 80, 81, *92*, 149, 151, 176, *184, 189, 192, 194*, 250, 252, 260, 267, 274, *301, 302*
Mukherjee, A. B., 155, 161, *189*
Mukherji, B., 291, *301*
Mulay, A. S., 123, *135*
Murphy, W. R., 122, *133*
Musella, A., 120, *131*
Myers, D. K., 264, 291, *301*
Myers, T. L., 277, *305*
Myrbäck, K., 27, *47, 48*

N

Nadal, C., 100, 108, *133*
Nagai, M., 291, *302*
Nakamura, T., 259, *298*
Nakao, K., 26, 36, *48*, 155, 161, *190*
Nakazato, H., 170, 180, *184*
Nandi, U. S., 36, *47*
Natarajan, A. T., 266, *302*
Neal, G. E., 121, *128*
Neil, G. L., 290, *301*
Neiman, P. E., 159, 160, *189, 190*
Nemoto, T., 291, *301*
Neresheimer, E. R., 68, *92*
Nettesheim, P., 122, *133*
Neu, R. L., 77, *92*
Neumeyer, J. L., 70, *92*
Newberne, P. M., 117, *134*
Newberry, W. M., 153, 173, *193*
Ngu, V. A., 121, *133*
Nias, A. H. W., 276, *301*
Nichol, C. A., 263, *298*
Nichols, W. H., 266, *301*
Niehaus, W. G., 120, *136*
Nilsson, J. R., 59, *64*
Nishida, M., 27, *47*
Nordman, C. T., 140, 142, *185, 194*
Norman, A., 147, *192*
Noteboom, W. D., 77, 80, 81, *92*
Norseth, T., 28, *47*
Novelli, G. D., 118, 119, *129*
Novogrodsky, A., 140, 142, 149, 150, 154, *190*
Nowell, P. C., 140, *190*

O

Oakman, N. J., 122, 123, *133*
O'Brien, P. J., 27, *47*
Ochoa, Jr., M., 225, *245*
Ockey, C. H., 275, *301*
O'Connor, A., 239, *243*
Odmark, G., 266, *299, 301*
Oehlert, W., 100, 101, 102, 116, 122, *128, 133*
Oettgen, H. F., 238, *245*, 291, *301*
Oftebro, R., 223, 228, *243*
Ogawa, M., 291, *295*
Ohkita, T., 239, *243*
Old, L. J., 238, *245*
Olin, E. J., 259, 281, 285, *300*
Olivera, B. M., 36, *47*
Oliverio, V. T., *245*
Olmsted, J. B., 15, 17, *23*

Ono, T., 155, 161, *190*
Oppenheim, J. J., 139, 140, 141, 142, 166, 171, *182, 188, 190*
Ord, M. G., 155, 161, *184*
Orfei, E., 126, *135*
Osgood, E. E., 140, *192*
Osman, O. H., 80, *91*
Ota, K., 239, *243*
Overgaard-Hansen, K., 256, 257, *301*

P

Pachman, L. M., 156, *190*
Padilla, G. M., 50, 56, 57, 59, 60, 61, 62, *64*
Pahnke, W. N., 78, *93*
Painter, J. T., 77, *92*
Painter, R. B., 273, *302*
Papermaster, B. W., 138, *190*
Pardee, A. B., 152, *190*
Parenti, F., 171, *190*
Parker, C. W., 153, 154, 173, *193*
Parker, V., 278, *304*
Parkhouse, R. M. E., 172, *190*
Parks, Jr., R. E., 256, *293*
Parnas, H., 160, *182*
Pars, H. G., 70, *91*
Paschkis, K. E., 120, *133*
Passow, H., 26, *47*
Pastan, I., 155, *190*
Patt, H. M., 288, *299*
Paul, A., 101, *130*
Paul, W. E., 153, *193*
Pauly, J. L., 142, *190*
Pazderka, J., 124, *135*
Pearmain, G., 141, *190*
Pearson, P., 170, *194*
Pechet, G., 103, *132*
Pegoraro, L., 173, *191*
Pelc, S. R., 220, *243, 244,* 250, 287, *298*
Penman, M., 60, *65,* 160, 165, *191, 194*
Penman, S., 60, *65* 160, 165, *191, 194,* 257, *303*
Peraino, C., 115, 124, *133*
Perez, A. G., 222, 230, 231, *244,* 276, 278, *299*
Perez-Tamayo, R., 122, *133*
Perlman, R. L., 155, *190*
Perlmann, P., 139, 142, 148, *182, 186*
Perry, L. D., 100, 109, *133*

Perry, R. P., 164, *183*
Peters, J. H., 152, 158, 159, *191*
Petersen, D. F., 62, *65,* 223, 230, *246,* 252, 260, 267, 271, 272, *302, 305*
Petersen, R. O., 205, *215,* 222, 230, *242*
Peterson, A. R., 274, *302*
Peterson, D. W., 82, *91*
Peterson, J. E., 116, *129, 133*
Peterson, R. D. A., 149, *188*
Pettengill, O. S., 268, *296*
Pettis, P., 81, *93*
Pfeifer, U., 105, *135*
Philips, F. S., 114, 120, *132, 134,* 223, 229, 238, *245,* 259, 280, 281, *300*
Piatigorsky, J., 170, *191*
Pick, E. P., 68, *92*
Pilgrim, C., 108, *127*
Platonow, N., 28, *48*
Plotz, P. H., 139, 171, *191*
Pogo, B. G. T., 148, 155, 157, 161, 167, *191*
Polgar, P. R., 150, 156, 181, *191*
Pollard, H., 14, *23*
Pomerat, C. M., 77, *92*
Popper, H., 124, 125, *127*
Pospišil, M., 125, *135*
Post, J., 97, 98, 99, 100, 117, *133,* 222, 223, 224, 225, 226, 227, 228, 229, 231, 232, 234, 235, 236, 237, 239, 240, 241, *244, 245,* 277, *302*
Potter, V. R., 97, 101, 102, *127*
Powell, A. E., 140, 149, *191*
Powers, H. O., 77, *92*
Powers, W. E., 279, *295*
Prescott, D. M., 147, *182,* 272, *294*
Price, D. J., 265, *302*
Price, R. I. M., 107, *135*
Priest, J. H., 275, *302*
Priest, R. E., 275, *302*
Prodi, G., 114, *131*
Prusoff, W. H., 235, *243, 245,* 259, *296*
Puck, T. T., 14, *23,* 223, 230, *245, 246,* 260, 261, 267, *302*
Pukhalskaya, E. C., 126, *133*

Q

Quastel, M. R., 151, 152, 153, 154, 179, *191*

AUTHOR INDEX

Quastler, H., 201, 213, 217, 220, 245

R

Raab, K. H., 122, 123, *133*
Rabes, H., 102, 106, 107, 116, 122, *133*
Rabinowitz, Y., 139, 173, *191*
Räsänen, R. T., 126, *134*
Raff, M. C., 139, 142, *192*
Raff, R. A., 170, *192*
Raick, A. N., 115, *134*
Raina, A., 71, 75, *92, 93*
Rainford, N., 126, *131*
Ramachandramurthy, P., 140, *193*
Ramel, C., 28, 29, 33, *48*
Rand, J., 74, 76, *93*
Randall, L., 170, *188*
Rannestad, J., 37, *48*
Rao, R. N., 266, *302*
Rasmussen, R. E., 273, *302*
Ratner, M., 81, *91*
Rauth, A. M., 262, 264, 287, *296, 302*
Razdan, R. K., 70, *91*
Reddy, J., 115, 116, 117, *134, 135*
Reed, B. L., 80, *93*
Regan, J. D., 262, 270, *302*
Reichard, P., 256, 258, *300, 302*
Reimann, E. M., 154, *192*
Reineke, L. M., 259, 285, *300*
Reisfeld, R., 141, *182*
Renaud, F. L., 14, *23*, 39, *48*
Resch, A., 269, *302*
Revel, J. P., 161, *185*
Rich, M. A., 258, 274, *296*
Richardson, L. C., 275, *301*
Riddick, D. H., 173, *192*
Rigas, D. A., 140, *192*
Rigler, R., 155, 156, 161, *187, 192*
Ringo, D. L., 15, *23*
Ripps, C., 171, *192*
Ritchie, A. C., 115, *134*
Rixon, R. H., 269, *302*
Rizzo, A. J., 122, 123, *134*, 231, *245*
Roath, S., 141, 171, *184*
Robbins, E., 275, 278, *302*
Robbins, J. H., 141, *188*
Roberts, D., 121, *133*
Roberts, K. B., 122, 123, *134*

Robison, G. A., 153, *192*
Rodriguez, V., 289, 291, *297*
Roeder, R. G., 60, *64,* 164, *188, 192*
Röschenthaler, R., 72, 74, 76, *92*
Rogentine, G. N., 142, *190*
Rogers, A. E., 103, 117, *132, 134*
Ronca, G., 27, *48*
Rosbush, M., 160, *191*
Rosen, O. M., 154, *184*
Rosenau, W., 140, *185*
Rosenbaum, J. L., 15, 17, *23*, 37, 39, *48*
Rosenberg, P., 79, *93*
Rosenkranz, H. S., 238, *245*
Rosenstreich, D. L., 139, 171, *192*
Rosenthal, A. S., 139, 153, 171, *192, 193*
Rossi, C. A., 27, *48*
Rossman, T., 81, *93*
Rothstein, A., 26, *47*
Rottman, F., 256, *302*
Rovera, G., 221, *243, 246*
Rowe, A. J., 14, *23*, 39, *48*
Rowe, D. S., *193*
Rubin, A. D., 148, 153, 157, 158, 164, 165, 166, *183, 185, 192*
Rubin, A. L., 142, *192*
Rubin, E., 116, *134*
Ruby, A., 17, 18, *23*
Rucker, R., 270, 275, *300*
Rudick, M. J., 36, *48,* 50, 54, 59, 60, *64,* 65
Rueckert, R. R., 176, *192,* 274, *302*
Rühl, H., 142, *187, 192*
Ruffin, N. E., 80, *93*
Russell, P., 139, 171, *193*
Rutter, W. J., 164, *188, 192*

S

Sacher, G., 209, *216*
Sadnik, I. L., 108, *127*
Saito, K., 28, 33, *48*
Saito, M., 28, 33, *48*
Sakamoto, K., 204, *216*
Sakiyama, A., 81, *93*
Sakuma, K., 122, 123, *134*
Salama, A. I., 49, *65*
Sandberg, A. A., 177, *187*
Sankaranarayanan, K., 270, 275, *300*
Sano, K., 291, *302*
Santavy, F., 121, *134*

Sartorelli, A. C., 233, 234, 238, *246,* 290, *302*
Sasaki, M., 79, *93*
Sasaki, M. S., 147, *192*
Sasovetz, D., 120, *131*
Sato, H., 15, *22*
Scaife, J. F., 262, 264, 269, 273, 276, *303*
Scanu, A. M., 14, *23*
Schabel, Jr., F. M., 6, 7, *10,* 261, 287, 288, 289, 290, 291, 292, *298, 300, 303, 304, 306*
Schachtschabel, D. O., 265, *303*
Schapira, L., 75, *93*
Scheidt, L. G., 269, 270, 274, 275, 279, 281, 283, *294*
Schenk, H., 265, 267, 269, *297*
Scherbaum, O., 82, *93*
Scherer, E., 107, 108, 116, *134*
Scherrer, K., 159, *192*
Schindler, R., 124, *134*
Schlicht, I., 124, 125, *134*
Schlumberger, J. R., 291, *300*
Schmid, R., 113, 126, *131, 132*
Schmidt, L. H., 290, *303*
Schneider, J. H., 120, *134*
Schneider, M., 289, 291, *300*
Scholar, E. M., 256, *293*
Scholze, P., 116, *133*
Schreibman, R., 140, 141, *186*
Schroeder, T., 20, *23*
Schroeter, D., 269, *302*
Schubert, D., 277, *303*
Schulte-Hermann, R., 124, 125, *134*
Schwartz, F. J., 100, 109, *133*
Schwartz, H. S., 117, 118, 120, 122, *134*
Schwartz, M. K., 238, *245*
Schwarzenberg, L., 289, 291, *300*
Scott, R. B., 231, *246*
Seed, J. C., 121, 122, *131*
Seevers, M. H., 68, *93*
Seidlová-Mašinová, V., 121, *134*
Seidman, I., 111, *134*
Selawry, H. S., 168, *192*
Sell, S., 142, 153, 176, *188, 193*
Selvig, S. E., 170, *192*
Setlow, R. B., 262, *302*
Sevastyanova, M. V., 268, 275, *297*
Shagoury, R. A., 70, *92*
Shani, A., 70, *92*
Shaw, M. W., 265, 279, *296*

Shealy, Y. F., 256, 257, 259, 282, *301*
Sheffer, J., 14, *23*
Shelanski, M. L., 14, 15, 16, 17, 18, 21, *21, 22, 23*
Sherman, F. G., 213, *217,* 220, *245*
Shirakawa, S., 223, 224, 233, 235, *244, 278,* 288, *299, 303*
Shohet, S. B., 151, *189*
Shortman, K., 139, 171, 173, *188, 193*
Shultice, R. W., 111, *129*
Siev, M., 257, *303*
Sigdestad, C. P., 195, 196, 200, 201, 202, 204, 209, 210, 213, *216*
Sigel, B., 107, 110, *135*
Silagi, S., 278, *303*
Silber, R., 149, *182,* 258, *297*
Silliman, R. P., 287, *303*
Simek, J., 122, 123, 124, 125, *135*
Simon, E. H., 277, *294*
Simon, E. J., 71, 72, 73, 74, 75, 76, 77, 79, 81, *92, 93, 94*
Simpson, L., 282, *294*
Simpson-Herren, L., 261, 287, 288, 290, 291, 292, *298, 300, 303*
Sinclair, R., 260, *303*
Sinclair, W. K., 197, 209, *217,* 222, 224, 238, *246*
Singer, A. J., 68, *93*
Sisken, J. E., 108, 121, *131,* 222, 226, 238, *243, 246*
Sisken, J. S., 268, *303*
Skerfving, S., 28, *48*
Skinner, A., 150, 152, *189*
Skipper, H. E., 6, 7, *10,* 261, 287, 288, 289, 290, *300, 303, 304*
Skora, I. A., 79, *91*
Slusarek, L., 15, 17, *22*
Smith, E. E., 140, 149, *185*
Smith, J. W., 153, 154, 173, *193*
Smolenskaya, I. N., 268, 275, *297*
Snow, G. A., 2, *10*
Socher, S. H., 269, *304*
Sodergren, J. E., 117, 118, 122, *134*
Somers, C. E., 275, *298*
Sorenson, G. D., 268, *296*
Sorm, F., 281, *305*
Souto, M., 97, *129*
Speake, R. N., 20, *22*
Special Report, 26, *48*
Speetzen, R., 116, *132*

Spicer, E., 141, *188*
Spohr, G., 159, *192*
Spooner, B. S., 20, 21, *23*
Sprent, J., 175, *193*
Stacey, K. A., 280, *293*
Staffeldt, E., 115, 124, *133*
Stanworth, D. R., 142, *186*
Starr, J. L., 168, *192*
Steel, G. G., 213, *215*, 241, 246, 288, *304*
Steffen, J., *245*
Steffen, J. A., 177, *193*
Stein, G. S., 221, *246*
Stein, H., 152, 158, 159, 173, *185*
Stein, J., 68, *92*
Steinberg, S. S., 223, 229, *245*
Steiner, A. L., 153, 154, 173, *193*
Steiner, D. F., 124, *136*
Stenzel, K. H., 142, *192*
Stephens, R. E., 14, 16, 19, *22, 23*, 85, *93*
Sternberg, S. S., 117, 118, 120, 122, *134*, 259, 280, 281, *300*
Stevens, C. E., 195, *216*
Stewart, C. C., 147, *193*
Stewart, J. E., 278, *304*
Stich, H. F., 113, 114, *132*
Stock, J. A., 223, 225, 228, 230, 231, 238, *246*
Stockert, J. C., 266, *297*
Stöcker, E., 97, 98, 99, 100, 105, 107, 113, 115, *135*
Stohlman, F., 147, 148, 149, *189*
Stolzmann, W. M., 177, *193*
Stone, L. E., 277, *296*
Strehler, B. L., 3, *10*
Strittmatter, P., 27, *48*
Strobo, J. D., 153, *193*
Stubblefield, E., 221, *244*
Studzinski, G. P., 176, *193*, 230, *244*, 260, 268, 271, 272, 275, 276, *296, 299, 304*
Sturm, H. A., 231, *242*
Süss, R., 231, *242*
Suhadolnik, R. J., 256, *295*
Sullivan, R. J., 119, *135*, 223, 231, 238, *246*
Sumner, J. B., 27, *48*
Sumwalt, M., 68, *92*
Sun, S. C., 123, 124, *131*
Surur, J. M., 97, *129*
Suskind, R. R., 142, *190*
Sutherland, E. W., 153, *192*, 256, *295*
Sutton-Gilbert, H., 199, *216*
Svoboda, D., 115, 116, 117, *134, 135*
Swaffield, M. N., 99, 101, 102, 103, 104, 106, *128*
Sylvester, R. F., 228, *243*
Symeonidis, A., 123, *135*
Szabó, K., 103, *132*
Szybalski, W., 235, *243,* 258, 262, 276, 277, *296*

T

Takahashi, T., 140, *193*
Takai, S.-I., 119, *128*
Takaku, F., 155, 161, *190*
Talal, N., 139, 171, *191*
Tallal, L., 238, *245*
Tannock, I. F., 203, 209, *216*, 288, *303, 304*
Taskinen, E., 126, *134*
Taylor, D., 126, *131*
Taylor, E. L., 20, *23*
Taylor, E. W., 14, 15, 17, 20, *21, 22, 23*
Taylor, G., 171, *184*
Taylor, J. H., 274, 277, *304, 305*
Teebor, G. W., 111, *134*
Teir, H., 122, *131*
Terayama, H., 122, 123, *134,* 155, 161, *190*
Terry, W. D., 142, *190*
Terskikh, V. V., 288, *297*
Thatcher, C. J., 262, *305*
Theologides, A., 118, *135*
Thom, R., 124, 125, *134*
Thomas, G. H., 264, *296*
Thomas, P., 2, 5, *10*
Thomson, J. L., 222, 231, *246*
Thomson, J. R., 261, *304*
Thor, D. E., 173, *193*
Thrasher, J. D., 30, 31, 32, 33, 34, 35, 36, 37, 38, 39, 40, 41, 46, *46, 48*, 213, *217*, 271, *298*
Till, J. E., 261, 274, *305*
Timasheff, S. N., 17, *23*
Timson, J., 265, *302*
Tingle, L. E., 41, 43, 44, 45, *47*
Tiollais, P., 170, *193*
Tissières, A., 170, *194*
Tjio, J., 78, *93*
Tobey, R. A., 62, *65*, 220, 223, 230, *246*, 252, 260, 267, 271, 272, *302, 305*

Todaro, G. J., 220, 221, *246*
Todo, A., 291, *295*
Tokuyasu, K., 161, *193*
Tolmach, L. J., 266, 278, *305*
Tongier, Jr., M., 270, *296*
Torelli, U. L., 159, 160, *194*
Tormey, D. C., 149, *194*
Trader, M. W., 261, *303, 304*
Trams, E. G., 123, *135*
Traut, R. R., 170, *194*
Troll, W., 155, 161, *186*
Trosko, J. E., 258, 269, *299*
Truxová, G., 291, *298*
Tsien, H. C., 50, *65*
Ts'o, P. O. P., 256, *305*
Tubiana, M., 288, 291, *297, 305*
Tuczek, H. V., 102, 106, 107, *133*
Tung, J., 274, *304*
Turner, K. J., 171, *185*
Turner, W. B., 20, *22*
Tushinski, R., 160, 170, 180, *184*
Tyler, R. W., 138, *184*
Tyrer, D. D., 290, 291, *305*

U

Ullberg, S., 28, *47*
Umeda, M., 28, 33, *48*
Umeda, T., 119, *128*
Usui, S., 81, *93*

V

Vail, M. H., 222, 223, 225, 227, *246, 247*, 264, 265, 267, 270, 276, 286, *305*
Valenti, C., 270, *293*
Valeriote, F. A., *243*, 261, 265, 270, 273, 279, 282, 283, 289, *295*
Balleron, A. J., 288, *297*
Van der Gaag, H., 262, *295*
Vandevoorde, J. P., 265, 267, *305*
Van Lancker, J. L., 119, 121, *130, 132*
Van Praag, D., 71, 72, 73, 74, *93*
Van't Hof, J., 268, *299*
Van Vroonhoven, T. J., 120, *135*
Vas, M. R., 141, *182*
Vassort, F., 288, *297*
Vaughan, M., 170, 180, *184*
Venditti, J. M., 290, 291, *298, 305*
Ventilla, M., 16, 17, *22, 23*

Verbin, R. S., 119, *135*, 222, 223, 231, 238, *246*
Verbo, S., 140, 141, *186*
Vermund, H., 235, *243, 246*
Vesely, J., 281, *305*
Vich, Z., 291, *298*
Vilcek, J., 81, *93*
Volini, F., 126, *135*
Von Der Decken, A., 111, *135*
von Haam, E., 115, *131*
Vos, O., 176, *182*
Voynow, R., 170, *188*

W

Wakisaka, G., 259, *298*
Walker, I. G., 262, *305*
Wall, R., 160, 170, 180, *184*
Wang, C. H., 156, *189*
Wang, J. C., 36, *47*
Wanntorp, H., 26, *47*
Waqar, M. A., 257, 259, 284, *305*
Waravdekar, V. S., 291, *298*
Ward, P. A., 147, 173, *194*
Warwick, G. P., 108, 117, *128, 136*
Wasicky, R., 68, *92*
Watkins, W. M., 149, *194*
Wattiaux, J. M., 50, *65*
Webb, T. E., 122, 123, *133, 134*, 231, *245*
Weber, T., 140, *194*
Weinberg, F., 164, *188*
Weinberg, R., 257, *303*
Weinberger, S., 238, *246*
Weinbren, K., 122, 123, *136*
Weisberger, A. S., 12, *22*
Weisenberg, R. C., 14, 15, 17, *21, 23*
Weiss, B G., 266, 278, *305*
Weissman, S. M., 159, 160, *194*
Weissmann, G., 151, 154, 155, 161, *186*
Welch, A. D., 259, *296*
Welsh, P. D., 141, 172, *182, 183*
Wessels, N. K., 20, 21, *23*
Whang, J., 166, *190*
Whang-Peng, J., 140, *188*
Wheeler, G. P., 222, 223, 225, 227, *246, 247*, 264, 265, 267, 270, 276, 286, *305*
Wheelock, E. F., 147, 173, *194*
Whitecar, J. P., 289, 291, *297*
Whitfield, J. F., 269, *302*

Whitmore, G. F., 199, *215,* 228, 234, *243, 244,* 259, 260, 262, 273, 274, 280, *297, 298, 305, 306*
Widholm, J., 36, *47*
Wiebel, F., 288, *306*
Wieser, O., 231, *242*
Wigler, P. W., 258, 271, *300*
Wigzell, H., 149, *194*
Wilcox, W. S., 261, 270, 289, *303, 304, 306*
Wiley, P. F., 290, *301*
Wilhite, B. A., 173, *191*
Wilkes, E., 238, *246,* 268, *303*
Wilkie, D., 49, 63, *64, 65*
Wilkoff, L. J., 6, 7, *10,* 223, 231 *247,* 261, 270, 274, 279, 287, *304, 306*
Willems, M., 60, *65,* 165, *194*
Williams, D. C., 121, *128*
Williams, N., 139, 171, *193*
Williams, N. E., 37, *48*
Williamson, N., 141, *188*
Wilson, J. W., 125, *136*
Wilson, L., 15, 17, 18, *22*
Wilson, M. J., 224, 233, 234, *244*
Winshell, E. B., 238, *245*
Wisniewski, H., 18, *22*
Withers, H. R., 196, 203, 209, *216, 217*
Witman, G. B., 15, 17, *23*
Wolberg, W. H., 291, *299*
Wolff, J. A., 228, *243*
Wolstencroft, R. A., 143, 147, 173, *189, 194*
Wolstenholme, G. E. W., 68, *93*
Wong, P., 173, *191*
Woodside, A. M., 168, 173, *189*
Wrba, H., 116, 122, *132, 133*
Wrenn, J. T., 20, *23*

Wright, P., 152, 153, 179, *191*
Wurster, N., 75, 76, 81, *93*

X

Xeros, N., 176, *194,* 265, 267, 271, *306*

Y

Yagoda, A., 291, *301*
Yakulis, V. J., 171, *186*
Yamada, K. M., 20, 21, *23*
Yang, S.-J., 271, *306*
Yarbro, J. W., 118, 120, *135, 136,* 238, *247*
Yesner, I., 249, *182*
Yielding, K. L., 27, *47,* 287, 291, *297*
York, J. L., 257, *306*
Yoshikawa-Fukada, M., 89, *93*
Yoshino, Y., 26, 36, *48*
Young, C. W., 227, 230, *247*
Young, R. J., 170, *194*
Young, R. S. K., 279, *306*
Younger, L. R., 124, *136*

Z

Zagerman, J., 120, *133*
Zajdela, F., 100, 108, *133*
Zak, F. G., 116, *134*
Zeldis, L. J., 161, *193*
Zetterberg, A., 265, *303*
Zeuthen, E., 82, *93, 94*
Zimmerman, A. M., 14, 15, *22,* 78, 79, 80, 85, *92, 93, 94,* 126, *136*
Zimmerman, S., 78, *94*
Zubrod, C. G., *245*
Zucker-Franklin, D., 81, *94*
Zwaveling, A., 2, 5, *10*

Subject Index

A

Acetylaminofluorene (AAF), hepatocarcinogen, 115
N-Acetyl galactosamine
 inhibitor of PHA stimulation, 149–150
Actinomycin D
 cancer therapy action, 6–8
 DNA inhibitor, 117–122
 effect on cell cycle, 222–238
  crypt cells, 201–205
 inhibitor of DNA, 12–13
 mode and site of action, 4–8
Aflatoxin B, hepatocarcinogen, 117
Alkaloids, antitumor drugs, 222–238
Alkylating agents, 225–238
Allium cepa, root tips, mercuric compound effect on, 28–31
N-Allylnormorphine, effect on *Amoeba proteus,* 79
Aminopterin, leukemia treatment, 228–230
5-Aminouracil
 inhibitor of DNA and RNA synthesis, 286–287
 mitotoxic effect of, 258–293
Amoeba proteus, drug studies in, 77–80
Antibiotics, *see also* specific antibiotic, 2, 215–238
 effect on cell cycle, 222–238
 mode and site of action, 4–8
Antifols, 228–230
Antigens, lymphocyte activation and, 141

Antiimmunoglobulin sera, mitogenic action, 142–143
Antilymphocyte sera, mitogenic action, 142–143
Antimycin, mode and site of action, 4–8
Antitumor drugs, 219–240
β-D-Arabinofuranosyladenine, mitotoxic effect of, 256–293
1-β-D-Arabinofuranosylcytosine, antitumor activity, 233–237
L-Asparaginase
 anticancer effects, 238
 effect on cell cycle, 222–238
ATP
 inhibition by adrenergic drugs, 49–65
 narcotic drug action on, 75–76
5-Azacytidine
 incorporation into DNA, 285
 mitotoxic effects of, 259–293
Azaserine, mode and site of action, 4–8

B

Bacteria, penicillin effect on, 4–8
Breast tumor, chemotherapy, 240–242
5-Bromodeoxyuridine
 incorporation into DNA, 285
 mitotoxic effects of, 258–293
Busulfan, effect on cell cycle, 225–228

C

Caffeine, mitotoxic effect, 256–293

SUBJECT INDEX

Cancer, *see also* Tumor
  chemotherapy
    cell cycle and, 6–8
    drugs used in, 4–8
    purines as agents, 287–293
    pyrimidine analogs as agents, 287–293
Carbon tetrachloride, hepatotoxin, 112–113
Cations
  heavy metal
    affinity to sulfhydryl group, 26–28
    association constants, 26–27
Cell(s), *see also* specific cells
  antibiotic effect on, 4–8
  colchicine action mechanism on, 14–17
  cytotoxin tolerance, 200–205
  density, desmethylimipramine effect on, 49–61
  dividing, mercuric compounds effect on, 25–48
  eukaryotic
    division and growth, drug effect on, 76–79
    drug action on, 76–82
    RNA inhibition, nature of, 87–90
    tolerance of drugs, 79–80
  hallucinogenic drug effects on, 6–10
  interaction, drugs, 11–23
  intestinal
    cytotoxin tolerance, 200–205
    proliferation, 201–213
    renewal, control of, 205–213
  irradiation and survival, 200–205
  narcotic drugs and, 8–10
  prokaryotic
    division and growth, drug effect on, 70–71
    drug action on, 70–76
    RNA inhibition, nature of, 87–90
    tolerance of drugs, 71–72
  tumor, chemotherapy of, 6–8
  vinblastine action on, 17–20
Cell cycle
  adrenergic drugs and, 49–65
  chemotherapy and, 1–8
  drug action onto, 11–23
  drug analysis indicator, 8–10
  effect of
    antitumor drugs on, 219–240

    hallucinogenic drugs on, 67–94
    narcotic drugs on, 67–94
    nucleosides on, 249–293
    purines on, 249–293
    pyrimidines on, 249–293
  hepatic, 95–125
    biochemistry of, 109
    hormones and, 122–124
    inhibition of, 117–122
  inhibition of, 11–23
    cytochalasins, 20–23
  lymphocytes, measurement, 177
  methylmercuric chloride and, 33–37
  phase specificity, cancer chemotherapeutic drugs and, 6–8
  radiation studies, 195–215
Chemotherapy
  cancer, 6–8, 265–293
  cell cycle and, 1–8
  objective of, 2–8
Chlorambucil, effects on cell cycle, 225–228
Chloramphenicol, mode and site of action, 4–8
1,3-bis(2-Chloroethyl)-1-nitrosourea (BCNU), cancer therapy action, 6–8
Chlorimipramine, effect on cell cycle, 49–50
Chlortetracycline, hepatocarcinogen, 118–119
Chromosomes
  antibiotics effect on, 4–8
  fragmentation, mercuric compound effect on, 29–47
Cilia, regeneration, inhibition by methylmercuric chloride, 37–40
Colchicine
  action mechanism, 14–17
  effect on crypt cells, 210–213
    microtubules, 14–17
    proteins, 13–17
  inhibitor of mitosis, 13–17
Colicin, inhibitor of phosphorylation, 12–13
Concanavalin A
  activity
    inhibition of, 149–150
    lymphocyte activation and, 140

Copper, affinity to sulfhydryl group, 26–28
Cortisone, hepatic cell cycle and, 122–124
Crypt progenitor cell (CPC), nature of, 196–200
3′,5′-Cyclic adenosine monophosphate (cAMP)
　mitotoxic effect of, 257–293
　role in cell cycle, lymphocytes, 153–155
Cycloheximide
　effect on cell cycle, 222–238
　mode and site of action, 4–8
Cyclophosphamide
　cancer therapy action, 6–8
　effect on cell cycle, 225–228
Cytochalasin(s)
　effect on
　　cell cycle, 20–23
　　microfilaments, 20–23
　　inhibitors of mitosis, 20–23
Cytochalasin B, cancer therapy action, 6–8
Cytodynamics, intestinal, 195–215
Cytosine arabinoside, effect on cell cycle, 222–238
Cytotoxins, intestinal tolerance to, 200–205

D

Daunomycin (rubidomycin), antitumor agent, 231
Death, causes of, 3
3′-Deoxyadenosine, incorporation into RNA, 286
Deoxyadenosines, mitotoxic effect of, 256–293
2′-Deoxyguanosine, mitotoxic effect of, 257–293
Desmethylimipramine (DMI)
　effect on
　　cell cycle, 49–65
　　DNA synthesis, 54–63
　　RNA synthesis, 50–63
　experimental studies, cell cycle, 50–61
Diethylnitrosamine (DEN), hepatocarcinogen, 115–116
Dimethylaminoazobenzene (DAB) hepatocarcinogen, 113–114
Dimethylbenzanthracene (DMBA), hepatocarcinogen, 114–115

DMI, experimental studies in $T.$ $pyriformis$, 50–61
DNA
　antibiotics and, 4–8
　synthesis
　　desmethylimipramine effect on, 54–63
　　drugs affecting it, 176–177
　　inhibition of, 117–122, 222–238
　　　hepatocarcinogens, 113–117
　　　hormones, 122–124
　　　narcotic drugs, 85–90
　　　purines and pyrimidines, 261–293
　　lymphocytes, 148–177
　　mercuric compounds and, 31–33
　　methylmercuric chloride and, 33–37
　　narcotic drug action on, 72–76
　transcription inhibition, 12–13
DNA polymerase
　activity inhibition, 119–122
　activity of, interference, 283
　lymphocytes, protein synthesis, 168–171
DON, mode and site of action, 4–8
Drugs, see also specific drug
　adrenergic, cell cycle and, 49–65
　cancer chemotherapy and, 4–8, 265–293
　cell cycle, inhibition studies, 11–23
　definition, 1, 11
　effect on
　　cell cycle, 261–293
　　hepatic cell proliferation, 95–125
　　liver, 9–10
　　lymphocyte DNA synthesis, 176–177
　hallucinogenic
　　action on cell cycle, 67–94
　　chemistry of, 69–70
　　effect on
　　　cells, 8–10
　　　　eukaryotic, 76–82
　　　　prokaryotic, 70–76
　　　　transport, 75–76
　　experimental studies on $T.$ $pyriformis$, 76–93
　　structural diagrams, 69
　inhibitors of
　　cell cycle, 11–23
　　mitosis, 12–23
　metabolism, residual liver, 111–112

narcotic
  action on cell cycle, 67–94
  ATP and, 75
  chemistry of, 69–70
  effect on
    eukaryotic cells, 76–82
    prokaryotic cells, 70–76
    transport, 75–76
  experimental studies on *T. pyriformis,* 76–93
  oxygen consumption and, 75
  structural diagrams, 69
  tolerance, 79–80
  radiation studies, 195–215

E

Egg(s), fertilized, methylmercuric chloride distribution in, 40–41
Enzyme(s), *see also* specific enzyme
  activity, organomercurials and, 27–47
  effect on cell cycle, 222–238
  interactions, drugs, 11–23
  lymphocytes, protein synthesis, 172–175
Erythromycin, mode and site of action, 4–8
Escherichia coli (E. coli), drug action on, 70–90
Ethylenediaminetetraacetic acid (EDTA), phytohemagglutinin inhibitor, 140

F

p-Fluorophenylalanine, antitumor effects, 238
5-Fluorouracil
  cancer therapy action, 6–8
  hepatocarcinogen, 120–121
  mitotoxic effect of 258–293
Folic acid antagonists, antitumor agents, 228–230

G

Glucosteroids, antitumor agents, 231–233
Glutamate dehydrogenase, activity increase by organomercurials, 27–28
Glycoprotein, lymphocytes, 150–151
Growth hormone, hepatic cell cycle and, 122–124

H

HeLa cells
  drug studies with, 76–82
  mercuric compound effect on, 28–31
Hepatic cell proliferation
  abbreviations, 95–96
  drug's general effect on, 112–125
  hormones and, 122–124
  kinetics of, 97–110
  liver, 97–100
  regulation, 110–111
Hepatocarcinogens, 113–117
Hepatocytes
  necrosis, 112–113
  normal, cyclic status of, 109–110
  proliferation and, 97–110
Hepatoma, chemotherapy, 240–242
Hepatotoxins, 112–113
Hormones, hepatic cell cycle and, 122–124
Hydrocortisone, effect on cell cycle, 222–238
Hydroxyurea
  cancer therapy action, 6–8
  cytotoxicity, kinetics of, 209–213
  effect on
    cell cycle, 222–238
    intestinal crypt cells, 197–200
  hepatocarcinogen, 120

I

Ileum
  chemotherapy, 240–242
  drug effect on, 225–238
  hydrocortisone and, 232–233
  methotrexate and, 228–230
Immunoglobulins, lymphocytes, protein synthesis, 171–172
Infection(s), bacterial, death by, 3
Inhibitors, action on cell cycle, macromolecular assembly, 11–23
Intestinal cytodynamics, drug radiation studies, 195–215
Iododeoxyuridine
  antitumor activity, 233–237
  effect on cell cycle, 222–238
5-Iododeoxyuridine
  incorporation into DNA, 285
  mitotoxic effects of, 259–293

Ions, drug effects on, 12–23
Irradiation, cell, and survival, 201–205
Isoproterenol, hepatic cell cycle and, 125–126

J

Jejunum, effect of
  cytotoxins on, 205–213
  radiation on, 205–213

K

Kinetics, hepatic cell proliferation, 97–110

L

Leukemia, drugs and, 227–238
Levallorphan, action on cell cycle, 69–94
  effect on
    nucleic acids synthesis, 72–76
    protein synthesis, 72–76
    T. pyriformis, 85
Levorphanol, action on cell cycle, 69–94
  effect on
    RNA synthesis, 72–76
    T. pyriformis, 85
Ligands, heavy metal binding, 26–28
Lipids, synthesis, stimulation by narcotic drugs, 85–90
Liver
  drug effects on, 9–10
  hepatic cell proliferation, 97–100
  residual, drug metabolism by, 111–112
Lymphocytes
  activation, 137–183
  binding sites, number, 152–153
  biochemical alterations, 150–153
  blastogenic factors, 143–144
  DNA synthesis, 148–177
  growth
    biological activity, 147–148
    mitogen-induced, 145–148
    PHA and, 145–147
    pokeweed and, 145–147
  glycoprotein metabolism, 150–151
  interactions with mitogens, 149–150
  mitogenic factors, 143–144
  mitogens and, 137–182
  morphological changes, 145–147
  peripheral blood, 138–139
  phospholipid metabolism, 150–151
  pinocytosis, 151
  protein synthesis, 166–175
  RNA metabolism, 156–166
  transformation, 148–177
  transport phenomena, 151–152
Lythechinus pictus, fertilized eggs, methylmercuric chloride distribution in, 40–41

M

Magnesium (Mg^{2+}), DNA-binding, desmethylimipramine effect on, 59–61
Malaria, treatment of, 2
Manganese (Mn^{2+}), DNA-binding, desmethylimipramine effect on, 59–61
Melphalan, effect on cell cycle, 225–228
Mepacrine, 2
Mercaptides, formation of, 26–28
6-Mercaptopurine
  cancer therapy action, 6–8
  mitotoxic effect of, 256–293
Mercuric chloride(s)
  effect on
    generation time of T. pyriformis, 32–33
  T. pyriformis structure, 41–47
Mercuric compounds, see also Organomercurials
  DNA synthesis and, 31–33
  effect on dividing cells, 25–48
  generation time and, 31–33
  protein synthesis and, 31–33
  toxicity to T. pyriformis, 30–47
Mercury, affinity to sulfhydryl group, 26–28
Metals, see also specific metal cations
  heavy, 25–48
    affinity to sulfhydryl group, 26–28
    effect on cell cycle, 25–48
Methotrexate
  effect on cell cycle, 223–238
  hepatocarcinogen, 120–121
  leukemia treatment, 228–230
  metabolic effects, 228–230
  mode and site of action, 4–8
Methyl α-D-mannoside, concanavalin A inhibitor, 149–150

SUBJECT INDEX

Methylcholanthrene, hepatic cell cycle and, 125
Methylmercuric chloride
 cell cycle and, 33–37
 DNA synthesis and, 33–37
 inhibitor of cilia regeneration, 37–40
 protein synthesis and, 33–37
Methylmercury, 26–48
Methylmercury hydroxide, effect on chromosome fragmentation, 29
6-(Methylthio)purine ribonucleoside, mitotoxic effect of, 257–293
Microtubules
 colchicine effect on, 14–17
 structures of, 18–20
 vinblastine effect on, 17–20
Mitogens
 bacterial, 141
 binding sites to lymphocytes, 149–150
 effect on
  lymphocyte growth, 145–148
  mitotic cycle, 137–182
 interactions with lymphocytes, 149–150
 lymphocytes and, 137–182
 nature of, 139–144
Mitomycin, mode and site of action, 4–8
Mitomycin C, effect on cell cycle, 222–238
Mitosis
 inhibition of, 12–23
 inhibition by
  colchicine, 13–17
  cytochalasins, 20–23
  vinblastine, 17–20
Mitotic cycle
 effect of mitogens on, 137–182
 lymphocytes
  cyclic AMP and, 153–155
  DNA synthesis, 175–177
  energy metabolism, 156
  nuclear alterations, 155–156
  protein synthesis, 166–175
  proteolysis, 155
  RNA metabolism, 156–166
 protein synthesis, enzymes and, 172–175
Mitotoxins
 classification, 251
 effect on cell cycle, 255–263
Morphine
 action on cell cycle, 69–93
 effect on *Amoeba proteus*, 79
 experimental studies, on *T. pyriformis*, 82–85
Mustards, effect on cell cycle, 222–238

N

Narcotics, effect on cells, 8–10
Necrosis, 112–113
Neosalvarsan, 2
Nitrogen mustard, effect on cell cycle, 225–228
Nitrosurea, effect on cell cycle, 222–238
Nogalomycin, hepatocarcinogen, 118
Nucleic acids, *see also* DNA and RNA
 anabolism, mitotoxin effect on, 254–259
 sites of inhibition, by mitotoxins, 254–259
 synthesis, narcotic drug effect on, 72–76
 T. pyriformis, desmethylimipramine effect on, 49–63
Nucleosides, 249–293
 biochemical mechanism as mitotoxins, 252–255
 cytotoxic effects of, 260–263
 mitotoxins, tabulation of effects, 268–281

O

Oligomycin, mode and site of action, 4–8
Opium alkaloids, 69–70, *see also* specific alkaloid
Organomercurials, *see also* Mercuric compounds
 effect on
  cell dividing, 26–48
  enzyme activity, 26–47
  c-mitotic effect of, 28–31
  toxicity, 27–47

P

Penicillin, 2–8
 mode and site of action, 4–8
PHA, *see* Phytohemagglutinin
Phenobarbital, hepatic cell cycle and, 124–125

Phenylmercuric acetate, effect on generation time of *T. pyriformis,* 33
Phenylmercury hydroxide, effect on chromosome fragmentation, 29
Phospholipids, lymphocytes, 150–151
Phosphorylase, activity inhibition by organomercurials, 27–28
Phosphorylation
  inhibition of, 12–13
  interferences, 4–8
Phytohemagglutinin (PHA)
  effect on lymphocyte growth, 145–147
  inhibition of stimulation, 149–150
  lymphocyte activation and, 140–148
Pinocytosis, lymphocytes, 151
Pokeweed mitogen, 141
  effect on lymphocyte growth, 145–147
Proliferation, hepatic cell, 95–125
Proteins
  interactions, drugs, 11–23
  synthesis,
    inhibition of, 119–122, 222–238
    inhibition by
      narcotic drugs, 85–90
      puromycin, 286–287
    lymphocytes, 166–175
    mercuric compounds and, 31–33
    methylmercuric chloride and, 33–37
    narcotic drug action on, 72–76
Proteolysis, mitotic cycle, 155
Purines, 249–293
  antitumor agents, 233–237
  biochemical mechanism as mitotoxins, 252–255
  cancer chemotherapy and, 287–293
  classification as mitotoxins, 251
  cytotoxic effects of, 260–263
  incorporation into DNA and RNA, 284–286
  mitotoxins, tabulation of effects, 268–281
Purine ribonucleotides synthesis, drugs interfering with, 263–287
Puromycin
  hepatocarcinogen, 119
  inhibitor of protein synthesis, 286–287
  mitotoxic effect of, 257–293
  mode and site of action, 4–8
Puromycin D, effect on cell cycle, 223–238

Pyrimidines
  biochemical mechanism as mitotoxins, 252–255
  cancer chemotherapy and, 287–293
  classification as mitotoxins, 251
  cytotoxic effects of, 260–263
  effect on cell cycle, 222–238, 249–293
  incorporation into DNA and RNA, 284–286
  mitotoxins, tabulation of effects, 268–281
Pyrimidine ribonucleotides, synthesis, drugs interfering with, 263–287
Pyrrolizidine alkaloids, hepatocarcinogens, 116–117

R

Rabbit, antiserum, lymphocyte activation and, 142–143
Radiation, intestinal crypt cells, 195–215
Ribonucleoside diphosphates, reduction of, interference, 282–283
Ribosomes, inhibitors of, 4–8, 12–13
Rifampicin, inhibitor of RNA polymerase, 12–13
Rifamycin, mode and site of action, 4–8
RNA, antibiotics and, 4–8
  desmethylimipramine effect on, 50–63
  metabolism, mitotic cycle, 156–166
  synthesis
    inhibition of, 117–122, 222–238
    inhibition by
      narcotic drugs, 85–90
    nature of, 87–90
    purines and pyrimidines, 261–293
    narcotic drug action on, 72–76
RNA polymerase, inhibition of, 12–13

S

Salts, inorganic, as mitogens, 142
Sarcoma, antitumor agents, 240–242
Serotonin, hepatic cell cycle and, 126
Serum, lymphocyte activation and, 142–143
Silver, affinity to sulfhydryl group, 26–28
Sodium periodate, mitogenic activity, 142
Spleen
  drug effect on, 225–238

SUBJECT INDEX

hydrocortisone and, 231–233
methotrexate and, 228–230
Staphylococcus aureus (S. aureus), drug action on, 70–90
Steroids, effect on cell cycle, 222–238
Streptolysin, S, 141
Streptomyces, antitumor antibiotics from, 230–231
Streptomycin
 inhibitor of ribosomal functions, 12–13
 mode and site of action, 4–8
Sugars
 inhibitors of
  concanavalin A, 149–150
  PHA stimulation, 149–150
Sulfanilamide, mode and site of action, 4–8
Sulfhydryl group, heavy metal affinity to, 26–28
Suramin, 2

T

Tetracyclines, mode and site of action, 4–8
Δ^9-Tetrahydrocannabinol (Δ^9-THC)
 action on cell cycle, 69–94
 experimental studies, on *T. pyriformis*, 82–85
Tetrahymena, 76–94, *see also* *Tetrahymena pyriformis*
Tetrahymena pyriformis, 27–47
 cell division, 82–85
 cilia regeneration in, 31–33
 desmethylimipramine effect on, 49–61
 drug studies with, 8–10, 76–93
 drug tolerance, 82–85
 growth rate, 82–85
 mercuric toxicity studies in, 30
 methylmercuric chloride and, 33–37
 structure, mercuric chloride effect on, 41–47
Thioacetamide (TA), hepatocarcinogen, 115

6-Thiodeoxyguanosine, mitotoxic effect of, 257–293
6-Thioguanine
 incorporation into DNA and RNA, 285–286
 mitotoxic effect of, 256–293
Thio-TEPA, effect on cell cycle, 225–228
Thymidine, mitotoxic effects of, 258–293
Thymidine kinase, activity inhibition, 118–119
Thymidylate synthetase, activity of, interference, 283
Tolerance, narcotic drugs, 79–80
Toxicity
 mercuric compounds, 26–48
 organomercurial compounds, 27–47
Trypanosomiasis, treatment of, 2
Trypsin, activity inhibition by organomercurials, 27–28
Tubulin, 13–23, *see also* Microtubules
Tumor, *see also* Cancer
 chemotherapy, 219–240, 265–293

U

Urease, activity inhibition by organomercurials, 27–28

V

Valinomycin, effect on ions, monovalent, 12–13
Vinblastine
 action mechanism of, 17–20
 antitumor drug, 221–238
 cancer therapy action, 6–8
 inhibitor of mitosis, 17–20
Vinca alkaloids, *see* specific drug
Vincristine
 antitumor drug, 221–238
 cancer therapy action, 6–8

X

X-Rays, cell survival after, 201–205